Cancer Treatment and Research

Volume 166

Series editor

Steven T. Rosen, Duarte, CA, USA

More information about this series at http://www.springer.com/series/5808

Chad A. Mirkin · Thomas J. Meade
Sarah Hurst Petrosko · Alexander H. Stegh
Editors

Nanotechnology-Based Precision Tools for the Detection and Treatment of Cancer

Springer

Editors
Chad A. Mirkin
Department of Chemistry and International Institute for Nanotechnology
Northwestern University
Evanston, IL
USA

Sarah Hurst Petrosko
Department of Chemistry and International Institute for Nanotechnology
Northwestern University
Evanston, IL
USA

Thomas J. Meade
Department of Chemistry and International Institute for Nanotechnology
Northwestern University
Evanston, IL
USA

Alexander H. Stegh
Department of Neurology, Robert H. Lurie Comprehensive Cancer Center and International Institute for Nanotechnology
Northwestern University
Chicago, IL
USA

ISSN 0927-3042
Cancer Treatment and Research
ISBN 978-3-319-36450-6 ISBN 978-3-319-16555-4 (eBook)
DOI 10.1007/978-3-319-16555-4

Springer Cham Heidelberg New York Dordrecht London

Softcover reprint of the hardcover 1st edition 2015

Printed on acid-free paper

Springer International Publishing AG Switzerland is part of Springer Science+Business Media (www.springer.com)

Chad A. Mirkin · Thomas J. Meade
Sarah Hurst Petrosko · Alexander H. Stegh
Editors

Nanotechnology-Based Precision Tools for the Detection and Treatment of Cancer

Springer

Editors
Chad A. Mirkin
Department of Chemistry and International Institute for Nanotechnology
Northwestern University
Evanston, IL
USA

Sarah Hurst Petrosko
Department of Chemistry and International Institute for Nanotechnology
Northwestern University
Evanston, IL
USA

Thomas J. Meade
Department of Chemistry and International Institute for Nanotechnology
Northwestern University
Evanston, IL
USA

Alexander H. Stegh
Department of Neurology, Robert H. Lurie Comprehensive Cancer Center and International Institute for Nanotechnology
Northwestern University
Chicago, IL
USA

ISSN 0927-3042
Cancer Treatment and Research
ISBN 978-3-319-36450-6 ISBN 978-3-319-16555-4 (eBook)
DOI 10.1007/978-3-319-16555-4

Springer Cham Heidelberg New York Dordrecht London

Softcover reprint of the hardcover 1st edition 2015

Printed on acid-free paper

Springer International Publishing AG Switzerland is part of Springer Science+Business Media (www.springer.com)

Preface

Significant progress in the fundamental understanding of cancer as well as its detection and treatment has been made by both the research and clinical communities over the past several decades. As a result, the death rates of many cancers are on the decline. However, despite tremendous efforts by many, much still remains to be understood about this highly heterogeneous and complex disease, from why cancers progress so differently in each individual patient to why each patient reacts differently to cancer treatments. In order to effectively treat every patient, advanced precision solutions are needed that can discern and contend with such heterogeneities, even those at the cellular and genetic levels. In order to give the patient the best chance for survival, their cancer must be detected early, when it has not yet metastasized, and treated using personalized options that are highly effective, noninvasive, biocompatible, and targeted and that do not cause significant unwanted immediate or long-term side effects. It would be ideal if such precision therapeutics could induce cancer regression and at the same time track its effect on the disease state in real-time. Many of the current solutions fall short of achieving many of these requirements, leaving room for additional research and innovation by scientists and physicians alike.

The field of nanoscience and technology is offering up myriad tools and materials that have the potential to dramatically impact cancer research, diagnostics, and treatment. The chemical and physical properties of nanostructures are highly dependent upon their size, shape, and composition. Therefore, the architecture of nanostructures can be tuned during synthesis or via post-synthetic modification techniques to produce materials with the desired properties for a given application, biomedical or otherwise. Indeed, nanoconstructs are highly modular, allowing them to be designed and synthesized with multiple functionalities in mind. This means that a single nanoconstruct can be used as a modality for cancer therapy, detection, and/or bioimaging tasks simultaneously. In addition, the small size of nanoparticles, which puts them on the same length scale as many biological structures, grants them privileged access to biological systems and tumor microenvironments, often resulting in unique and potentially useful interactions with biological structures. These and other factors have enabled nanostructures to be the cornerstones of new technologies and processes that surpass traditional ones used for the study, detection, and treatment of cancer in terms of their capabilities and efficacies.

A diverse array of nanostructures that have found application within the field of biomedicine, specifically cancer research, detection, and treatment, are highlighted in this book. Many of these nanostructures possess both inorganic and organic or biological components; the properties of such structures are a synergistic combination of each and hence many possess theranostic (combined therapy and detection/imaging) abilities. For example, spherical nucleic acids (SNAs), which are highlighted in Chapters "Nanoflares as Probes for Cancer Diagnostics" and "Therapeutic Applications of Spherical Nucleic Acids", are made by templating a shell of highly oriented oligonucleotides on the surface of an organic or inorganic nanoparticle (e.g., gold, silver, iron oxide, liposomes). These nanomaterials, which can be made from one or more different types of oligonucleotides and modified with fluorophores and other tracking entities, are revolutionizing aspects of the intracellular detection and gene regulation arenas. Magnetic nanostructures (MNS, Chapter "Theranostic Magnetic Nanostructures (MNS) for Cancer"), which can possess cores comprised of iron, nickel, zinc, or cobalt, and nanodiamonds (NDs, Chapter "Nanodiamond-Based Chemotherapy and Imaging"), which have carbon-based cores, have been coated with a variety of small or polymeric molecules and successfully used in drug delivery and magnetic resonance imaging (MRI), among other areas. Lipid-based nanostructures (Chapter "Theranostic Lipid Nanoparticles for Cancer Medicine") have found use as imaging agents in computerized tomography (CT) and positron emission tomography (PET), and they can also be used in photothermal (PT) and photodynamic (PD) therapy applications. High-density lipoprotein (HDL)-like nanostructures can be used to precisely target lymphoma cells and to deliver a variety of therapeutic cargos, including small molecule drugs and siRNA, in a highly specific manner (Chapter "Synthetic High-Density Lipoprotein-Like Nanoparticles as Cancer Therapy"). Nanoparticles of gold or iron are being used as radiosensitizers (Chapter "Radiosensitization and Nanoparticles") to enhance the effects of radiation on tumor cells via DNA damage and hyperthermia. Finally, the porosities of hybrid particles that are comprised of metal-organic frameworks (MOF) and polysilane moieties have been used as carriers for cancer imaging and therapeutic agents (Chapter "Hybrid Nanoparticles for Cancer Imaging and Therapy").

Many of these nanomaterials have found a unique place in biology and medicine, and some are available in the marketplace. Spherical nucleic acid (SNA) nanoconstructs (Chapters "Nanoflares as Probes for Cancer Diagnostics" and "Therapeutic Applications of Spherical Nucleic Acids") were first invented in 1996 and have since been commercialized extensively. Indeed, there are now over 1,800 products and a robust pipeline of therapeutic lead compounds that exist based upon SNAs. For example, SmartFlares™, commercialized by EMD Millipore and AuraSense, LLC, are changing the way circulating tumor cells (CTCs) are tracked and studied by providing the only way to sort live cells based on intracellular genetic and small molecule markers (Chapter "Nanoflares as Probes for Cancer Diagnostics"); SNAs are also important lead structures for the treatment of cancers, including those of the brain (glioblastoma multiforme) and skin (Chapter "Therapeutic Applications of Spherical Nucleic Acids"). Likewise, nanodiamonds

(Chapter "Nanodiamond-Based Chemotherapy and Imaging"), due to their unique surface properties, are becoming a popular platform for theranostic and chemotherapeutic applications. These structures are currently being validated in nonhuman primate/large animal studies and the first in-human clinical trials are being planned.

However, despite the almost limitless potential of nanostructures and the significant progress that has been made thus far, many questions still remain. For instance, intense research is currently being undertaken to understand how nanoparticles interact with biological environments, including cancer cells, from how they pass through the tumor microenvironment (Chapter "Exploring the Tumor Microenvironment with Nanoparticles") and enter cancer cells to how they escape from the endosome and are ultimately exocytosed (Chapter "How Nanoparticles Interact with Cancer Cells"). One major consideration in this process is the nature and formation of the "protein corona" (Chapter "Engineering the Nanoparticle-Protein Interface for Cancer Therapeutics"), a protein accumulation layer that forms on the surface of colloidal particles in biological environments, including blood and serum. This corona can change the effective structure of the nanoconstruct, and accordingly its behavior and interactions with biological systems important in cancer research and treatment. So that the effect of each individual architectural parameter on the system can be isolated, it is important to develop materials that are truly "calibration-quality" (Chapter "Calibration-Quality Cancer Nanotherapeutics"). The particle replication in non-wetting templates (PRINT®) process is one method that can be used to fabricate polymer-based nanoparticles with independent control over each particle parameter for this purpose.

Given the progress that has been made in the past ten years, we are optimistic that the next decade will bring about more exciting advances in the area of cancer nanotechnology. A growing number of nanostructures and processes will complete the often lengthy and complicated US Food and Drug Administration (FDA) approval phase (Chapter "Cancer Nanotherapeutics in Clinical Trials") and enter the clinic where they can be used to save lives and contribute positively to humanity.

Evanston, IL
January 2015

Chad A. Mirkin
Thomas J. Meade
Sarah Hurst Petrosko
Alexander H. Stegh

Contents

Nanoflares as Probes for Cancer Diagnostics

Pratik S. Randeria, William E. Briley, Alyssa B. Chinen, Chenxia M. Guan, Sarah Hurst Petrosko and Chad A. Mirkin

Abstract

Patients whose cancer is detected early are much more likely to have a positive prognosis and outcome. Nanoflares hold promise as a practical diagnostic platform for the early detection of cancer markers in living cells. These probes are based on spherical nucleic acid (SNAs) and are typically composed of gold nanoparticle cores and densely packed and highly oriented oligonucleotide shells; these sequences are complementary to specific mRNA targets and are hybridized to fluorophore-labeled reporter strands. Nanoflares take advantage of the highly efficient fluorescence quenching properties of gold, the rapid cellular

Pratik S. Randeria and William E. Briley contributed equally.

P.S. Randeria
Department of Biomedical Engineering, Northwestern University, Sheridan Road, 2145, Evanston, IL 60208, USA

P.S. Randeria · W.E. Briley · A.B. Chinen · C.M. Guan · S.H. Petrosko · C.A. Mirkin (✉)
International Institute for Nanotechnology, Northwestern University, Sheridan Road, 2145, Evanston, IL 60208, USA
e-mail: chadnano@northwestern.edu

W.E. Briley
Interdepartmental Biological Sciences Program, Northwestern University, Tech Drive, 2205, Evanston, IL 60208, USA

A.B. Chinen · S.H. Petrosko · C.A. Mirkin
Department of Chemistry, Northwestern University, Sheridan Road, 2145, Evanston, IL 60208, USA

C.M. Guan
Department of Chemical and Biological Engineering, Northwestern University, Sheridan Road, 2145, Evanston, IL 60208, USA

C.A. Mirkin et al. (eds.), *Nanotechnology-Based Precision Tools for the Detection and Treatment of Cancer*, Cancer Treatment and Research 166,
DOI 10.1007/978-3-319-16555-4_1

uptake of SNAs that occurs without the use of transfection agents, and the enzymatic stability of such constructs to report a highly sensitive and specific signal in the presence of intracellular target mRNA. In this chapter, we will focus on the synthesis, characterization, and diagnostic applications of nanoflares as they relate to cancer markers.

Keywords
Gold nanoparticles · Spherical nucleic acids · Nanoflares · Cancer diagnostics · Early stage cancer detection

Contents

1 Introduction

For the past several decades, tremendous efforts have been made by many to battle cancer, one of the leading causes of death in the United States and around the world. Fortunately, as a result, the death rates for all cancers among men and women of all major racial and ethnic groups are on the decline (rates for both sexes combined decreased by 1.5 % per year from 2001 through 2010) [1]. The reductions in the death rates of four types of cancers (lung and bronchus, colon and rectum, female breast, and prostate) account for more than two-thirds of this overall drop. Notably, regular screening is routinely performed for all of these types of cancers among high-risk populations, which appears to contribute significantly to the falling numbers. According to one statistic, the five-year relative survival rate for an older female patient is 92 % if their breast cancer is diagnosed early, while it is still in a localized stage. However, the same patient's survival rate drops precipitously to 80 and 20 % if the cancer spreads to either regional or distant lymph nodes, tissues, and organs, respectively. Similar trends have been observed for other common cancer types. As a whole, the evidence points toward the fact that the earlier and more accurately a patient's cancer can be detected and diagnosed, the more likely they are to survive.

Despite the obvious need for early and accurate cancer detection and diagnosis, intense research in both the laboratory and the clinic is still needed to make this goal a reality. Cancer is a highly heterogeneous disease [2–4]; hundreds of cancers exist, and the levels of the disease indicators can vary widely from person-to-person. Some cells within a tumor population may express high levels of dangerous oncogenes, while others may express lower levels or none at all. Many of the current diagnostic tools, including imaging modalities such as mammography and computerized tomography (CT) and laboratory tests such as polymerase chain reaction (PCR) and enzyme-linked immunosorbent assay (ELISA), interrogate bulk samples; thus, many of these vital, subtle differences may not be able to be detected. These analytical techniques are not designed to deal with sample heterogeneity. Novel methods are needed to quickly and reliably detect ultra-low concentrations of cancer markers, such as oncogenic RNA and proteins, and/or small numbers of cancer cells (e.g., circulating tumor cells, cancer stem cells) in ways that are minimally invasive to the patient so that accurate diagnoses of highly varied cancer disease states can be made. Such tools might also be valuable from a research standpoint as they could be used to learn more about the fundamental chemistry and biology of cancer in ways that are impossible with current techniques.

Many nanotechnology-based systems, including those based on the detection of genetic markers of cancer, are being developed and explored as a way to meet these needs [5, 6]. Cancerous cells are endowed with invasive properties through the unsilencing or overexpression of known oncogenes. Therefore, the ideal nanoparticle probe would be one that enables the direct detection of gene expression at low levels (i.e., low enough that they can be produced by a small number of malignant cells in the early stages of the disease) in a quantitative fashion. Such constructs would be able to enter any cell and interrogate intracellular genetic material without altering the cellular environment (i.e., they could be used with live cells). The ideal probe would recognize intracellular targets with very high specificity such that false positives would be minimized, and it would enable single-cell resolution in order to accurately analyze heterogeneous bulk samples. Further, such nanoprobes would be stable against degradation in biological environments and would not cause cytotoxic effects, immunogenic reactions, or unwanted side effects. It would also be advantageous if the construct could be used in the context of both single- and multi-gene detection and potentially in theranostic applications that combine detection and therapeutic schemes. One type of nanoprobe—the nanoflare [7], which is based upon spherical nucleic acids (SNAs) [8]—is proving to fulfill many of these requirements.

2 Spherical Nucleic Acids: Synthesis and Properties

SNAs are three-dimensional conjugates consisting of densely functionalized, highly oriented nucleic acids covalently attached to the surface of a nanoparticle (Fig. 1) [8]. The core serves two purposes: (1) it imparts the conjugate with novel chemical and physical properties, and (2) it acts as a scaffold for assembling and orienting the oligonucleotides into a dense arrangement. The nucleic acid shell confers upon the SNA many of its functional properties, including high cellular uptake in over 60 tested mammalian cell lines [9], resistance to enzymatic degradation [10], and action without apparent cytotoxicity [11] or unintended immunogenicity [12]. In addition, SNAs have significantly higher binding constants than linear nucleic acids of the same sequence [13]. These structures are highly modular: the core composition and

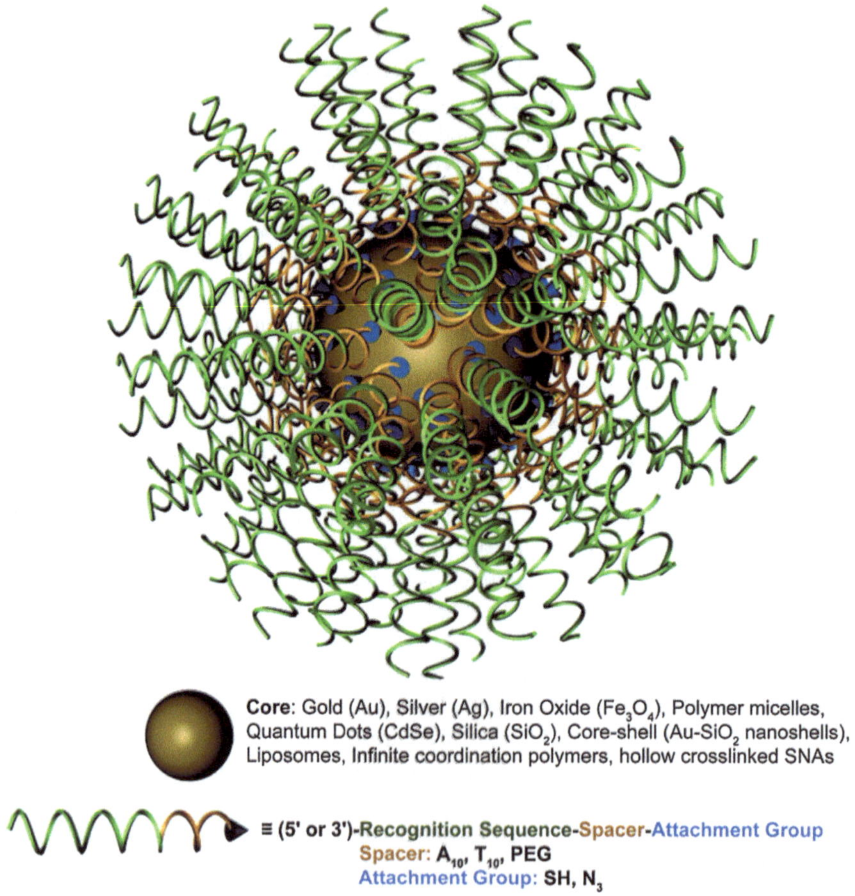

Fig. 1 Structure of a spherical nucleic acid (SNA)—gold nanoparticle conjugate. (Reproduced with permission from [8])

core size as well as the nucleic acid class and sequence can be tailored for the desired application. Because of their modularity and biocompatibility, SNAs are a promising candidate for gene detection and regulation in live cells.

2.1 Synthesis of SNAs

We will first describe how these constructs are synthesized. SNAs can be prepared with inorganic cores (silver [14], gold [15], iron oxide [16], quantum dots [17], silica [18], infinite coordination polymers [19]) or be coreless (cross-linked alkyne polymers [20], liposomes [21]) in nature. However, SNAs with gold particle cores are most commonly used in gene detection assays involving nanoflares because gold is inert and has the ability to quench fluorescent molecules in a distance-dependent manner [9]. Citrate-capped gold colloids can be synthesized using the Frens method, where chloroauric acid ($HAuCl_4$) is reduced by sodium citrate [22]. This well-established, solution-based method allows for the facile synthesis of highly uniform gold colloids of a specific size ranging from 5 to 150 nm in diameter [22]. The particle acts as a template for the subsequent attachment of nucleic acids and thus plays an important role in determining the final size of the overall nanostructure. Indeed, a core size must be chosen such that the SNA will be able to enter cells and function appropriately once inside. To date, the most commonly used SNAs for diagnostic applications have typically consisted of a 10–15 nm gold cores [9] with thiol [23] or cyclic disulfide [24] attachments.

The oligonucleotides that make up the nucleic acid shell are typically 25–40 bases in length (7–12 nm) and can be composed of single- and double-stranded DNA [10], short-interfering RNA (siRNA) [25], micro-RNA (miRNA) [26], RNA/DNA hybrids [27], and modified nucleic acids such as peptide nucleic acids (PNA) [28] and locked nucleic acids (LNA) [29]. Each oligonucleotide anchored to the core is made up of three distinct regions: (1) an alkylthiol [23] or cyclic disulfide [24] chemical tethering group that can be used to link the oligonucleotides to the gold nanoparticle's surface, (2) a recognition element (usually 15–25 base pairs in length) that is complementary to the target biomolecule of interest, and (3) a spacer region that can be modulated in length to tune the distance between the recognition element and the nanoparticle surface. Because of the ability of nucleobases to interact with the gold surface, it is necessary to extend the recognition element away from the particle surface, giving it more free volume to interact with incoming strands due to the curvature of the nanoparticle [30]. Poly-adenine (poly-A) or oligo-ethylene glycol (OEG) are commonly utilized as spacers [31]. Poly-thymine (poly-T) spacers are usually avoided for intracellular applications because of their ability to interact with the poly-A tails of a variety of mRNA, diminishing the sequence specificity of the SNAs.

To create a dense nucleic acid shell, the alkylthiol modified oligonucleotides are first mixed in solution with the citrate-capped gold particles (Fig. 2). Thiol moieties have a high affinity for gold; thus, the thiolated oligonucleotides displace the citrate

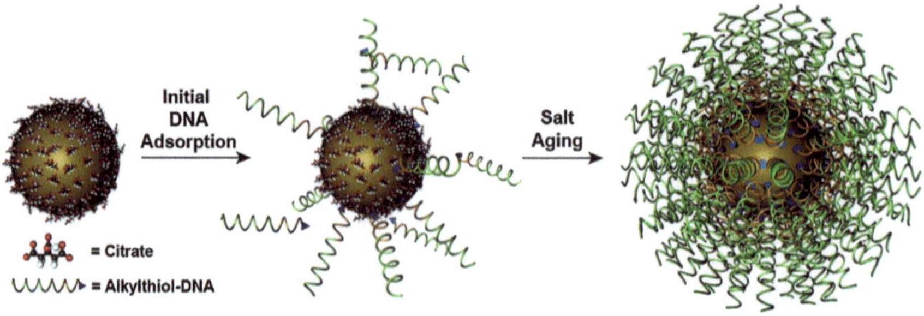

Fig. 2 Synthesis of SNA conjugates. Citrate-stabilized particles are incubated with alkylthiol-functionalized oligonucleotides in water to form a low-density monolayer. By incubating the nanoparticles in aqueous solutions with successively higher concentrations of salt (typically, 0.15–1.0 M), a high-density spherical nucleic acid shell is formed. (Reproduced and modified with permission from [8])

ions and adsorb onto the gold surface. To facilitate the formation of a dense monolayer of oligonucleotides, monovalent counterions (such as sodium ions) need to be slowly added into the mixture to screen the negative charges of the phosphate backbones. One can control the number of strands that adsorb to the gold surface (from 50 to 250 oligonucleotides per 15 nm particle) by varying the amount of added sodium ions from 0.05 to 1.0 M and the spacer type to achieve a dense multivalent nanostructure [31]. We will discuss the importance of tuning the oligonucleotide density in the context of cellular entry as well as endogenous nucleic acid binding in later sections. The three-dimensional architecture of the nucleic acids around the gold core confers some unique and biologically attractive properties, which make SNAs desirable for high sensitivity intracellular detection assays. These properties will be described in the next section.

2.2 Properties that Make SNAs Ideal as Intracellular Diagnostic Probes

Unlike linear nucleic acids, SNAs have been shown to be capable of rapidly entering many cell lines without complexation to carrier moieties, despite their dense polyanionic shell (Fig. 3) [32, 9, 33]. Generally, negatively charged nucleic acids require cationic moieties, such as lipoplexes, peptides, or viruses, to traverse through the negatively charged cellular membrane. These transfection agents can be harmful to cells and result in off-target effects, which may alter the expression of a variety of genes in an uncontrollable manner [34]. In many detection assays, such perturbations are significant enough to result in false-positive signals [34]. One does not encounter this issue when using SNAs, as they do not require additional carriers. The dense, highly oriented array of nucleic acids on the surface of SNAs are recognized by class A scavenger receptors (SR-A), allowing for their uptake into cells in high quantities (on the order of 10^6/cell) as single entities via lipid

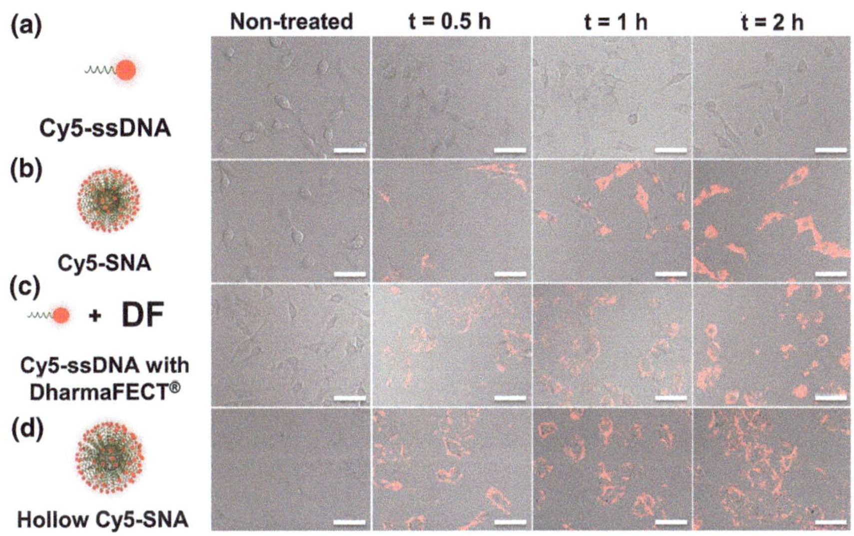

Fig. 3 SNA uptake into cells. The three-dimensional architecture of the SNA allows it to be rapidly endocytosed via scavenger receptor-mediated endocytosis. Linear nucleic acids of the same sequence are unable to enter cells without the use of transfection agents. (Reproduced and modified with permission from [32]

raft-dependent, caveolae-mediated endocytosis [32]. We confirmed that this was the primary mechanism of SNA uptake by demonstrating that uptake was reduced by over 90 % when the SR-A levels of cells that normally took up SNAs in high quantities were depleted via conventional viral vector-based siRNA knockdown [32]. Further, our group found that the cellular uptake of SNAs scales with the density of oligonucleotides adsorbed to the nanoparticle core; a higher number of oligonucleotides loaded on the particle typically increase SNA entry into cells [35].

Upon entering cells, SNAs resist enzymatic degradation by nucleases [10] and treated cells exhibit no apparent toxicity [9] or off-target effects [12]. Each of these properties is necessary for the development of a nucleic acid detection platform that can be used in live cells. Nature has engineered intracellular machinery, such as nucleases to recognize and degrade exogenous DNA and RNA, rendering them unable to perform their intended function to protect cells from foreign organisms. Unlike free oligonucleotides, the SNA's dense nucleic acid shell sterically resists the binding of nucleases to individual oligonucleotides [10]. We have shown that the salt cloud associated with SNAs that screen the negative charges of neighboring strands also inhibits enzyme activity [36]. Cellular cytotoxicity is not observed after 72 h of treatment with SNAs [20], and the whole-genome expression profiles of cells treated with SNAs did not show significant up- or down-regulation of genes compared to untreated cells [37]. However, elevated cytotoxicity was observed with linear nucleic acids of an identical sequence that were transfected into cells at therapeutically necessary doses using a cationic lipoplex (DharmaFECT 1); changes in 427 unintended genes were also observed (Fig. 4).

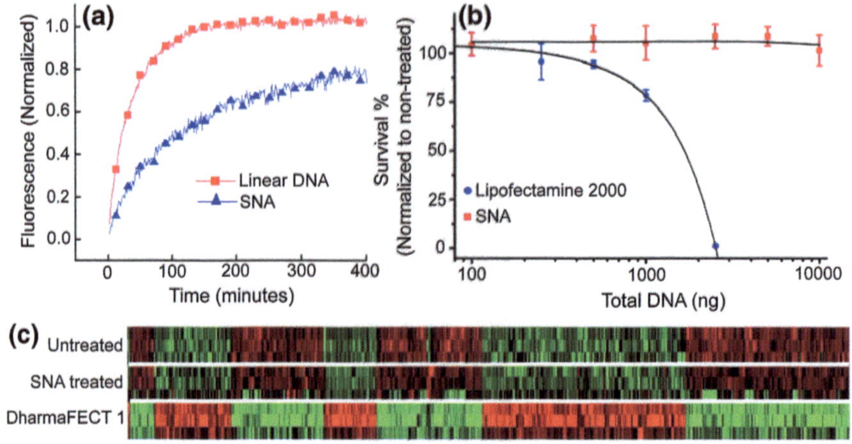

Fig. 4 Intracellular properties of SNAs. Studies demonstrated that SNAs resist enzymatic degradation (**a**), cause little cytotoxicity (**b**), and exhibit limited interference with normal cellular behavior (**c**) compared to linear nucleic acids. (Reproduced and modified with permission from [20, 36, 37])

Once inside cells, an ideal intracellular probe for early stage cancer diagnosis must also be able to detect gene expression at low levels. SNAs have significantly higher binding constants for free complementary oligonucleotide strands than free DNA of the same sequence (over 100 times higher) [13], indicating that these structures are able to be utilized for sensitive diagnostic assays (Nanosphere, Inc., Northbrook, IL). This enhanced binding strength manifests itself as an increase in the melting temperature, T_m, the temperature at which half of the complementary strands in solution are dehybridized, as well as a decrease in the full width at half maximum (FWHM) of the melting transition (Fig. 5). In one proposed mechanism, the enhanced binding strength was found to be directly related to the high density of

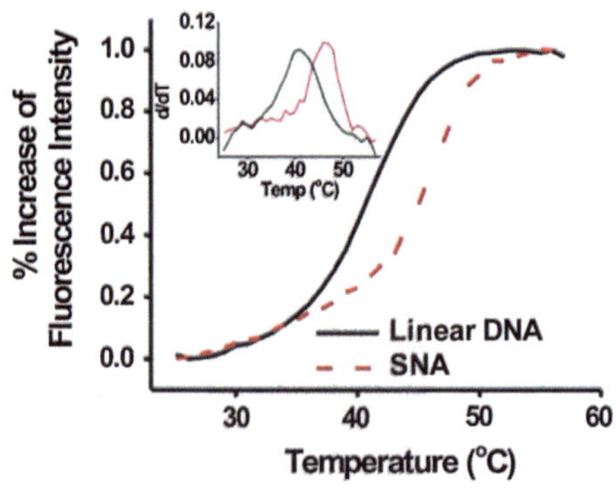

Fig. 5 Complementary strands in solution have a higher binding constant when bound to SNAs versus free linear nucleic acids of the same sequence. (Reproduced with permission from [13])

DNA strands on and high local salt concentration around the nanoparticle surface rather than the absolute amount of bound DNA [38]. Further, the high oligonucleotide density on the SNA restricts the configurational microstates that can be explored by individual strands and increases the binding constant of complementary nucleic acids even more [38]. The increase in binding strength exhibited by the SNAs in comparison to the free molecular probes directly translates into lower limits of detection for target sequences, and it allows for the differentiation of fully complementary and mismatched DNA targets that contain a single-nucleotide polymorphism [39].

3 The Nanoflare: A Platform for Gene Detection and Regulation in Live Cells

The unique characteristics of the SNA have allowed this platform to thrive in numerous applications in biology and medicine. As a therapeutic agent, the nanoconjugate can be delivered intravenously [40] or as a topical agent [37] without toxicity or immunogenicity, enter most cells tested to date efficiently and in high quantities [35], and regulate the expression of targeted genes; they have even been found to cross the blood–brain and blood–tumor barriers [40]. Even so, the SNA platform has shown the most immediate success as a gene detection and diagnostic agent in an architecture known as the nanoflare. The nanoflare is an SNA-gold nanoparticle conjugate functionalized with ssDNA or DNA/LNA hybrid sequences, which are designed to be complementary to genes of interest (Fig. 6). These strands are referred to as the recognition or antisense strands. A short internal complementary strand containing a terminal fluorophore is hybridized onto the antisense strand. When bound in close proximity to the gold particle, the fluorescence of the fluorophore is quenched. However, in the presence of an mRNA target complementary to the antisense sequence, the target forms a longer, more stable duplex and displaces the shorter flare strand. Displacement of the fluorophore from the nanoparticle surface results in a "turn on" of a fluorescence signal proportional to the concentration of the target transcripts. It should be noted that at higher

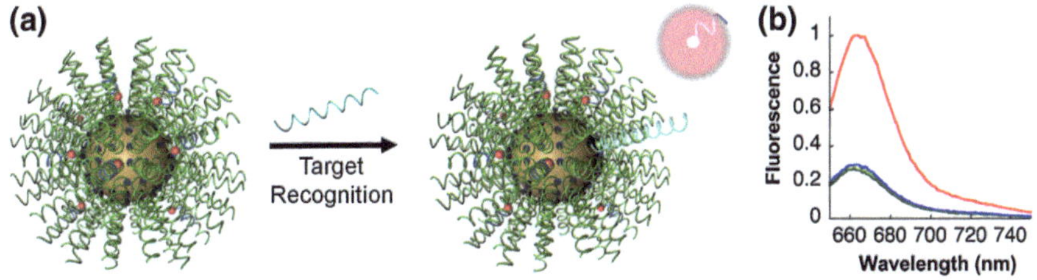

Fig. 6 **a** Schematic representation of sequence-specific recognition of target DNA using nanoflares. **b** Fluorescence response of nanoflares alone (*green*), and nanoflares upon recognition of target DNA (*red*) compared to noncomplementary DNA (*blue*). (Reproduced and modified with permission from [7])

concentrations, nanoflares have also been shown to bind to target mRNA and inhibit translation of the corresponding protein [11]. The unique architecture of the SNA allows the nanoflare to enter live cells and detect mRNA expression without the significant disruption of cellular homeostasis, a result difficult to achieve with other fluorescence-based RNA detection assays, such as those based on fluorescent in situ hybridization (FISH) [41] or molecular beacons [42]. Furthermore, the increased binding thermodynamics and selectivity of complement hybridization exhibited by the SNA enable the nanoflare to attain extremely sensitive and specific detection of polynucleotide targets [43]; so much so that in vitro the nanoflare has been able to distinguish complementary targets from those with single nucleotide polymorphisms (SNPs) [11].

4 Detection of Intracellular Oncogenes

In 2007, the nanoflare was first introduced as a promising platform for mRNA detection in living cells. In a proof-of-concept experiment, nanoflares were utilized to detect intracellular mRNA targets with single-cell resolution and without perturbing cell function. Such results cannot be attained with conventional mRNA quantification techniques, such as real-time polymerase chain reaction (RT-PCR) [44], and the nanoflare enables the profiling of cells based on their genetic content, while preserving their viability for further analysis (*vide infra*). Since the completion of this work, nanoflares have been commercialized by EMD Millipore and sold under the trade name SmartFlare™, with over 1700 genetically unique versions of these constructs currently available in over 230 countries [45].

4.1 Fluorescence Response to Target Oligonucleotides in Extracellular Conditions

To investigate the ability of nanoflares to detect oncogenes in a breast cancer cell model, a type of nanoflare-targeting survivin, an anti-apoptotic gene that is upregulated in a range of cancer types, was designed [7]. Prior to their use in cells, nanoflares were tested in extracellular conditions with synthetic DNA targets to confirm the sequence specificity of the release of the fluorophore-labeled DNA flare strands upon target recognition and binding. Specifically, upon incubation with target DNA, the Cy5 fluorescence of survivin nanoflares was enhanced 3.8-fold. In contrast, the fluorescence signal did not change in the presence of a noncomplementary DNA strand (Fig. 6). Taken together, these results demonstrate that nanoflares efficiently signal the presence of target oligonucleotide in a sequence-specific manner. This capability is vital to the use of nanoflares as intracellular mRNA detection probes.

4.2 Uptake and Fluorescence Response in Cell Culture

Based on their ability to enter cells without additional transfection reagents, nanoflares can be used in cell culture models. This was first demonstrated in SKBR3 cells, a human breast cancer model that expresses high quantities of the survivin mRNA transcript [46]. As a control, noncomplementary nanoflares that did not recognize the survivin transcript were designed to have similar background fluorescence, melting properties, and signaling ability as the targeting probe. Cultured SKBR3 cells were treated with either survivin-targeting or noncomplementary nanoflares, and imaged by confocal microscopy (Fig. 7). A control cell line, C166 mouse endothelial cells [7], that does not express human survivin, was also treated with nanoflares and imaged. The fluorescence of SKBR3 cells treated with survivin nanoflares was significantly higher than those treated with noncomplementary nanoflares. In addition, there was no distinguishable difference in the fluorescence of C166 cells treated with either survivin or noncomplementary nanoflares. These results demonstrate the use of nanoflares to qualitatively profile cells for the expression of an important oncogene, survivin.

In order to quantify the intracellular fluorescence signal from nanoflares, analytical flow cytometry was used. In addition to signal quantification, flow cytometry allows for the collection of data from a larger population of cells than practically attainable using microscopy. Cells that were treated with nanoflares were observed as a uniform population of fluorescent cells, consistent with the high cell penetration (>99 %) that is observed with SNAs (Fig. 7) [10]. The SKBR3 cells treated with survivin-targeting nanoflares were 2.5 times more fluorescent than those treated with noncomplementary nanoflares. In addition, as seen in the microscopy analysis, C166 cells treated with either survivin-targeting or noncomplementary nanoflares displayed low fluorescence. These results demonstrate the nanoflare's capability of distinguishing cancerous cell populations based on the expression of an mRNA target of interest.

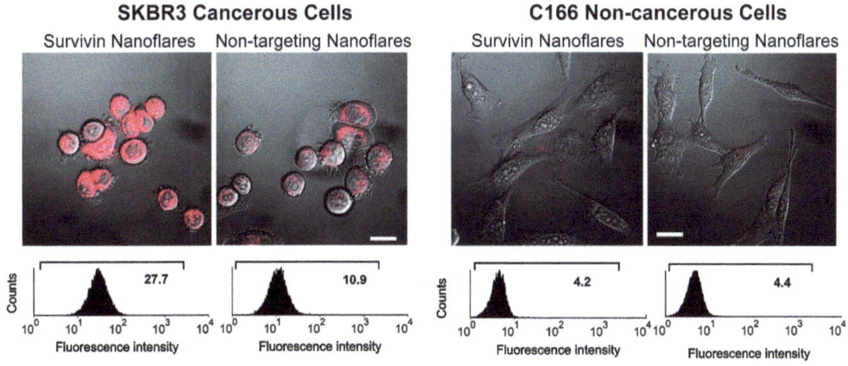

Fig. 7 Fluorescence microscopy (*top*) and flow cytometry data (*bottom*) for SKBR3 (breast cancer, *left two*) and C166 (healthy mouse endothelial, *right two*) cells following treatment with survivin or noncomplementary nanoflares. (Reproduced and modified with permission from [7])

5 Concomitant Detection of Multiple Targets: Multiplexed Nanoflares

Cancers are highly heterogeneous and intricate, and in such disease states, multiple genes can be up- and/or down-regulated concurrently. The original nanoflare architecture only allowed for the detection of a single mRNA's expression level. To target more than one gene, multiple, individual nanoflares would need to be designed and synthesized, significantly increasing the complexity of the system. Thus, to make the nanoflare more relevant as a diagnostic probe for diseases like cancer, its architecture was modified such that a single probe could be used to detect multiple mRNA expression profiles simultaneously. This modified architecture has been termed the multiplexed nanoflare. While our group pioneered the idea of the multiplexed nanoflare by showing that a single probe could be used to identify two independent mRNA expression profiles [47], other groups have since shown that three and four transcripts can also be detected simultaneously [48, 49].

While the original nanoflares were made by anchoring a single type of recognition strand to the gold core, the multiplexed nanoflares are synthesized by functionalizing multiple antisense sequences onto the same particle. As a proof-

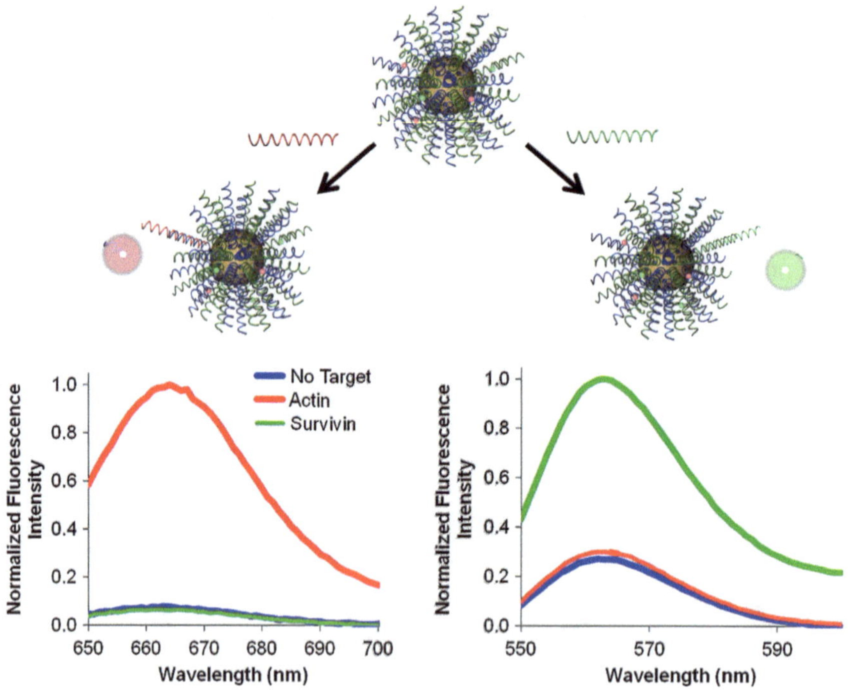

Fig. 8 The addition of DNA targets complementary to each flare (*green* and *red*) to multiplexed nanoflares, individually, caused the increase in fluorescence intensity of each flare on the Cy5 and Cy3 channels, respectively. (Reproduced and modified with permission from [47])

of-concept, our group developed multiplexed nanoflares that can recognize actin and survivin mRNA sequences (Fig. 8). Specifically, equimolar concentrations of actin and survivin recognition sequences were functionalized onto the particles and then hybridized with flare strands bearing Cy3 and Cy5 fluorophores, respectively. Therefore, in the presence of survivin mRNA, the probes are designed to release the Cy3 flares, and in the presence of actin, the probes are designed to release the Cy5 flares. To test the specificity of flare release, actin and survivin target DNA were incubated with the multiplexed nanoflares for 1 h at 37 °C, and Cy3 and Cy5 emission fluorescence spectra were collected. The Cy5 channel showed an increase in fluorescence only in the presence of actin target DNA, and the Cy3 channel showed an increase only in the presence of the survivin target, confirming that the flares are released in a sequence-specific manner (Fig. 8).

5.1 Confocal Measurement of Survivin and Actin Gene Expression in Live Cells

To determine if these constructs could distinguish the cellular expression levels of actin and survivin simultaneously, multiplexed nanoflares were added to HeLa cells, a cell line known to contain high levels of both actin and survivin, that were pretreated with either actin siRNA or survivin siRNA (Fig. 9). This way, one set of cells would have a high level of survivin mRNA present and low level of actin mRNA, while the other set of cells would have the opposite ratio. A decrease in Cy3 fluorescence (Fig. 9 red, left panel) was seen when the survivin expression was knocked down using survivin-targeted siRNA as measured by flow cytometry and confocal microscopy; the fluorescence in the Cy5 actin (Fig. 9 green, left panel) channel remained constant. A decrease in Cy5 fluorescence intensity (Fig. 9 green, right panel) was observed when the cells were treated with actin-targeted siRNA and multiplexed nanoflares, a significant change was not seen in the Cy3 survivin (Fig. 9 red, right panel) channel. These data demonstrated that the multiplexed

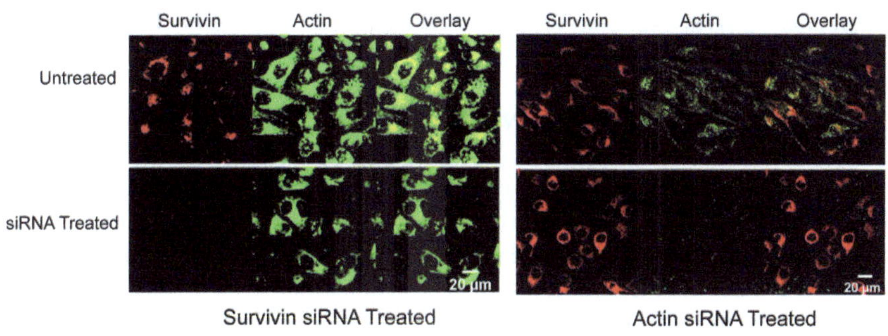

Fig. 9 The fluorescence intensity of cells treated with multiplexed nanoflares changes as a function of intracellular gene expression, as seen by confocal microscopy. (Reproduced and modified with permission from [47])

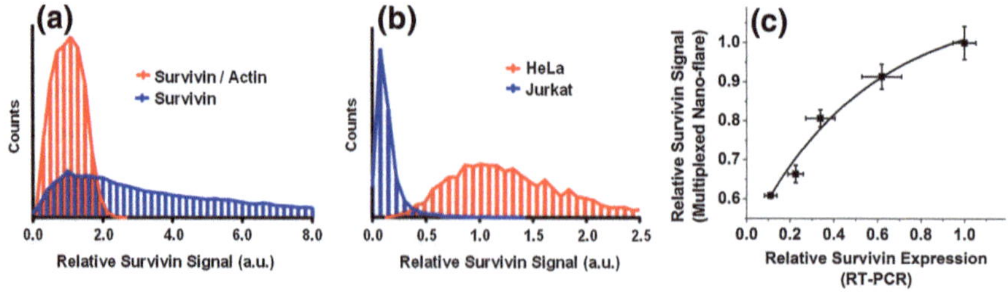

Fig. 10 Quantitative and qualitative determination of mRNA expression in live cells. **a** The detection of intracellular survivin mRNA was performed with traditional nanoflares and multiplexed nanoflares. The ratio of survivin to actin was expressed for each system. The cell-associated fluorescence intensities were measured in HeLa cells using flow cytometry and presented as histograms. **b** In a similar experiment, the multiplexed nanoflares were used to compare the ratio of survivin to actin expression in HeLa and Jurkat cells. **c** The expression of survivin mRNA was measured as a function of survivin siRNA delivered by both the multiplexed nanoflare and RT-PCR. (Reproduced and modified with permission from [47])

nanoflares were able to enter live cells and release two different types of flares with a high level of specificity.

5.2 Increase in Signal-to-Noise Ratio

Due to cell-to-cell variations in nanoflare uptake, the fluorescence intensities of different cells treated with the same single-target nanoflare probes typically have a wide distribution. However, when the expression of a widely abundant control gene is also monitored as a reference gene in the same cell through the use of multiplexed nanoflares, the extent of cell-to-cell variability can be compensated for. Specifically, when the fluorescence of survivin-Cy3 was normalized to that of actin-Cy5, the standard deviation in the fluorescence signal as monitored by flow cytometry decreased tenfold (Fig. 10a). This increase in the signal-to-noise ratio considerably improves the sensitivity of multiplexed nanoflares, and it has allowed for cells with differential survivin expression to be sorted using flow cytometry (Fig. 10b).

5.3 Quantification of mRNA Expression Using Multiplexed Nanoflares

The sensitivity of multiplexed nanoflares can also be applied to quantify relative mRNA expression. Multiplexed nanoflares were applied to cells with decreasing amounts of survivin expression (as a result of transfecting increasing amounts of survivin-targeting siRNA). A decrease in relative survivin expression, as determined by RT-PCR, strongly correlates with decreasing nanoflare fluorescence as determined by flow cytometry analysis (Fig. 10c). This study enforces the idea that multiplexed nanoflares can quantitatively determine relative gene expression inside live cells, while also allowing one to differentiate between cell populations based on their genetic profiles.

6 Detection and Isolation of Live Circulating Tumor Cells from Whole Blood

Since nanoflares exhibit high cellular uptake, enhanced resistance toward nuclease degradation, and do not cause observable cytotoxicity, they are uniquely suited for the detection and isolation of circulating tumor cells (CTCs). Alternative methods to detect CTCs rely on the recognition of certain cell surface proteins, including EpCAM, thus they are ineffective in cell populations that do not express significant levels of such proteins [50, 51]. In addition, these methods are not able to distinguish single-cell genetic profiles amongst heterogeneous cancer cell populations, limiting their ability to accurately predict metastatic potential. To overcome these challenges, nanoflares were designed to target markers of the epithelial-to-mesenchymal transition (EMT), an integral part of cancer metastasis, and were used for the capture of live circulating breast cancer cells [52]. This approach provides an unprecedented opportunity to isolate cancer stem cells based on the presence of genetic markers and may improve cancer diagnosis and prognosis.

6.1 Nanoflares Targeting Markers of the Epithelial-to-Mesenchymal Transition

Vimentin, an intermediate filament protein, and fibronectin, an extracellular matrix protein, are often expressed in cancerous cells undergoing the EMT. Thus, nanoflares were designed to target these two oncogenes, and then tested in a model metastatic breast cancer cell line, MDA-MB-231, using analytical flow cytometry. As a control, healthy epithelial mammary cells, HMLE, were used. Low fluorescence was observed in HMLE cells treated with the noncomplementary control, vimentin, and fibronectin nanoflares. In contrast, the fibronectin nanoflare-treated MDA-MB-231 cells were 6 times more fluorescent and the vimentin nanoflare-treated MDA-MB-231 cells were 8 times more fluorescent compared to those treated with the noncomplementary control (Fig. 11).

6.2 Recovery Yield of Model Circulating Tumor Cells Isolated from Whole Blood

Based on the ability of vimentin and fibronectin nanoflares to distinguish epithelial cells from metastatic cancer cells, we next sought to identify and isolate circulating tumor cells. mCherry cDNA was expressed in MDA-MB-231 cells to provide an orthogonal method of tracking the cells, and a known number of these cells was spiked into human whole blood from a healthy volunteer. Samples of this blood were

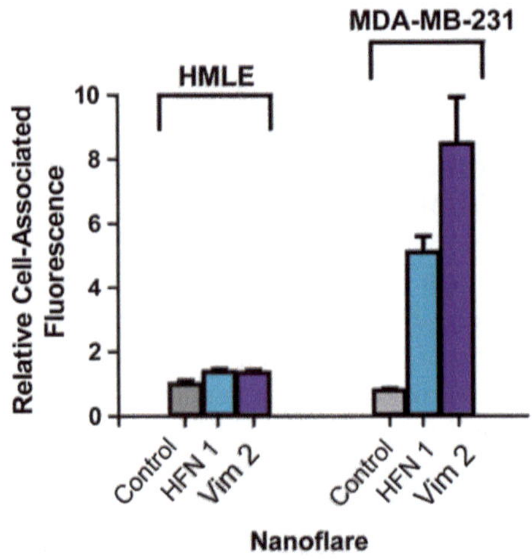

Fig. 11 Vimentin-targeting and fibronectin-targeting nanoflares tested in noncancerous epithelial mammary cells, HMLE, and metastatic breast cancer cells, MDA-MB-231. (Reproduced with permission from [52])

treated with vimentin, fibronectin, or noncomplementary control nanoflares. Red blood cells and peripheral blood mononuclear cells were depleted using a CD45 immunomagnetic separation and a Ficoll gradient, and the samples were resuspended in a cell culture medium (Fig. 12). The samples were then analyzed by flow

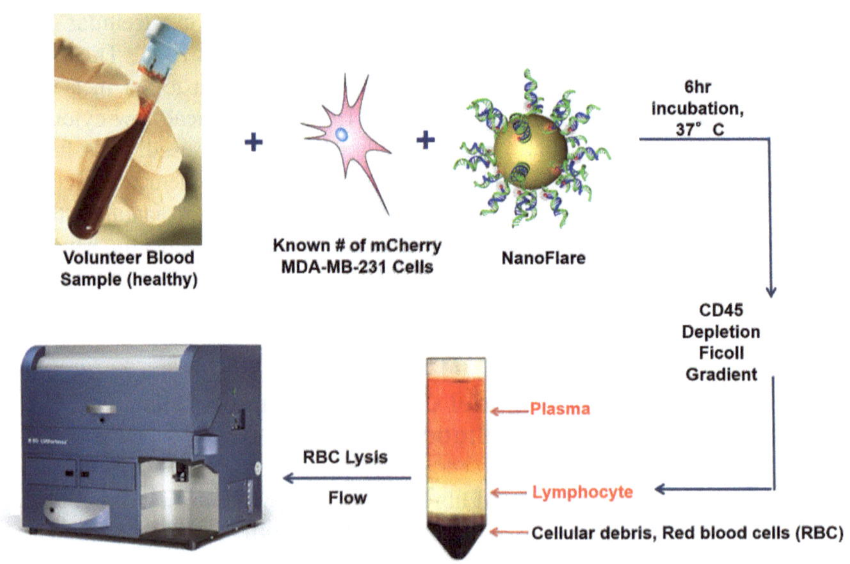

Fig. 12 Schematic representation of isolation of model circulating tumor cells doped into human whole blood

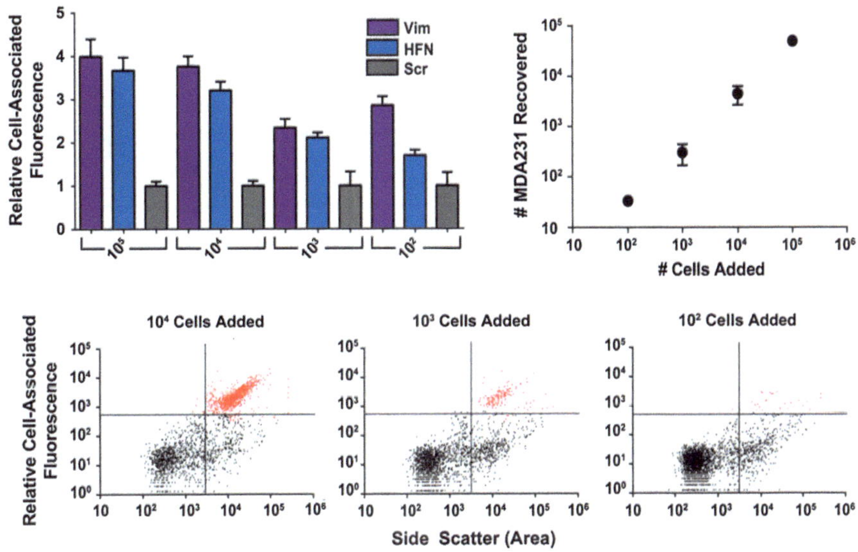

Fig. 13 Recovery of model circulating tumor cells doped into human whole blood. (Reproduced with permission from [52])

cytometry for both mCherry and Cy5 nanoflare fluorescence to determine the recovery yield of MDA-MB-231 cells that were doped into the blood samples (Fig. 13). In samples treated with the vimentin or fibronectin nanoflares, over 99 % of cells showing high mCherry fluorescence also showed a strong nanoflare fluorescence signal. Additionally, the vimentin nanoflare and fibronectin nanoflare provided a 4- and 3.5-fold fluorescent enhancement, respectively, over the non-complementary scrambled nanoflare. On average, treatment with nanoflares yielded a recovery rate of approximately 68 ± 14 % of the mCherry MDA-MB-231 cells that were added, which is consistent with commercially available methods of circulating tumor cell isolation that employ a densitometric enrichment step, such as a Ficoll gradient [53]. Importantly, nanoflares were able to detect as few as 100 mCherry MDA-MB-231 cells that were doped into the whole blood samples. These results demonstrate the ability of the nanoflares to survey the metastatic potential of cells in the blood stream, and nanoflares provide a means for clinicians to track therapeutic efficacy throughout a treatment plan on an individual patient basis.

6.3 Isolation and Characterization of Circulating Tumor Cells from a Mouse Xenograft Model of Human Metastatic Breast Cancer

An orthotopic model of triple-negative breast cancer with widespread metastases to several organs was used to test the ability of the nanoflare to detect circulating tumor cells in a murine model. Blood samples were obtained 6 weeks following

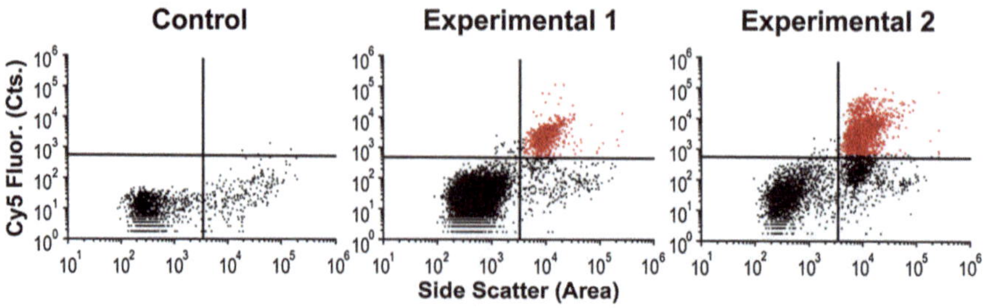

Fig. 14 Isolation of circulating tumor cells from an orthotopic mouse model of triple-negative breast cancer. The cancerous MDA-MB-231 cells are represented by the *red dots* and the noncancerous cells are represented by the *black dots*. (Reproduced with permission from [52])

tumor inoculation with mCherry-labeled MDA-MB-231 cells and were treated with vimentin, fibronectin, or noncomplementary scrambled nanoflares. Samples were processed to remove red blood cells and peripheral blood mononuclear cells and analyzed for mCherry and nanoflare fluorescence using flow cytometry (Fig. 14). Samples treated with vimentin or fibronectin nanoflares exhibited high mCherry fluorescence and also showed a significant nanoflare fluorescence signal, with greater than 90 % of the mCherry MDA-MB-231 cells also exhibiting strong nanoflare fluorescence. The vimentin nanoflare and fibronectin nanoflare showed a 1.5- and 1.25-fold fluorescent enhancement relative to a noncomplementary scrambled nanoflare, respectively. Based on this, the nanoflares were able to isolate circulating mCherry MDA-MB-231 cells derived from an orthotopic murine model of metastatic triple-negative breast cancer based on genetic markers, reinforcing the potential use of nanoflares in cancer diagnostics and patient prognostics. This advance may be valuable in the design of better cancer treatment plans that are adjusted after observing the changes in the metastatic potential of cells from patient blood in response to therapy.

6.4 Isolation and Characterization of Seeded Recurrent Cells in Human Blood

The nontoxic nature of the nanoflare provides a unique opportunity to isolate live circulating tumor cells, which could allow further analysis of the cancer population. To investigate this capability, cells were doped into human blood, isolated using nanoflares, and further cultured. A GFP-expressing recurrent cell line isolated from tumors in a HER2 mouse model of breast cancer was chosen. Cells retrieved using nanoflares were GFP positive, and formed mammospheres (Fig. 15), spherical clusters formed by cancer stem cells, indicating that cells originally spiked into the blood sample were successfully retrieved and that they remained viable after isolation. These results suggest that it may also be possible to isolate and further culture live circulating tumor cells from human patients, providing the opportunity to study cancer cell homogeneity and its relation to patient outcomes.

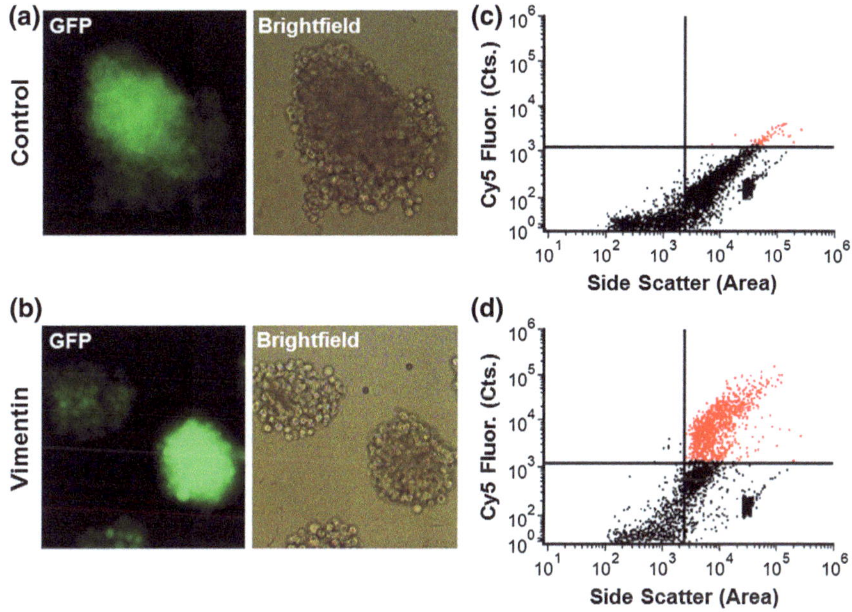

Fig. 15 **a**, **b** Culture of recovered recurrent HER2 tumor cells isolated from whole blood resulted in the formation of mammospheres. **c**, **d** These cells were isolated using either a noncomplementary control nanoflare or a vimentin-targeting nanoflare, with nanoflare fluorescence significantly enhanced for samples treated with vimentin nanoflares. (Reproduced with permission from [52])

7 Conclusion

This chapter highlights the remarkable progress that has been made in the application and translation of the SNA nanoconstruct and the SNA-based nanoflare to research and clinical settings. In many ways, the nanoflare is an ideal probe, enabling the highly sensitive and selective intracellular detection of target genes with little undesired impact on the cellular environment. Thus, they offer tremendous potential in cancer research and treatment, among other fields. Indeed, in the seven short years since their inception, nanoflares have formed the basis for the founding of one company, AuraSense (Skokie, IL) and its commercialization of thousands of SmartFlare™ constructs (through partnership with EMD Millipore) for a variety of research purposes. Although herein the nanoflare's usage in animal and human samples involving breast cancer was emphasized, because of the tailorability of the SNA structure and its unique chemical and physical properties, such nanostructures could theoretically be used to detect and analyze any disease with a genetic basis, including most forms of cancer. Indeed, the discovery of nanoflares and their subsequent development for use in such applications have marked a paradigm shift from detecting protein markers on CTCs to nucleic acid markers within CTCs. Future work centers on continuing to explore this possibility, carrying the nanoflare through to clinical trials and using the construct to elucidate fundamental aspects of cancer biology and genetics. Our hope is that the rapid implementation of the nanoflare architecture enables the lives of many to be improved or saved.

References

1. Edwards BK, Noone A-M, Mariotto AB, Simard EP, Boscoe FP, Henley SJ, Jemal A, Cho H, Anderson RN, Kohler BA, Eheman CR, Ward EM (2014) Annual Report to the Nation on the status of cancer, 1975–2010, featuring prevalence of comorbidity and impact on survival among persons with lung, colorectal, breast, or prostate cancer. Cancer 120:1290–1314
2. Almendro V, Marusyk A, Polyak K (2013) Cellular heterogeneity and molecular evolution in cancer. Annu Rev Pathol 8:277–302
3. Burrell RA, McGranahan N, Bartek J, Swanton C (2013) The causes and consequences of genetic heterogeneity in cancer evolution. Nature 501:338–345
4. Meacham CE, Morrison SJ (2013) Tumour heterogeneity and cancer cell plasticity. Nature 501:328–337
5. Briley WE, Halo TL, Randeria PS, Alhasan AH, Auyeung E, Hurst SJ, Mirkin CA (2012) Biochemistry and biomedical applications of spherical nucleic acids (SNAs). In: Nanomaterials for biomedicine. American Chemical Society, Washington, DC
6. Elsabahy M, Wooley KL (2012) Design of polymeric nanoparticles for biomedical delivery applications. Chem Soc Rev 41:2545–2561
7. Seferos DS, Giljohann DA, Hill HD, Prigodich AE, Mirkin CA (2007) Nano-flares: probes for transfection and mRNA detection in living cells. J Am Chem Soc 129:15477–15479
8. Cutler JI, Auyeung E, Mirkin CA (2012) Spherical nucleic acids. J Am Chem Soc 134:1376–1391
9. Giljohann DA, Seferos DS, Daniel WL, Massich MD, Patel PC, Mirkin CA (2010) Gold nanoparticles for biology and medicine. Angew Chem Int Ed 49:3280–3294
10. Rosi NL, Giljohann DA, Thaxton CS, Lytton-Jean AKR, Han MS, Mirkin CA (2006) Oligonucleotide-modified gold nanoparticles for intracellular gene regulation. Science 312:1027–1030
11. Prigodich AE, Seferos DS, Massich MD, Giljohann DA, Lane BC, Mirkin CA (2009) Nano-flares for mRNA regulation and detection. ACS Nano 3:2147–2152
12. Massich MD, Giljohann DA, Seferos DS, Ludlow LE, Horvath CM, Mirkin CA (2009) Regulating immune response using polyvalent nucleic acid—gold nanoparticle conjugates. Mol Pharm 6:1934–1940
13. Lytton-Jean AKR, Mirkin CA (2005) A thermodynamic investigation into the binding properties of DNA functionalized gold nanoparticle probes and molecular fluorophore probes. J Am Chem Soc 127:12754–12755
14. Lee J-S, Lytton-Jean AKR, Hurst SJ, Mirkin CA (2007) Silver nanoparticle—oligonucleotide conjugates based on DNA with triple cyclic disulfide moieties. Nano Lett 7:2112–2115
15. Mirkin CA, Letsinger RL, Mucic RC, Storhoff JJ (1996) A DNA-based method for rationally assembling nanoparticles into macroscopic materials. Nature 382:607–609
16. Cutler JI, Zheng D, Xu X, Giljohann DA, Mirkin CA (2010) Polyvalent oligonucleotide iron oxide nanoparticle "Click" conjugates. Nano Lett 10:1477–1480
17. Zhang C, Macfarlane RJ, Young KL, Choi CHJ, Hao L, Auyeung E, Liu G, Zhou X, Mirkin CA (2013) A general approach to DNA-programmable atom equivalents. Nat Mater 12:741–746

18. Young KL, Scott AW, Hao L, Mirkin SE, Liu G, Mirkin CA (2012) Hollow spherical nucleic acids for intracellular gene regulation based upon biocompatible silica shells. Nano Lett 12:3867–3871
19. Calabrese CM, Merkel TJ, Briley WE, Randeria PS, Narayan SP, Rouge JL, Walker DA, Scott AW, Mirkin CA (2014) Biocompatible infinite-coordination-polymer nanoparticle-nucleic-acid conjugates for antisense gene regulation. Angew Chem Int Ed Engl 54(2):476–480
20. Cutler JI, Zhang K, Zheng D, Auyeung E, Prigodich AE, Mirkin CA (2011) Polyvalent nucleic acid nanostructures. J Am Chem Soc 133:9254–9257
21. Banga RJ, Chernyak N, Narayan SP, Nguyen ST, Mirkin CA (2014) Liposomal spherical nucleic acids. J Am Chem Soc 136:9866–9869
22. Frens G (1973) Controlled nucleation for regulation of particle-size in monodisperse gold suspensions. Nat Phys Sci 241:20–22
23. Brust M, Schiffrin DJ, Bethell D, Kiely CJ (1995) Novel gold-dithiol nano-networks with non-metallic electronic properties. Adv Mater 7:795–797
24. Letsinger RL, Elghanian R, Viswanadham G, Mirkin CA (2000) Use of a steroid cyclic disulfide anchor in constructing gold nanoparticle—oligonucleotide conjugates. Bioconjug Chem 11:289–291
25. Giljohann DA, Seferos DS, Prigodich AE, Patel PC, Mirkin CA (2009) Gene regulation with polyvalent siRNA—nanoparticle conjugates. J Am Chem Soc 131:2072–2073
26. Hao L, Patel PC, Alhasan AH, Giljohann DA, Mirkin CA (2011) Nucleic acid-gold nanoparticle conjugates as mimics of microRNA. Small 7:3158–3162
27. Rouge JL, Hao L, Wu XA, Briley WE, Mirkin CA (2014) Spherical nucleic acids as a divergent platform for synthesizing RNA–nanoparticle conjugates through enzymatic ligation. ACS Nano 8:8837–8843
28. Patel PC, Giljohann DA, Seferos DS, Mirkin CA (2008) Peptide antisense nanoparticles. Proc Natl Acad Sci 105:17222–17226
29. Seferos DS, Giljohann DA, Rosi NL, Mirkin CA (2007) Locked nucleic acid-nanoparticle conjugates. ChemBioChem 8:1230–1232
30. Hill HD, Millstone JE, Banholzer MJ, Mirkin CA (2009) The role radius of curvature plays in thiolated oligonucleotide loading on gold nanoparticles. ACS Nano 3:418–424
31. Hurst SJ, Lytton-Jean AKR, Mirkin CA (2006) Maximizing DNA loading on a range of gold nanoparticle sizes. Anal Chem 78:8313–8318
32. Choi CHJ, Hao L, Narayan SP, Auyeung E, Mirkin CA (2013) Mechanism for the endocytosis of spherical nucleic acid nanoparticle conjugates. Proc Natl Acad Sci 110:7625–7630
33. Patel PC, Giljohann DA, Daniel WL, Zheng D, Prigodich AE, Mirkin CA (2010) Scavenger receptors mediate cellular uptake of polyvalent oligonucleotide-functionalized gold nanoparticles. Bioconjug Chem 21:2250–2256
34. Nguyen J, Szoka FC (2012) Nucleic acid delivery: the missing pieces of the puzzle? Acc Chem Res 45:1153–1162
35. Giljohann DA, Seferos DS, Patel PC, Millstone JE, Rosi NL, Mirkin CA (2007) Oligonucleotide loading determines cellular uptake of DNA-modified gold nanoparticles. Nano Lett 7:3818–3821
36. Seferos DS, Prigodich AE, Giljohann DA, Patel PC, Mirkin CA (2008) Polyvalent DNA nanoparticle conjugates stabilize nucleic acids. Nano Lett 9:308–311
37. Zheng D, Giljohann DA, Chen DL, Massich MD, Wang X-Q, Iordanov H, Mirkin CA, Paller AS (2012) Topical delivery of siRNA-based spherical nucleic acid nanoparticle conjugates for gene regulation. Proc Natl Acad Sci 109:11975–11980
38. Long H, Kudlay A, Schatz GC (2006) Molecular dynamics studies of ion distributions for DNA duplexes and DNA clusters: salt effects and connection to DNA melting. J Phys Chem B 110:2918–2926
39. Elghanian R, Storhoff JJ, Mucic RC, Letsinger RL, Mirkin CA (1997) Selective colorimetric detection of polynucleotides based on the distance-dependent optical properties of gold nanoparticles. Science 277:1078–1081

40. Jensen SA, Day ES, Ko CH, Hurley LA, Luciano JP, Kouri FM, Merkel TJ, Luthi AJ, Patel PC, Cutler JI, Daniel WL, Scott AW, Rotz MW, Meade TJ, Giljohann DA, Mirkin CA, Stegh AH (2013) Spherical nucleic acid nanoparticle conjugates as an RNAi-based therapy for glioblastoma. Sci Transl Med 5:209ra152
41. Gall JG, Pardue ML (1969) Formation and detection of RNA-DNA hybrid molecules in cytological preparations. Proc Natl Acad Sci USA 63:378–383
42. Santangelo PJ, Nix B, Tsourkas A, Bao G (2004) Dual FRET molecular beacons for mRNA detection in living cells. Nucleic Acids Res 32:e57
43. Prigodich AE, Lee O-S, Daniel WL, Seferos DS, Schatz GC, Mirkin CA (2010) Tailoring DNA structure to increase target hybridization kinetics on surfaces. J Am Chem Soc 132:16296
44. Nolan T, Hands RE, Bustin SA (2006) Quantification of mRNA using real-time RT-PCR. Nat Protoc 1:1559–1582
45. EMD Millipore (2014) SmartFlare RNA detection probes. http://www.emdmillipore.com/smartflare
46. Peng X-H, Cao Z-H, Xia J-T, Carlson GW, Lewis MM, Wood WC, Yang L (2005) Real-time detection of gene expression in cancer cells using molecular beacon imaging: new strategies for cancer research. Cancer Res 65:1909–1917
47. Prigodich AE, Randeria PS, Briley WE, Kim NJ, Daniel WL, Giljohann DA, Mirkin CA (2012) Multiplexed nanoflares: mRNA detection in live cells. Anal Chem 84:2062–2066
48. Li N, Chang C, Pan W, Tang B (2012) A multicolor nanoprobe for detection and imaging of tumor-related mRNAs in living cells. Angew Chem Int Ed 51:7426–7430
49. Pan W, Zhang T, Yang H, Diao W, Li N, Tang B (2013) Multiplexed detection and imaging of intracellular mRNAs using a four-color nanoprobe. Anal Chem 85:10581–10588
50. Riethdorf S, Fritsche H, Müller V, Rau T, Schindlbeck C, Rack B, Janni W, Coith C, Beck K, Jänicke F, Jackson S, Gornet T, Cristofanilli M, Pantel K (2007) Detection of circulating tumor cells in peripheral blood of patients with metastatic breast cancer: a validation study of the cell search system. Clin Cancer Res 13:920–928
51. Yu M, Stott S, Toner M, Maheswaran S, Haber DA (2011) Circulating tumor cells: approaches to isolation and characterization. J Cell Biol 192:373–382
52. Halo TL, McMahon KM, Angeloni NL, Xu Y, Wang W, Chinen AB, Malin D, Strekalova E, Cryns VL, Cheng C, Mirkin CA, Thaxton CS (2014) NanoFlares for the detection, isolation, and culture of live tumor cells from human blood. Proc Natl Acad Sci USA 111:17104–17109
53. Noble PB, Cutts JH (1967) Separation of blood leukocytes by Ficoll gradient. Can Vet J 8:110–111

Therapeutic Applications of Spherical Nucleic Acids

Stacey N. Barnaby, Timothy L. Sita, Sarah Hurst Petrosko, Alexander H. Stegh and Chad A. Mirkin

Abstract

Spherical nucleic acids (SNAs) represent an emerging class of nanoparticle-based therapeutics. SNAs consist of densely functionalized and highly oriented oligonucleotides on the surface of a nanoparticle which can either be inorganic (such as gold or platinum) or hollow (such as liposomal or silica-based). The spherical architecture of the oligonucleotide shell confers unique advantages over traditional nucleic acid delivery methods, including entry into nearly all cells independent of transfection agents and resistance to nuclease degradation. Furthermore, SNAs can penetrate biological barriers, including the blood–brain and blood–tumor barriers as well as the epidermis, and have demonstrated

Stacey N. Barnaby and Timothy L. Sita contributed equally to this work.

S.N. Barnaby · S.H. Petrosko · C.A. Mirkin (✉)
Department of Chemistry, Northwestern University,
2145 Sheridan Road, Evanston, IL 60208, USA
e-mail: chadnano@northwestern.edu

S.N. Barnaby · T.L. Sita · S.H. Petrosko · A.H. Stegh (✉) · C.A. Mirkin
International Institute for Nanotechnology, Northwestern University,
2145 Sheridan Road, Evanston, IL 60208, USA
e-mail: a-stegh@northwestern.edu

T.L. Sita · C.A. Mirkin
Interdepartmental Biological Sciences Program, Northwestern University,
2205 Tech Drive, Evanston, IL 60208, USA

A.H. Stegh
Ken and Ruth Davee Department of Neurology, The Northwestern Brain Tumor Institute,
the Robert H. Lurie Comprehensive Cancer Center, Northwestern University,
303 East Superior, Chicago, IL 60611, USA

C.A. Mirkin et al. (eds.), *Nanotechnology-Based Precision Tools for the Detection and Treatment of Cancer*, Cancer Treatment and Research 166,
DOI 10.1007/978-3-319-16555-4_2

efficacy in several murine disease models in the absence of significant adverse side effects. In this chapter, we will focus on the applications of SNAs in cancer therapy as well as discuss multimodal SNAs for drug delivery and imaging.

Keywords
Spherical nucleic acids · SNAs · siRNA · Nanoparticles · Cancer · Therapeutics

Contents

1 Introduction

Nucleic acid-based therapeutic agents typically consist of short strands of oligonucleotides that are capable of performing gene regulatory functions. The DNA and RNA comprising such therapeutics can be customized to selectively silence any mutated or deregulated genes, thus offering tremendous potential as tools for precision medicine, where patient-specific treatments are designed to address the genetic basis underlying an individual patient's disease [1, 2].

Nucleic acid-based therapeutics can be primarily divided into two categories: those that are comprised of double-stranded RNA (dsRNA) molecules that function via the RNA interference (RNAi) pathway and those that consist of single-stranded DNA molecules that act as antisense oligonucleotides (ASO). Both types can interfere with mRNA molecules to silence protein expression, and many of these structures have been extensively investigated in clinical trials for cancer, hereditary disorders, heart disease, inflammatory conditions, and viral infections [3–6]. The mechanisms of action of these two types of oligonucleotide-based therapeutics differ as described briefly below (for additional information see [7] (for RNAi-based gene silencing) and [8] (for ASO-based silencing)).

1.1 Antisense Oligonucleotides (ASOs)

ASOs can inhibit protein translation via either a steric blockade of translation or the recruitment of the endonuclease RNase H [8, 9]. The former method involves sequence-specific binding of ASOs to target mRNA in the cytoplasm, thereby preventing ribosomal translation of the mRNA (Fig. 1a). The latter mechanism involves RNase H-dependent cleavage, in which RNase H recognizes an RNA–DNA heteroduplex, selectively cleaves the RNA strand, and releases the intact DNA strand [10]. The intact DNA strand can then engage additional target mRNAs and recruit RNase H, enhancing its potency.

1.2 The RNA Interference (RNAi) Pathway

Fire and Mello first reported the discovery of RNAi-based gene silencing in *Caenorhabditis elegans* in 1998. This discovery later earned them the Nobel Prize in Physiology or Medicine in 2006 [11]. The RNAi pathway was then established in mammalian cells by Tuschl and colleagues, prompting rigorous development of RNAi-based therapeutics to battle diseases previously considered "undruggable" by traditional pharmaceutical approaches (i.e., small molecules and biotherapeutic antibodies) [12–15]. The RNAi pathway is activated by the presence of dsRNA in the cytoplasm (Fig. 1b) [16–18]. Dicer, a cytoplasmic endoribonuclease, cleaves longer dsRNA into small interfering RNA (siRNA) or microRNA (miRNA) segments, which are typically 21–23 nucleotides in length. The "antisense" or "guide" strands of siRNA or miRNA segments are then recognized and loaded into the RNA-induced

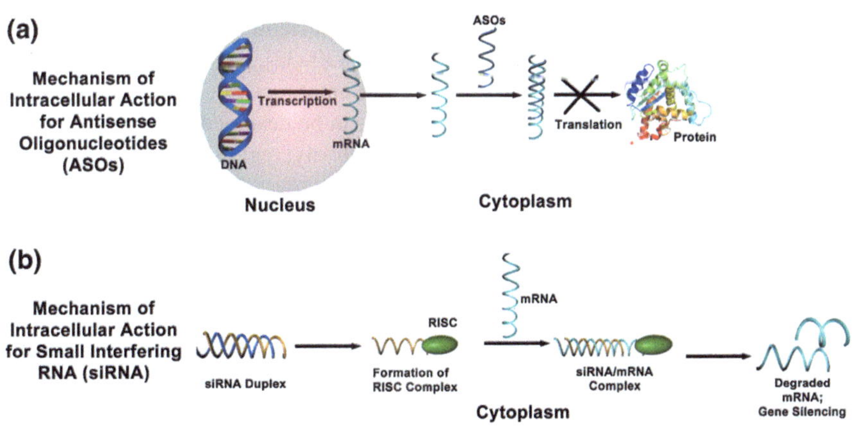

Fig. 1 Different mechanisms of intracellular action of antisense oligonucleotides (ASOs) and small interfering RNAs (siRNAs). **a** Mechanism of action for ASOs, which bind complementary mRNA and cause inhibition of translation or recruit RNase H to cleave the RNA moiety within an RNA–DNA duplex; **b** Mechanism of action for siRNAs, which includes formation of the RNA-induced silencing complex (RISC) and subsequent degradation of target mRNA

silencing complex (RISC), while the "sense" or "passenger strands" are degraded. This activates the RISC complex, leading to Watson-Crick base pairing of complementary target mRNA. Once the target mRNA is bound, its expression can be modified by distinct mechanisms, depending on the biological context. In siRNA-based RNAi, Argonaute 2 (Ago2, an RNA endonuclease in the RISC complex) subsequently cleaves the target mRNA, thereby inhibiting its translation. The RISC complex is recycled and thus cleaves mRNA continuously, resulting in persistent knockdown lasting between 3 and 7 days in dividing cells and up to 3–4 weeks in nondividing cells [19]. In miRNA-based RNAi, multiple mechanisms of silencing are possible, including repression of protein translation and/or deadenylation of target mRNA subsequently leading to its degradation [7, 20].

1.3 Challenges for Oligonucleotide Drug Delivery

Despite these cellular mechanisms that allow for highly specific therapeutic manipulation of genetic expression, a number of barriers to effective delivery are encountered when nucleic acids are systemically injected, limiting their utility in vivo. Unmodified oligonucleotides experience rapid renal clearance, are subject to cleavage by RNases and DNases in serum, and display inefficient uptake by target tissues. Additionally, unmodified oligonucleotides do not efficiently cross the cell membranes and have been shown to trigger a cellular immune response [21–23]. These in vivo barriers to oligonucleotide delivery have slowed the translation of nucleic acid-based therapeutics to the clinic and mandated the use of oligonucleotide-carrier systems, such as polymers/polyplexes [24, 25], dendrimers [26], and lipids [27, 28]. Notably, each of these systems has their own safety concerns and delivery limitations [29]. Nanomaterial-based systems have emerged as potential therapeutic agents for delivering oligonucleotides to cells, and early experiments have shown their great promise [30–32]. Among nanomaterials, spherical nucleic acids (SNAs; Fig. 2) represent an attractive class of single-entity agents, where therapeutic oligonucleotides that can be designed and synthesized to function through either the RNAi or antisense pathway; and these structures are the focus of this chapter.

1.4 Spherical Nucleic Acids (SNAs)

SNAs, typically composed of densely functionalized and highly oriented nucleic acids on nanoparticle cores, represent an emerging class of therapeutics for diseases, including many forms of cancer, because they are capable of overcoming the limitations of traditional oligonucleotide delivery methods and provide an alternative path to gene regulation (Fig. 3) [33, 34]. SNAs are single-entity agents that exhibit unique chemical and physical properties in biological environments. They are readily taken up by almost any type of cell (over 60 tested to date) in high quantities without the use of ancillary transfection reagents ($>10^6$ nanoparticles per

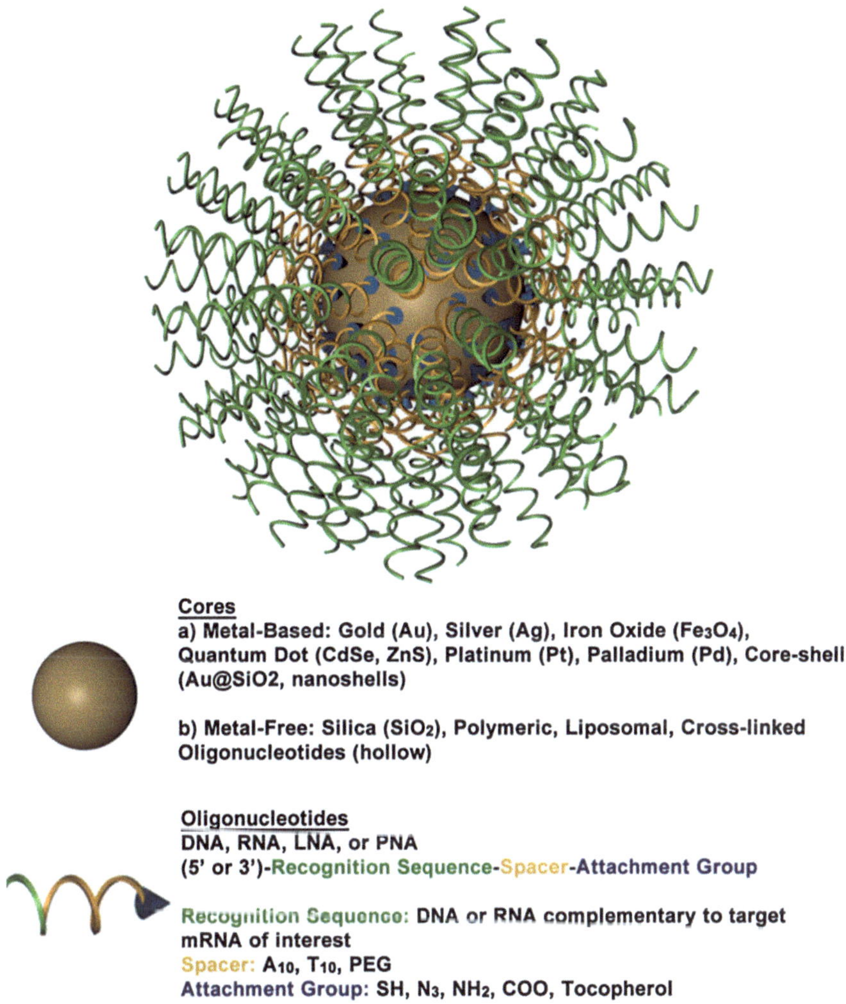

Fig. 2 A 3D drawing of a spherical nucleic acid (SNA). SNAs consist of densely functionalized and highly oriented nucleic acids on the surface of a nanoparticle. Because the properties of SNAs are derived from the shell of nucleic acids, many different cores can be used, such as metal nanoparticles (Au, Pt, etc.), liposomes, and polymers. SNAs can even be core-free. Adapted with permission from Cutler et al. [33]. Copyright 2012. American Chemical Society

cell) [35] through caveolae-mediated endocytosis initiated by recognition through class A scavenger receptors (SR-A) [36, 37]. They elicit a minimal immune response (i.e., 25-fold reduced immune response compared to delivery by cationic carriers) [38, 39] and exhibit increased stability compared to free oligonucleotides in solution [40–43]. These properties stem from the dense shell of highly oriented nucleic acids presented at the surface of these structures [33, 44, 45]. The fact that SNAs facilely enter cells without causing a significant immune response makes

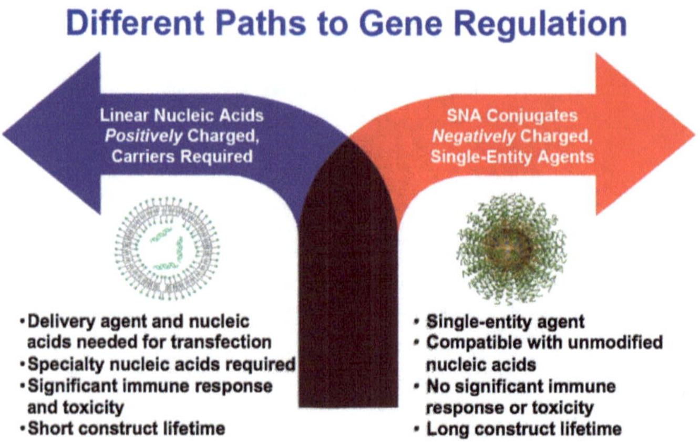

Fig. 3 SNAs offer a different paradigm for gene regulation, where negatively charged nucleic acids do not need to be precomplexed with synthetic positively charged carriers to enter cells and cause gene regulation. If the nucleic acids are densely oriented at the nanoscale, they enter cells in high numbers, exhibit nuclease resistance, show no apparent toxicity, and do not activate the innate immune response. Reproduced with permission from Cutler et al. [33]. Copyright 2012. American Chemical Society

them ideal for local delivery, such as through the skin; however, the fact that they are nonspecifically picked up by nearly all cells will need to be addressed for systemic applications.

The composition of SNAs is highly tailorable, making them an ideal therapeutic platform because they can be tuned to meet the needs of a given application. SNAs can be composed of a variety of oligonucleotides (e.g., DNA, siRNA, microRNA, peptide nucleic acid (PNA), or locked nucleic acid (LNA)) and a variety of different types of nanoparticle cores, such as gold (Au) [34], silver (Ag) [46], iron oxide (Fe_3O_4) [47, 48], quantum dots (CdSe, CdSe/ZnS) [49, 48], platinum [48], silica (SiO_2) [50], core-shell ($Au@SiO_2$) [50], and liposomes [51] typically ranging in size from 10 to 50 nm. Coreless versions of these structures can also be made that display the same useful properties as the core-filled structures, emphasizing the concept that the properties of SNAs stem from their densely functionalized and highly oriented nucleic acid shell and not from the nanoparticle core [51–54]. Some types of the hollow SNAs that have been synthesized thus far include those consisting of cross-linked oligonucleotides [53], DNA-block copolymer micelles [55, 56], infinite coordination polymers [52], metal organic frameworks [54], and liposomes [51]. Hollow SNAs, such as the liposomal SNAs, represent an exciting new class of metal-free SNAs that can be useful in gene regulation, and their potential is only beginning to be realized [51]. The liposomal SNAs have some exciting advantages over conventional liposomal structures, as the oligonucleotide cargo is arranged on the surface of the liposomal entity and thus stabilizes liposomes in the sub-100 nm range. Other synthetic advances in SNA development include the ability to attach RNA to DNA-based SNAs via enzymatic ligation to create RNA-DNA hybrid SNAs [57]. These structures can be used to regulate gene expression in a manner similar to other types of SNAs, and they are more cost-effective to synthesize than SNAs composed solely of RNA oligonucleotides. SNAs can also be backfilled with a variety of surface passivating molecules, such as polyethylene glycol (PEG) [40, 45] or oligoethylene glycol (OEG) [41, 45], which have been known to improve colloidal stability, increase circulation time, and reduce protein adsorption [58, 59].

1.5 Applications of SNAs in Cancer Research and Treatment

The ability to tune the oligonucleotide sequence of SNAs is extremely powerful in the development of SNAs as cancer therapeutics. For example, treatment with SNAs in vitro has resulted in gene knockdown of model targets, such as luciferase [41] and enhanced green fluorescent protein (eGFP) [42, 50], as well as targets involved in cancer cell growth and proliferation, such as HER2 (an oncogenic receptor tyrosine kinase (RTK) responsible for development and progression of cancers, in particular breast cancers) [51, 60], *Bcl2L12* (a GBM oncoprotein and potent inhibitor of effector caspases and p53) [45], and epidermal growth factor receptor (EGFR; a RTK that is important for maintaining epidermal homeostasis and a potent oncogene in several cancers when overexpressed or mutated [61] both in vitro and in vivo). Because the oligonucleotide sequence can be designed to target virtually any mRNA of interest, we have only begun to scratch the surface of the potential of SNAs as cancer therapeutics; SNAs could theoretically be used to target any disease with a genetic basis, including many forms of cancer. To illustrate this versatility, we will highlight three applications below in which SNAs are used to treat cancer. First, we will highlight how SNAs can be used in the treatment of glioblastoma multiforme (GBM), the most prevalent and aggressive form of primary central nervous system malignancies. Then, we will discuss how SNAs can be delivered topically to regulate EGFR in the treatment of hyperproliferative skin disorders and skin cancer. Finally, we explore SNAs as multifunctional therapeutic agents, where drug conjugation to the SNA results in the simultaneous delivery of oligonucleotides and drugs. These structures are being evaluated for the treatment of prostate and breast cancer.

2 SNAs for the Treatment of Glioblastoma Multiforme (GBM)

2.1 *Bcl2L12*-Targeting siRNA SNAs

Due to the unabated growth of GBM tumors and their extensive resistance to therapies, only 3–5 % of patients survive longer than 3 years postdiagnosis [62]. Most therapeutics tested to treat GBM do not penetrate the blood–brain barrier/blood–tumor barrier (BBB/BTB) and therefore are extremely ineffective [63]. However, given that SNAs are rapidly taken up by scavenger receptors, including those found on the surface of endothelial cells of the BBB/BTB, SNAs merited investigation as a therapeutic platform that could cross the BBB/BTB and pervasively penetrate glioma tissue [64, 65]. To preclinically evaluate SNAs for the treatment of GBM, Jensen et al. utilized an in vitro co-culture model of the human BBB, consisting of human primary brain microvascular endothelial cells (huBMECs) and human astrocytes [45]. The authors designed SNAs consisting of a 13 nm gold nanoparticle core conjugated to thiolated siRNA duplexes targeting the GBM oncogene *Bcl2L12*, an effector caspase and p53 inhibitor overexpressed in the vast majority (>90 %) of GBM tumors [66–70]. By labeling these SNAs with a fluorescent Cy5.5 dye and employing fluorescence microscopy, the authors

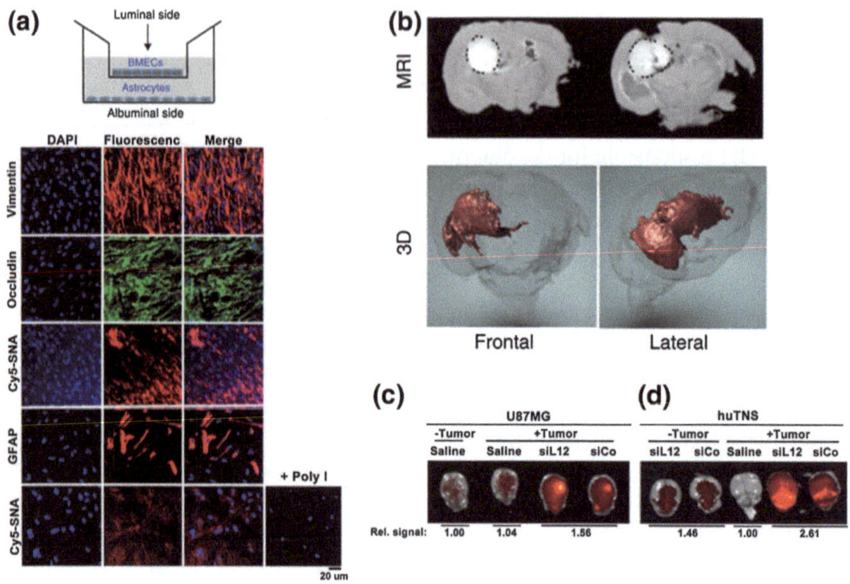

Fig. 4 *BcL2L12*-SNAs in treatment of glioblastoma multiforme. SNAs cross the BBB/BTB and selectively accumulate in glioma tissue. **a** Noncontact in vitro BBB model using a co-culture of human primary brain microvascular endothelial cells (huBMECs) and human astrocytes. Representative confocal fluorescence microscopy images demonstrate Cy5.5-SNA (*red*) distribution in endothelial and astrocytic cells. Endothelial and astrocytic cells stained positively for occludin (a marker for tight junctions) and glial fibrillary acidic protein (GFAP), respectively. The cytoplasm is stained with DAPI and the nuclei stained with anti-vimentin; **b** Magnetic resonance (MR) images of tumor-bearing mouse brains injected intracranially with SNAs-Gd^{III}. Two representative coronal sections imaged 24 h after SNAs-Gd^{III} injection (*upper panel*) show localization of SNAs-Gd^{III} within the intracerebral lesion. Gd^{III} *signal is white* and *outlined* with a *black dotted line*. Also shown is the corresponding three-dimensional (3D) reconstruction of MR images (Gd^{III} *signal in red*); **c** and **d** IVIS analysis of brains with or without human glioblastoma-astrocytoma (U87MG) (**c**) or human tumor neurospheres (huTNS) (**d**) tumors 48 h after systemic delivery of saline or Cy5.5-SNAs. SNA accumulation is indicated by increased fluorescence (*yellow*). Quantification of radiant efficiency is shown as relative signal amount under the images. SiCo = scrambled control sequences. Adapted with permission from Jensen et al. [45]. Copyright 2013. *Science Translational Medicine*

demonstrated that *Bcl2L12*-targeting SNAs (*L12*-SNAs) were able to undergo transcytosis through the huBMEC layer and enter human astrocytes (Fig. 4a) [71–73]. Consistent with the previous reports of SR-A-mediated uptake of SNAs [36, 37], this BBB-penetrating capacity was abolished when polyinosinic acid (Poly-I), which blocks SR-A-dependent SNA uptake and likely mediates transcytosis, was added prior to SNA treatment. Next, BBB and glioma tissue penetration was evaluated in vivo in both healthy and glioma-bearing mice. In conjunction with Cy-5 dye, gadolinium (Gd^{III}) was conjugated to SNAs to visualize and quantify their tissue penetration; the biodistribution of these SNAs was evaluated via inductively coupled plasma-mass spectrometry (ICP-MS), magnetic resonance imaging (MRI), and confocal fluorescence microscopy. Gd^{III}-SNA conjugates were prepared from alkyne-modified DNA thymine (dT) nucleotides and azide-labeled Gd^{III} complexes through click chemistry. Following local administration, both 3D reconstruction of MRI images and confocal fluorescence demonstrated extensive intratumoral dissemination by SNAs (Fig. 4b). ICP-MS further validated these results, showing a 10-fold higher accumulation of SNAs in tumor versus nontumor brain regions, possibly due to the enhanced permeation and retention (EPR) effect [74]. Following tail vein injection of Cy5.5-labeled *L12*-SNAs, in vivo imaging system (IVIS) quantification of radiant intensities showed a 1.8-fold higher accumulation of SNAs in GBM-xenograft-bearing mice compared to sham GBM-inoculated mice (Fig. 4c). Furthermore, systemically delivered *L12*-SNAs successfully neutralized *Bcl2L12* expression, increased intratumoral apoptosis, reduced tumor burden, and increased survival. Systemically administered *L12*-SNAs did not induce inflammatory cytokines, cause any changes in blood chemistry and complete blood counts, or elicit changes in histopathology compared to saline or control SNAs. With no evidence to date of toxicity and promising in vivo results thus far, *L12*-SNAs represent a promising construct for GBM treatment that is headed toward early clinical testing.

2.2 Delivery of Therapeutic miRNA Using miR-182 SNAs

miRNAs have been shown to be important regulators of GBM pathogenesis and therapeutic susceptibility [75]. Genomic studies have characterized miRNA-controlled signaling pathways, which include critical growth and survival pathways, such as receptor tyrosine kinase (RTK)-phosphoinositide 3-kinase (PI3K)-phosphatase and tensin homolog (PTEN), retinoblastoma (Rb), B-cell lymphoma 2 (Bcl-2), and tumor protein p53 (p53) signaling pathways [76, 77]. Given the global overexpression of *Bcl2L12*, its roles in the pathogenesis of GBM, and its involvement in therapy resistance, the Kessler, Peters, Mirkin, and Stegh labs sought to identify miRNAs that control the expression of *Bcl2L12* in GBM [78]. In silico studies of GBM samples from the multidimensional Cancer Genome Atlas (TCGA) dataset (http://cancergenome.nih.gov/dataportal/) were designed to discover miRNAs with expression levels negatively correlated with *Bcl2L12* mRNA levels (Cancer Genome Atlas Research [79]. From these studies, miR-182 was identified as a potential miRNA candidate that regulates *Bcl2L12* expression.

The authors demonstrated that miR-182 acts as a tumor suppressor in GBM by not only controlling the expression and activity of *Bcl2L12*, but also levels of the RTK c-Met and the transcription factor hypoxia-inducible factor 2α (HIF2α). To harness miR-182-related tumor suppressive functions as a therapeutic in vitro and in vivo, the authors synthesized SNAs functionalized with mature miR-182 sequences (182-SNAs). Treatment of glioma cells with 182-SNAs in vitro was shown to potently decrease *Bcl2L12* and c-Met protein levels compared to control SNA-treated cultures, while substantially increasing apoptotic responses and reducing cellular growth. The authors then evaluated 182-SNAs in mice bearing orthotopic GBM xenografts in vivo. In 182-SNA-treated mice compared to control SNA-treated mice, average tumor weights were reduced, and 182-SNA-treated mice experienced significantly prolonged survival relative to control SNA-treated mice.

Taken together, 182-SNAs were shown to effectively decrease *Bcl2L12* and c-Met protein levels, enhance apoptotic responses to chemotherapy, drastically reduce tumor burden, and extend survival of GBM-xenograft-bearing mice. Coupled with the absence of any observable side effects or toxicity, 182-SNAs represent a novel platform for delivering therapeutic miRNAs in GBM.

3 Topical Delivery of SNAs to Regulate Epidermal Growth Factor Receptor (EGFR)

In order to suppress cancer-causing genes in the skin using oligonucleotides, the oligonucleotide must pass the epidermal barrier. One of the greatest challenges associated with topical drug delivery is the design and synthesis of materials that can pass through this barrier [80–82]. In skin, topical delivery is the desired route for delivering agents that can regulate gene suppression because the skin is easily accessible and because topical delivery reduces the risk of systemic side effects. Therefore, an agent that can deliver oligonucleotides through the epidermal barrier with relatively little cytotoxicity would be ideal. Recently, it has been shown that SNAs can be delivered topically in a commercial moisturizer or phosphate buffered saline (PBS) solution to target EGFR, an important gene for epidermal homeostasis and potent oncogene that is frequently overexpressed or mutated in cancer [83, 84]. SNAs were able to penetrate through hairless mice and human skin equivalents without any clinical or histological evidence of toxicity [39]. The Mirkin and Paller labs first measured the uptake of SNAs in normal human keratinocytes (hKCs) because they are notoriously difficult to transfect [85]. ICP-MS revealed that SNA uptake is five times higher in hKCs as compared to HaCaT (spontaneously immortalized hKCs) or HeLa cells [39]. Although the underlying mechanism for the high cellular uptake in hKCs is still under investigation, these data show the potential of SNAs as a strong candidate for topical oligonucleotide delivery. When the uptake in hKCs was visualized by confocal microscopy using nontargeting SNAs with a Cy3 dye, strong fluorescence was seen in the cytoplasm of the cells and morphological differences were not seen between untreated hKCs and SNA-treated hKCs (Fig. 5a). To determine potential off-target effects of SNAs, genome-wide expression profiling revealed only seven up-regulated genes (none were downregulated); in contrast, 427 genes were up- or down-regulated in the case of lipid-based siRNA delivery using the DharmaFECT1™® transfection reagent. This is a promising result because SNA toxicity should be relatively low since there are virtually no off-target effects.

After confirming that SNAs enter hKCs and cause relatively little immune response, the potential for EGFR mRNA and protein knockdown was assessed. Western blot analysis showed that there was greater EGFR suppression by 0.3 nM siRNA delivered by SNA than 30 nM siRNA delivered by DharmaFECT1™ in hKCs (Fig. 5b). In fact, total suppression of EGFR, as seen by Western blot, was observed with incubation of 1.5 nM total siRNA delivered by SNAs, something that was not seen even at 30 nM siRNA delivered by DharmaFECT1™, thus demonstrating the potency of oligonucleotide delivery via SNAs.

The authors then investigated the potential of SNAs to penetrate mouse skin. In both SKH10E hairless mice and hair-bearing C57BL/6 J mice that were shaved 24 h before treatment, SNA penetration through the stratum corneum and into the epidermis and dermis was seen in as few as 3 h after a single SNA dose (Fig. 5c). Furthermore, after treating SKH1-E hairless mice daily for 3 days with nontargeting SNAs and monitoring the gold content in skin for 10 days post-treatment, it was found that only 2 % of the initial gold was still present in the skin (Fig. 5d). In summary, SNAs were shown to penetrate the skin of two different mice strains and subsequently clear the skin by 10 days post-treatment.

The next step was to investigate the efficacy of EGFP suppression in mouse skin. Western blot analysis showed that the protein expression of EGFP was nearly eliminated and the downstream phosphorylation of ERK was inhibited by 74 % from the application of SNAs in Aquaphor®; Aquaphor® alone, nontargeting SNAs, and free siRNA did not have an effect on EGFP expression (Fig. 5e). Furthermore, the decrease in EGFP expression was accompanied by a 74 % decrease in the phosphorylation of downstream extracellular signal-regulated kinases (ERK1/2), demonstrating the specificity of EGFR knockdown (total ERK1/2 expression remained constant). To confirm the phenotypic effect of EGFR knockdown on mouse skin, computerized morphometric analysis of histological sections of mouse skin treated with EGFR SNAs showed an almost 40 % reduction in thickness compared to the mouse skin treated with Aquaphor® alone or nontargeting SNAs (Fig. 5c). Similar results were seen with human skin equivalents. Taken together, these data demonstrate the ability of SNAs to knockdown a specific gene target in vivo, which causes a specific biological response with minimal off-target effects. This study lays the groundwork for SNAs to be utilized for topical oligonucleotide delivery with vast potential in the treatment of skin diseases and disorders, such as metastatic melanoma and psoriasis.

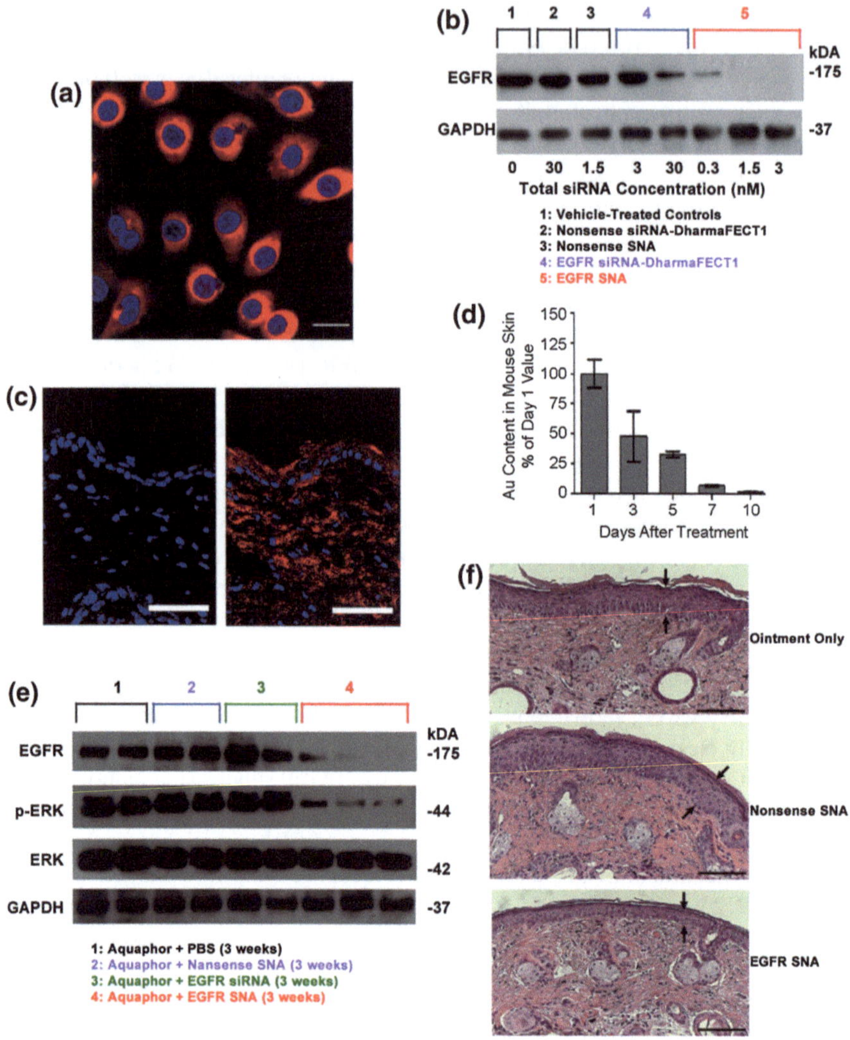

(a)
(b)
1 2 3 4 5
kDA
EGFR -175
GAPDH -37
0 30 1.5 3 30 0.3 1.5 3
Total siRNA Concentration (nM)
1: Vehicle-Treated Controls
2: Nonsense siRNA-DharmaFECT1
3: Nonsense SNA
4: EGFR siRNA-DharmaFECT1
5: EGFR SNA
(c)
(d)
Au Content in Mouse Skin % of Day 1 Value
150 125 100 75 50 25 0
1 3 5 7 10
Days After Treatment
(e)
1 2 3 4
kDA
EGFR -175
p-ERK -44
ERK -42
GAPDH -37
1: Aquaphor + PBS (3 weeks)
2: Aquaphor + Nansense SNA (3 weeks)
3: Aquaphor + EGFR siRNA (3 weeks)
4: Aquaphor + EGFR SNA (3 weeks)
(f)
Ointment Only
Nonsense SNA
EGFR SNA

◀ **Fig. 5** Topical delivery of SNAs through human and mouse skin. **a** Uptake of Cy3-labeled nonsense SNAs (*red*) in the cytoplasm of approximately 100 % of the primary human keratinocytes (hKCs) after 24 h incubation. The nuclei were stained *blue* with Hoechst 33343. Scale bar, 20 μm; **b** Western blot showing epidermal growth factor receptor (EGFR) protein levels for hKCs treated for 48 h. Note the greater suppression by 0.01 nM EGFR SNAs (equivalent to 0.3 nM siRNA) as compared with the 30 nM EGFR siRNA delivered with DharmaFECT1™ at 60 h; **c** Mouse (SKH1-E) skin treated topically with 1:1 Aquaphor® only (*left*) or with 50 nM Cy5-labeled (*red*) SNAs dispersed in the 1:1 Aquaphor® (*right*). The SNAs are seen in the cytoplasm of epidermal cells and the dermis 3 h after application. DAPI-stained nuclei in blue. Scale bars, 100 μm; **d** Mouse skin was treated daily for 3 days with nonsense SNAs and analyzed by ICP-MS for gold content. The gold content in mouse skin progressively decreases after cessation of topical treatment; 10 days after the final treatment, only 2 % of the original gold content remains ($n = 3$ at each time point); **e** The protein expression of EGFR was nearly eliminated in the EGFR SNA-treated group, whereas the downstream phosphorylation of ERK was inhibited by 74 %; total ERK expression remained constant; **f** The mean thickness of EGFR SNA-treated skin was 40 % less than that of control-treated skin ($P < 0.001$), as measured by computerized morphometry. Epidermal thickness was measured from the top of the stratum granulosum to the basement membrane (*arrows*) at three equidistant sites. Adapted with permission from Zheng et al. [39]. Copyright 2012. *Proceedings of the National Academy of Sciences of the United States*

4 Multifunctional SNAs

Thus far, we have discussed SNAs that enter cells and tumor tissues and perform a single function (e.g., regulate gene expression via the RNAi or antisense pathway). However, SNAs that can perform multiple therapeutic, diagnostic, targeting, and imaging functionalities simultaneously within a cell can also be synthesized. Using solid-phase DNA synthesis, modified phosphoramidites can be used to add specific chemical functional groups onto oligonucleotides; and these functional groups can be used as handles to attach drugs, small molecules, or antibodies of interest to these oligonucleotides. When these modified oligonucleotides are formulated as SNAs, the SNA structure allows for the cellular entry of the conjugate and the oligonucleotides allow for the regulation of gene expression, while the small molecule drug or contrast agent that is appended to the oligonucleotides that comprise the SNAs, for example, can be used as an additional therapeutic component or imaging modality, respectively.

4.1 SNA-Drug Conjugates for Drug Delivery

Cisplatin and carboplatin are widely regarded as effective treatments for testicular and ovarian cancers, and these drugs have also been utilized for the treatment of bladder, cervical, head and neck, esophageal, and small cell lung cancer [86, 87]. However, many of the synthetic delivery systems used for cisplatin and carboplatin are associated with systemic toxicity [88]. Because SNAs have been shown to enter cells in high quantities [35] without provoking a significant immune response [38], they are a promising platform for the delivery of these and other platinum (Pt) compounds. To this end, the Lippard and Mirkin labs collaborated to synthesize a DNA oligonucleotide with a terminal dodecyl amine to which the Pt (IV) pro drug (*c*,*c*,*t*-{$Pt(NH_3)_2Cl_2(OH)(O_2CCH_2CH_2CO_2H)$}) was conjugated via amide linkages (Fig. 6a) [89]. Platinum atomic absorption spectroscopy (AAS) showed that 98 % of the DNA amines on the SNA were conjugated to platinum. The mechanism of action for Pt (IV) complexes requires that they are reduced to cytotoxic Pt (II) in a reducing environment, such as inside cells or blood [90]. Electrochemical studies confirmed that the conjugation of a Pt (IV) payload to the SNA did not significantly alter the reduction potential of the complex, making it likely that the axial ligands of the Pt (IV) complex could be removed once it entered a reducing environment. It was also confirmed that the SNAs still entered HeLa cervical cancer cells with the Pt (IV) prodrugs attached (Fig. 6b), and specifically colocalized with microtubules

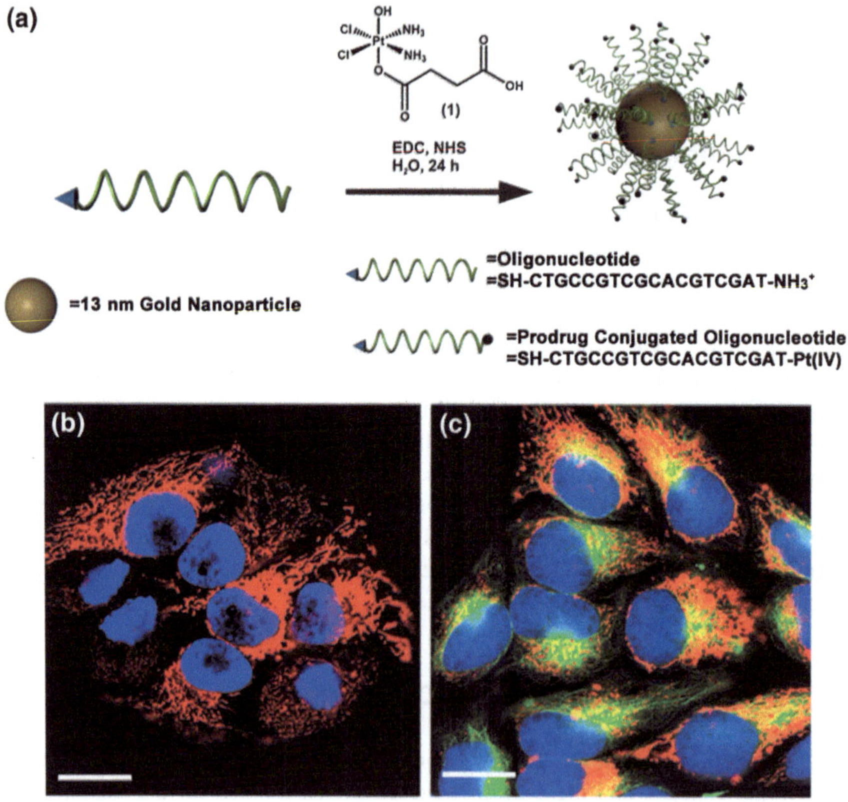

Fig. 6 Drug and oligonucleotide dual therapy. **a** Scheme for the synthesis of Pt (IV) terminated SNAs; **b** Live cell imaging of HeLa cells upon incubation with platinum-tethered Cy5-SNAs for 12 h; **c** Co-localization of the particles with the cytoplasmic microtubules. Hoechst 33342 was used for nuclear staining. Scale bars, 20 μm. Adapted with permission from Dhar et al. [89]. Copyright 2009. American Chemical Society

(Fig. 6c). Finally, the efficacy of the SNA-bound Pt (IV) prodrug was assessed in four different cell lines. For A549 lung epithelial cancer cells in particular, the conjugation of the Pt (IV) prodrug to the SNA resulted in superior killing efficiency, with a half maximal inhibitory concentration (IC_{50}) of 0.9 μM, compared to an IC_{50} of 11 μM for free cisplatin. Therefore, it appears that the Pt (IV) prodrug is more effective when conjugated to DNA on the SNA surface. Future work will include using the oligonucleotide, whether it is DNA or RNA, to knockdown gene expression, which, in conjunction with drug delivery, should render these constructs even more effective.

The Ho and Mirkin labs employed a similar strategy with the chemotherapeutic Paclitaxel [91]. Paclitaxel is used to treat cancers such as ovarian, breast, and nonsmall cell lung cancers [92, 93]. Paclitaxel is challenging to deliver because of its low aqueous solubility; cells also acquire chemoresistance toward this drug and it often causes harmful side effects [94]. Therefore, it was hypothesized that Paclitaxel's conjugation to SNAs may increase its aqueous solubility and perhaps decrease the associated side effects.

Zhang and co-workers conjugated a thiolated oligonucleotide containing a terminal Paclitaxel group to 13 nm gold nanoparticles (AuNPs) (Fig. 7a; compound 3) [91]. This step was accomplished by modifying Paclitaxel molecules with succinic anhydride groups to form a Paclitaxel carboxylic acid derivative (compound 1), which was conjugated to DNA oligonucleotides with a terminal amine group using EDC/sulfo-NHS chemistry (compound 2). The average number of Paclitaxel molecules per gold nanoparticle was measured to be 59 ± 8. It is interesting to note that while free Paclitaxel is not soluble in phosphate buffered saline (PBS) at a 5 μM concentration, the SNA-Paclitaxel conjugates remain well dispersed in PBS, as confirmed by dynamic light scattering (DLS) (Fig. 7b) and transmission electron microscopy (Fig. 7c). Free Paclitaxel has a maximum solubility of 0.4 μg/mL in aqueous solution [95] and the SNA-Paclitaxel conjugate exhibits a maximum solubility for Paclitaxel of 21.35 μg/mL, a greater than 50-fold enhancement in drug solubility. Confocal microscopy confirmed the internalization of fluorophore-labeled SNA-Paclitaxel conjugates in MCF7 human breast adenocarcinoma cells and MES-SA/Dx5 human uterine sarcoma cells after a 6 h incubation. A terminal deoxynucleotidyl transferase dUTP nick end-labeling (TUNEL) assay [96] determined that DNA fragmentation and apoptosis was induced by Paclitaxel. It was shown that Paclitaxel remained active when bound to SNAs and that the SNA-Paclitaxel conjugates have the potential to overcome Paclitaxel resistance in cells.

Furthermore, the ability of SNA-Paclitaxel conjugates to kill cancer cells derived from different types of cancer (MCF7 breast cancer cells, MES-SA/Dx5 multidrug resistant cells derived from uterine sarcoma, and SKOV-3 ovarian cancer cells) was assessed. Enhanced cytotoxicity was observed in all three cell lines for the SNA-Paclitaxel conjugate when compared with free Paclitaxel and the DNA-Paclitaxel conjugate. Therefore, the SNA formulation of Paclitaxel may overcome the cellular cross-resistance of chemotherapeutics in vitro. There are also significant advantages to using the SNA-Paclitaxel formulation compared to the free Paclitaxel and the DNA-Paclitaxel conjugate when looking at the IC_{50} (in nM Paclitaxel; Table 1). In

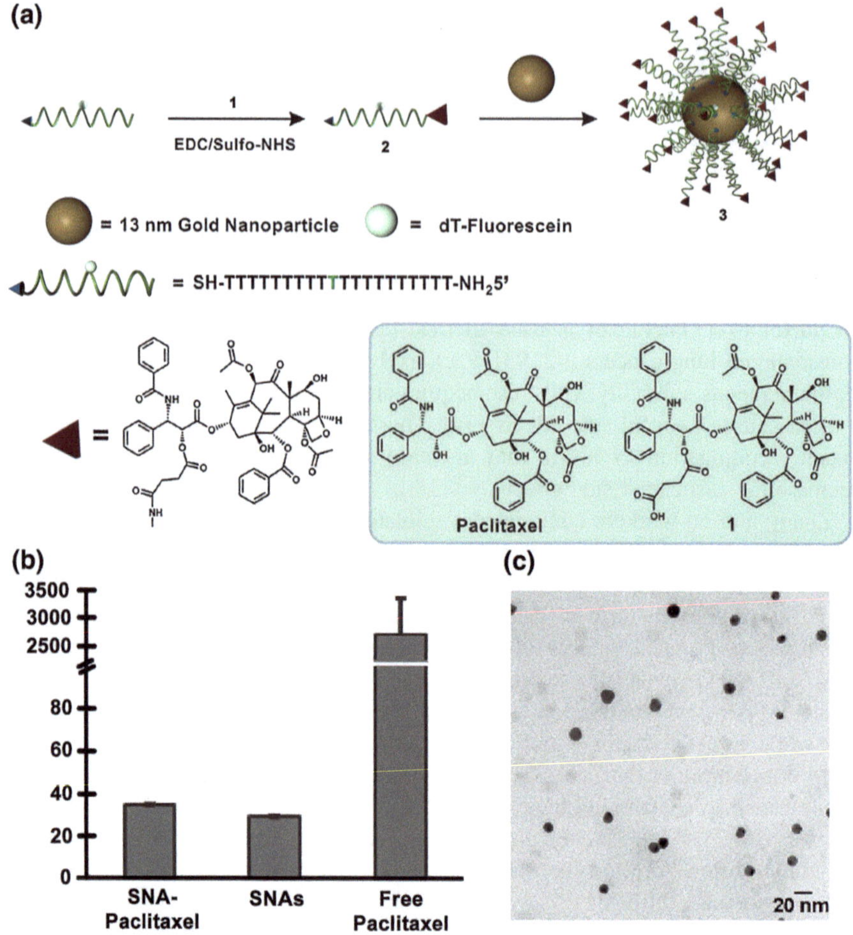

Fig. 7 Drug solubility and efficacy of paclitaxel increases upon covalent linkage with SNAs. **a** Synthesis of SNA-Paclitaxel conjugates; **b** Hydrodynamic sizes of SNA-Paclitaxel conjugates (3), SNAs, and Paclitaxel in PBS buffer ($n = 3$). The compounds were suspended in PBS buffer at the equivalent Paclitaxel concentration of 21.3 μg/mL (25 μM) for dynamic light scattering (DLS) measurement; **c** TEM image of SNA-Paclitaxel conjugates (3). Scale bar is 20 nm. Adapted with permission from Zhang et al. [91]. Copyright 2011. *ACS Nano*

MCF7 cells, for example, the IC_{50} decreases from above 1 μM and 193 nM for free Paclitaxel to 119.4 and 52.6 nM for SNA-Paclitaxel conjugates after 12 and 48 h, respectively. Therefore, conjugation to SNAs represents a potential new route to solubilize previously insoluble drugs and improve their biological response. Future work is still needed to better understand the mechanism of enhanced efficacy in this system, but the groundwork has been set for the use of SNAs as platforms to which other biologically relevant small molecules can be conjugated.

Table 1 IC_{50} of SNA-Paclitaxel conjugates (3), free Paclitaxel, and compound (1) after 12 and 48 h Incubation in MCF7 Breast cancer, SKOV-3 Ovarian Cancer, and MES-SA/Dx5 multidrug resistant cells

Cell line	Incubation time (h)	IC_{50} (nM Paclitaxel)		
		SNA-Paclitaxel	Paclitaxel	Paclitaxel carboxylic acid derivative
MCF7	12	119.4	>1000	>1000
	48	52.6	193.0	133.2
SKOV-3	12	4.3	175.6	>1000
	48	17.5	28.9	188.0
MES-SA/Dx5	12	118.0	>1000	>1000
	48	104.5	>1000	>1000

Adapted with permission from Zhang et al. [91]. Copyright 2011. *ACS Nano*

4.2 SNA-Antibody Conjugates for Cellular Targeting

While SNAs solve many challenges associated with the intracellular delivery of oligonucleotides, the fact that they allow entry into most cells in a nonspecific fashion could present a problem in the targeting of genes essential for normal organ homeostasis. To enhance SNA association with target cells, the Mirkin lab conjugated a monoclonal antibody (mAb) designed to bind to the human epithelial growth factor receptor 2 (HER2) to antisense DNA-SNAs to create HER2-targeting antisense DNA-SNAs [60]. HER2 is involved in signal transduction pathways leading to increased cellular growth and differentiation, and it is up-regulated in many epithelial cancers, including breast, ovarian, gastric, and salivary [97, 98]. The authors utilized copper (I) (Cu(I)) click chemistry to conjugate a HER2 antibody to antisense DNA, specifically linking an azide-functionalized HER2 mAb to DNA with a 3′ alkyne group (Fig. 8a).

Using ICP-MS, the authors evaluated the cell uptake of both HER2-targeted and nontargeted SNAs as a function of time in HER2 overexpressing SKOV-3 ovarian cancer cells (Fig. 8b). In the first 6 h of incubation, the HER2-targeted SNAs demonstrated a significantly higher rate of uptake compared to the nontargeted SNAs ($\sim$236 particles s^{-1} $cell^{-1}$ for HER2-targeted SNAs vs. $\sim$19 particles s^{-1} $cell^{-1}$ for nontargeted SNAs). However, as incubation time was extended to 24 h, the selectivity diminished, perhaps due to the cells' inability to replenish HER2 on their surfaces after mAb binding and endocytosis [99]. Furthermore, band density analysis by Western blot for HER2-targeted SNAs in SKOV-3 cells revealed that at certain concentrations, HER2 expression could no longer be detected (Fig. 8c). Thus, HER2-targeting SNAs were taken up by HER2 over expressing cells to a greater extent and at a faster initial rate compared to nontargeted particles. Additionally, HER2-targeting SNAs demonstrated potent antisense gene knockdown, requiring only pM amounts of SNAs to drastically silence HER2

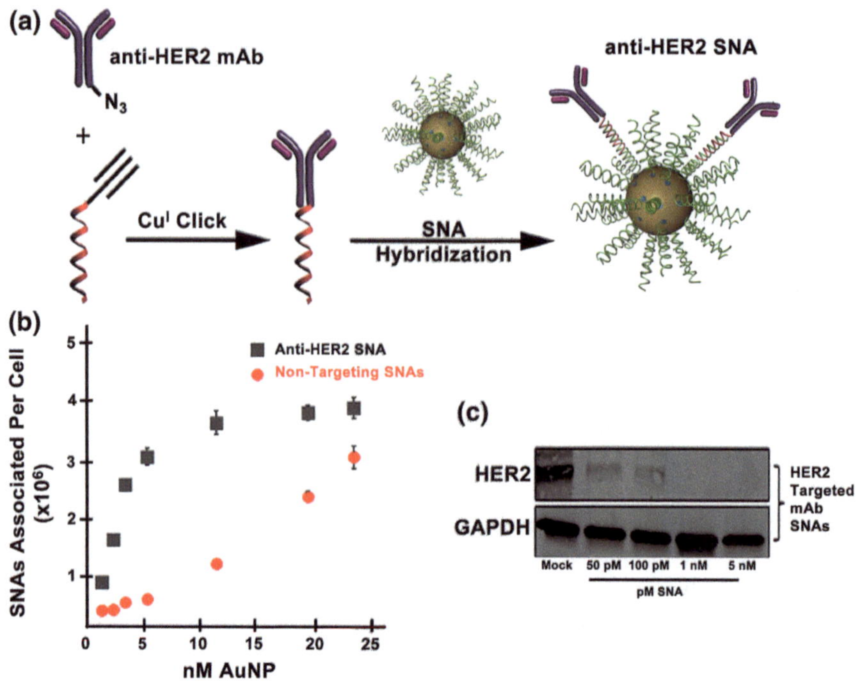

Fig. 8 Human epidermal growth factor receptor 2 (HER2)-targeted SNAs. **a** Synthesis of HER2-targeting SNAs; **b** Cell uptake of targeted versus nontargeted SNAs in SKOV-3 ovarian cancer cells as a function of time; **c** Western blot of HER2 expression in SKOV-3 cells after treatment with increasing concentrations of HER2-targeting SNAs. GAPDH is used as an internal reference. Adapted with permission from Zhang et al. [60]. Copyright 2012. American Chemical Society

expression. Hence, these SNA-antibody constructs are capable of increasing both the selectivity and efficacy of the SNA platform.

4.3 Gadolinium-Enriched SNAs for Cellular Imaging

The Meade and Mirkin labs utilized SNAs as a platform to synthesize bioactivated contrast agents that are capable of penetrating cells for use in cell-tracking experiments [100]. The most commonly used contrast agents, paramagnetic gadolinium (III) (Gd^{III}) complexes, are advantageous because Gd^{III} complexes have high relaxivities since they reduce the longitudinal relaxation time (T_1) of local water protons [101]. Song and co-workers sought to fill the need for a magnetic resonance imaging (MRI) agent with both high Gd^{III} loading (for enhanced contrast) and efficient cellular uptake (for imaging small cell populations) by conjugating Gd^{III} to SNAs [100]. The Gd^{III} complexes were attached through click chemistry to poly DNA thymine (poly dT) oligonucleotides, which contained five conjugation sites of hexylamino labeled DNA thymine (dT groups conjugated with a cross-linker, azidobutyrate-*N*-hydroxysuccinimidester) (Fig. 9a). Through ICP-MS, the Gd^{III} loading per SNA was calculated; there were 342 ± 1 Gd^{III} per 13 nm AuNP and 656 ± 20 Gd^{III} per 30 nm AuNP. The relaxation efficiencies of DOTA-Gd^{III}, DNA-Gd^{III}, and SNA-Gd^{III} (13 and 30 nm AuNPs) were measured by taking the slope of a plot of the measured $1/T_1$ as a function of Gd^{III} concentration (Table 2). The relaxivity at 37 °C in water at 60 MHz increased from 3.2 $mM^{-1}\ s^{-1}$ for DOTA-Gd^{III} to 8.7 $mM^{-1}\ s^{-1}$ for DNA-Gd^{III} to 16.9 $mM^{-1}\ s^{-1}$ for 13 nm SNA-Gd^{III} and 20.0 $mM^{-1}\ s^{-1}$ for 30 nm SNA-Gd^{III}. These results are consistent with the Soloman-Bloembergen-Morgan theory, which states that there is a concomitant decrease in rotational correlation time (τ_r) with an increase in r_1 [102–106]. T_1 weighted MR images in solution phantoms as well as NIH/3T3 mouse fibroblast cells show that SNA-Gd^{III} is significantly brighter than DOTA-Gd^{III} at multiple

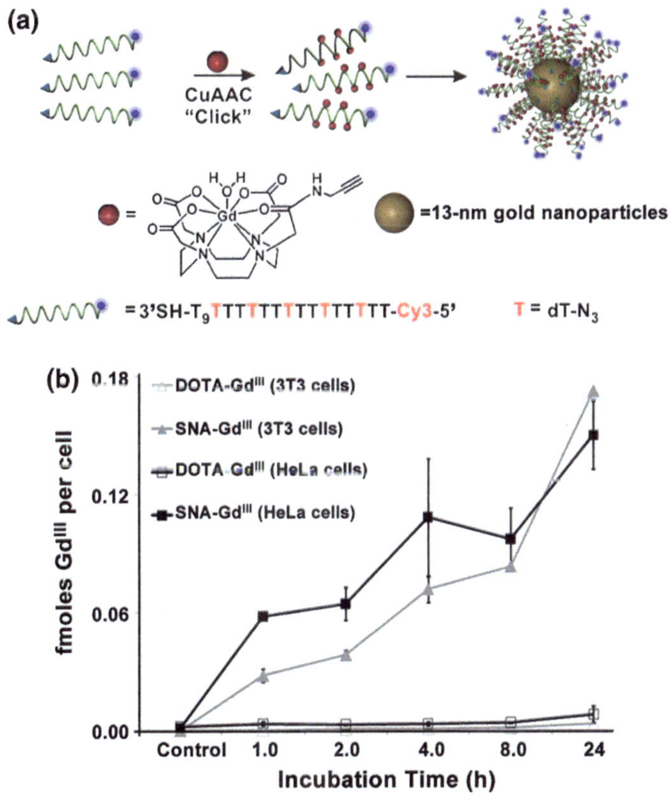

Fig. 9 Multimodal SNA-Gd^{III}. **a** Synthesis of Cy3-SNA-Gd^{III} conjugates using copper catalyzed azide alkyne cycloaddition (CuAAC); **b** Time-dependent cellular uptake of SNA-Gd^{III} conjugates compared to DOTA–Gd^{III} in NIH/3T3 mouse fibroblast and HeLa cervical cancer cells. Cells were incubated with 6.5 μM Gd^{III} for both contrast agents. Error bars represent ±1 standard deviation of the mean for duplicate experiments. Adapted with permission from Song et al. [100]. Copyright 2009. *Angewandte Chemie*

Table 2 Relaxivities (r) of Gd ^{III} complexes and conjugates at 60 and 600 MHz

	r_1, (mM^{-1} s^{-1})	
	60 MHz (1.41 T)	600 MHz (14.1 T)
DOTA-Gd^{III}	3.2[a]	2.2
DNA-Gd^{III}	8.7	NM
13 nm SNA-Gd^{III}/ionic	16.9	5.1
13 nm SNA-Gd^{III}/particle	5779	1275
30 nm SNA-Gd^{III}/ionic	20.0	NM
30 nm SNA-Gd^{III}/particle	13,120	NM

60 MHz measured in pure water at 37 °C and 600 MHz measured in cell media at 25 °C. NM = not measured

[a]Taken from [107]

Adapted with permission from Song et al. [100]. Copyright 2009. *Angewandte Chemie*

Gd^{III} concentrations. To confirm efficiency of cellular uptake, NIH/3T3 and HeLa cells were incubated with SNA-Gd^{III} or DOTA-Gd^{III} for increasing amounts of time. At all concentrations, the Gd^{III} uptake was more than 50-fold higher for SNA-Gd^{III} than for DOTA-Gd^{III} (Fig. 9b). Therefore, SNAs functionalized with Gd^{III} are outstanding contrast agents that freely enter cells.

Finally, the cell labeling efficiency was assessed using analytical flow cytometry. It was found that incubation of Cy3 SNA-Gd^{III} with NIH/3T3 cells results in 80 and 100 % of the cells being labeled after four and 24 h, respectively. Taken together, these data demonstrate the versatility of the SNA platform and show that conjugation of Gd^{III} to SNAs results in a multimodal, cell permeable MR contrast agent that is biocompatible, has a high Gd^{III} loading, and relatively high relaxivity. These Gd-SNA conjugates also show a greater than 50-fold increase in cell uptake compared to clinically available DOTA-Gd^{III}. When also designed and synthesized to target a gene of interest, such SNAs could be efficiently used as dual imaging-therapeutic moieties.

4.4 Self-Assembled, Multimodal SNAs for Cancer Therapy

The design of multimodal nanomaterials (i.e., nanoconjugates capable of simultaneously performing various functions including imaging, sensing, and drug delivery) is becoming increasingly desired. As such, the Tan lab sought to further expand the multimodal potential of SNAs by generating self-assembled, SNA-based constructs that combine fluorescent imaging, specific target cell recognition, and high drug loading capacity [108]. To achieve this, 30 nm streptavidin magnetic nanoparticles (MNPs) were conjugated with a biotin-DNA initiator strand that triggered a cascade of DNA-assisted hybridization reactions of monomer DNA sequences, resulting in the formation of a long DNA polymer (diameter after assembly $\sim$124 nm) as the nanoparticle shell (Fig. 10).

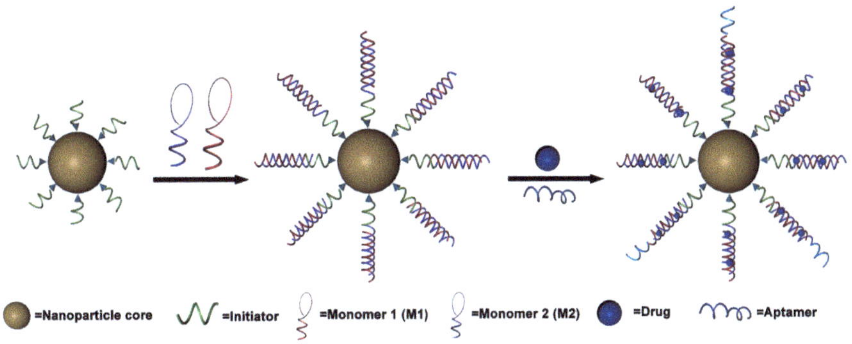

Fig. 10 Schematic representation of multimodal SNAs. Adapted with permission from Zheng et al. [108]. Copyright 2013

The monomer DNA sequences were designed to carry fluorophores for fluorescence imaging and targeting sequences for cell-specific recognition. One of the monomer DNA sequences contained an embedded aptamer AS1411, which forms a stable G-quadruplex structure that specifically targets nucleolin, a protein overexpressed in tumor cells [109]. AS1411 has been tested in phase II clinical trials for patients with relapsed or refractory acute myeloid leukemia and renal cell carcinoma [110]. In G-quadruplex form, AS1411 can bind 5, 10, 15, 20-tetrakis (4-N-methylpyridiniumyl)porphyrin (TMPyP4), a photodynamic therapy drug that is cytotoxic when exposed to light [111]. Hence, AS1411 enabled cell-specific targeting as well as drug loading capacity for these multimodal SNAs.

When nucleolin-expressing SKOV-3 ovarian epithelial adenocarcinoma cells and control, non-nucleolin-expressing HBE135 lung epithelial cells were incubated with FITC-labeled AS1411/MNP-SNAs and random, nontargeted MNP-SNAs, a significant fluorescence shift was found only with AS1411/MNP-SNAs in SKOV-3 cells, as assessed via flow cytometry. These data indicated that the binding capacity of the AS1411 aptamer remained intact after conjugation to the MNP-SNAs. Additionally, no significant fluorescence change was detected between the random MNP-SNAs and the AS1411/MNP-SNAs in HBE135 cells, indicating the specificity of the AS1411/MNP-SNAs for nucleolin.

Furthermore, a cytotoxicity assay was performed that compared TMPyP4 only to TMPyP4 bound on AS1411/MNP-SNAs (TMPyP4/AS1411/MNP-SNAs) in SKOV-3 and HBE135 cells. Cell death was induced by laser irradiation after treating with TMPyP4 only or TMPyP4/AS1411/MNP-SNAs for 10 min and assessed via flow cytometry monitoring of propidium iodide (PI)-labeled dead cells. The phototoxicity of the TMPyP4/AS1411/MNP-SNAs was significantly higher than TMPyP4 only in nucleolin-expressing SKOV-3 cells (IC_{50} of TMPyP4/AS1411/MNP-SNAs $\sim$0.15 μM vs. IC_{50} of TMPyP4 only $\sim$0.4 μM). Additionally, in non-nucleolin-expressing HBE135 cells, TMPyP4/AS1411/MNP-SNAs induced phototoxicity was significantly less than TMPyP4 only (IC_{50} of TMPyP4/AS1411/MNP-SNAs $\gg$3 μM vs. IC_{50} of TMPyP4 only $\sim$1.5 μM). These results indicated that TMPyP4/AS1411/MNP-SNAs enabled cell-specific drug entry and acted as cytotoxic agents to nucleolin-expressing cancer cells, while protecting non-nucleolin-expressing control cells relative to TMPyP4 alone.

Thus, TMPyP4/AS1411/MNP-SNAs were shown to be a promising, multifunctional construct, with the ability to target cancer cells specifically, act as cytotoxic payloads to cancer cells, and be fluorescently imaged. Given the tunable nature of the monomer DNA sequences, these self-assembled, multimodal SNA conjugates are adaptable and could potentially be expanded as a therapeutic platform for cancer.

5 Future Outlook

Spherical nucleic acids (SNAs) represent a model platform for oligonucleotide-based therapeutics. These novel materials enter cells in high quantities without the use of transfection agents and can be utilized to regulate gene expression without eliciting an immune response. They are also resistant to nuclease degradation and can be formulated as multifunctional materials when the nucleic acids on their surface are conjugated to entities such as small molecules, drugs, or antibodies. While it is advantageous that these materials enter almost every known cell line and can cross the blood–brain and blood–tumor barriers as well as the epidermis, their use does not come without its own challenges. The nonspecific uptake of SNAs by virtually all cells must be addressed in the context of systemic delivery, and the coupling of SNAs to targeting moieties, such as antibodies, are being explored to increase their specificity for certain cell types. While their nonspecific uptake necessitates thoughtful target selection for systemic applications, SNAs are ideal candidates for local administration. The challenges for this class of therapeutics, going forward, will be picking appropriate genetic targets and diseases to treat. The fact that the oligonucleotide sequence can be tuned to specifically target a disease-causing mRNA sequence, while sparing healthy mRNA sequences, has tremendous potential for increasing the therapeutic precision and minimizing off-target effects. The customizable nature of the oligonucleotides becomes even more important when one considers that the genetic basis of many diseases is constantly changing and that diseases present themselves differently in each patient. SNAs may allow for on-demand, personalized, or individualized therapeutic options to combat ever-evolving microorganisms and cancer diversity.

References

1. Burnett JC, Rossi JJ, Tiemann K (2011) Current progress of siRNA/shRNA therapeutics in clinical trials. Biotechnol J 6(9):1130–1146
2. Stegh AH (2013) Toward personalized cancer nanomedicine—past, present, and future. Integr Biol 5(1):48–65
3. Burnett JC, Rossi JJ (2012) RNA-based therapeutics: current progress and future prospects. Chem Biol 19(1):60–71

4. Davidson BL, McCray PB (2011) Current prospects for RNA interference-based therapies. Nat Rev Genet 12(5):329–340
5. Kanasty R, Dorkin JR, Vegas A, Anderson D (2013) Delivery materials for siRNA therapeutics. Nat Mater 12(11):967–977
6. Kim DH, Rossi JJ (2007) Strategies for silencing human disease using RNA interference. Nat Rev Genet 8(3):173–184
7. Carthew RW, Sontheimer EJ (2009) Origins and mechanisms of miRNAs and siRNAs. Cell 136(4):642–655
8. Bennett CF, Swayze EE (2010) RNA targeting therapeutics: molecular mechanisms of antisense oligonucleotides as a therapeutic platform. Annu Rev Pharmacol Toxicol 50:259–293
9. Magen I, Hornstein E (2014) Oligonucleotide-based therapy for neurodegenerative diseases. Brain Res 1584:116–128
10. Cerritelli SM, Crouch RJ (2009) Ribonuclease H: the enzymes in eukaryotes. FEBS J 276 (6):1494–1505
11. Fire A, Xu S, Montgomery MK, Kostas SA, Driver SE, Mello CC (1998) Potent and specific genetic interference by double-stranded RNA in Caenorhabditis elegans. Nature 391 (6669):806–811
12. Castanotto D, Rossi JJ (2009) The promises and pitfalls of RNA-interference-based therapeutics. Nature 457(7228):426–433
13. Elbashir SM, Harborth J, Lendeckel W, Yalcin A, Weber K, Tuschl T (2001) Duplexes of 21-nucleotide RNAs mediate RNA interference in cultured mammalian cells. Nature 411 (6836):494–498
14. Hannon GJ, Rossi JJ (2004) Unlocking the potential of the human genome with RNA interference. Nature 431(7006):371–378
15. Novina CD, Sharp PA (2004) The RNAi revolution. Nature 430(6996):161–164
16. Hannon GJ (2002) RNA interference. Nature 418(6894):244–251
17. McManus MT, Sharp PA (2002) Gene silencing in mammals by small interfering RNAs. Nat Rev Genet 3(10):737–747
18. Zamore PD, Tuschl T, Sharp PA, Bartel DP (2000) RNAi: double-stranded RNA directs the ATP-dependent cleavage of mRNA at 21 to 23 nucleotide intervals. Cell 101(1):25–33
19. Bartlett DW, Davis ME (2007) Effect of siRNA nuclease stability on the in vitro and in vivo kinetics of siRNA-mediated gene silencing. Biotechnol Bioeng 97(4):909–921
20. Fabian MR, Sonenberg N, Filipowicz W (2010) Regulation of mRNA translation and stability by microRNAs. Annu Rev Biochem 79:351–379
21. Kanasty RL, Whitehead KA, Vegas AJ, Anderson DG (2012) Action and reaction: the biological response to siRNA and Its delivery vehicles. Mol Ther 20(3):513–524
22. Schroeder A, Levins CG, Cortez C, Langer R, Anderson DG (2010) Lipid-based nanotherapeutics for siRNA delivery. J Intern Med 267(1):9–21
23. Whitehead KA, Langer R, Anderson DG (2010) Knocking down barriers: advances in siRNA delivery. Nat Rev Drug Discov 9(5):412
24. Fichter KM, Ingle NP, McLendon PM, Reineke TM (2013) Polymeric nucleic acid vehicles exploit active interorganelle trafficking mechanisms. ACS Nano 7(1):347–364
25. Nelson CE, Kintzing JR, Hanna A, Shannon JM, Gupta MK, Duvall CL (2013) Balancing cationic and hydrophobic content of PEGylated siRNA polyplexes enhances endosome escape, stability, blood circulation time, and bioactivity in vivo. ACS Nano 7(10):8870–8880
26. Patil ML, Zhang M, Taratula O, Garbuzenko OB, He H, Minko T (2009) Internally cationic polyamidoamine PAMAM-OH dendrimers for siRNA delivery: effect of the degree of quaternization and cancer targeting. Biomacromolecules 10(2):258–266
27. Alabi CA, Love KT, Sahay G, Yin H, Luly KM, Langer R, Anderson DG (2013) Multiparametric approach for the evaluation of lipid nanoparticles for siRNA delivery. Proc Natl Acad Sci USA 110(32):12881–12886

28. Rungta RL, Choi HB, Lin PJC, Ko RWY, Ashby D, Nair J, Manoharan M, Cullis PR, MacVicar BA (2013) Lipid nanoparticle delivery of siRNA to silence neuronal gene expression in the brain. Mol Ther Nucleic Acids 2:e136
29. Nayerossadat N, Maedeh T, Ali PA (2012) Viral and nonviral delivery systems for gene delivery. Adv Biomed Res 1(2):14
30. Bharali DJ, Klejbor I, Stachowiak EK, Dutta P, Roy I, Kaur N, Bergey EJ, Prasad PN, Stachowiak MK (2005) Organically modified silica nanoparticles: a nonviral vector for in vivo gene delivery and expression in the brain. Proc Natl Acad Sci USA 102 (32):11539–11544
31. Giljohann DA, Seferos DS, Daniel WL, Massich MD, Patel PC, Mirkin CA (2010) Gold nanoparticles for biology and medicine. Angew Chem Int Ed 49(19):3280–3294
32. Kneuer C, Sameti M, Bakowsky U, Schiestel T, Schirra H, Schmidt H, Lehr C-M (2000) A nonviral DNA delivery system based on surface modified silica-nanoparticles can efficiently transfect cells in vitro. Bioconjug Chem 11(6):926–932
33. Cutler JI, Auyeung E, Mirkin CA (2012) Spherical nucleic acids. J Am Chem Soc 134 (3):1376–1391
34. Mirkin CA, Letsinger RL, Mucic RC, Storhoff JJ (1996) A DNA-based method for rationally assembling nanoparticles into macroscopic materials. Nature 382(15):607–609
35. Giljohann DA, Seferos DS, Patel PC, Millstone JE, Rosi NL, Mirkin CA (2007) Oligonucleotide loading determines cellular uptake of DNA-modified gold nanoparticles. Nano Lett 7(12):3818–3821
36. Choi CHJ, Hao L, Narayan SP, Auyeung E, Mirkin CA (2013) Mechanism for the endocytosis of spherical nucleic acid nanoparticle conjugates. Proc Natl Acad Sci USA 110 (19):7625–7630
37. Patel PC, Giljohann DA, Daniel WL, Zheng D, Prigodich AE, Mirkin CA (2010) Scavenger receptors mediate cellular uptake of polyvalent oligonucleotide-functionalized gold nanoparticles. Bioconjug Chem 21(12):2250–2256
38. Massich MD, Giljohann DA, Seferos DS, Ludlow LE, Horvath CM, Mirkin CA (2009) Regulating immune response using polyvalent nucleic acid—gold nanoparticle conjugates. Mol Biopharm 6(6):1934–1940
39. Zheng D, Giljohann DA, Chen DL, Massich MD, Wang XQ, Iordanov H, Mirkin CA, Paller AS (2012) Topical delivery of siRNA-based spherical nucleic acid nanoparticle conjugates for gene regulation. Proc Natl Acad Sci USA 109(30):11975–11980
40. Barnaby SN, Lee A, Mirkin CA (2014) Probing the inherent stability of siRNA immobilized on nanoparticle constructs. Proc Natl Acad Sci USA 111(27):9739–9744
41. Giljohann DA, Seferos DS, Prigodich AE, Patel PC, Mirkin CA (2009) Gene regulation with polyvalent siRNA—nanoparticle conjugates. J Am Chem Soc 131(6):2072–2073
42. Rosi NL, Giljohann DA, Thaxton CS, Lytton-Jean AKR, Han MS, Mirkin CA (2006) Oligonucleotide-modified gold nanoparticles for intracellular gene regulation. Science 312 (5776):1027–1030
43. Seferos DS, Prigodich AE, Giljohann DA, Patel PC, Mirkin CA (2009) Polyvalent DNA nanoparticle conjugates stabilize nucleic acids. Nano Lett 9(1):308–311
44. Hao L, Patel PC, Alhasan AH, Giljohann DA, Mirkin CA (2011) Nucleic acid-gold nanoparticle conjugates as mimics of microRNA. Small 7(22):3158–3162
45. Jensen SA, Day ES, Ko CH, Hurley LA, Luciano JP, Kouri FM, Merkel TJ, Luthi AJ, Patel PC, Cutler JI, Daniel WL, Scott AW, Rotz MW, Meade TJ, Giljohann DA, Mirkin CA, Stegh AH (2013) Spherical nucleic acid nanoparticle conjugates as an RNAi-based therapy for glioblastoma. Sci Transl Med 5(209):209ra152
46. Lee J-S, Lytton-Jean AKR, Hurst SJ, Mirkin CA (2007) Silver nanoparticle—oligonucleotide conjugates based on DNA with triple cyclic disulfide moieties. Nano Lett 7(7):2112–2115
47. Cutler JI, Zheng D, Xu X, Giljohann DA, Mirkin CA (2010) Polyvalent oligonucleotide iron oxide nanoparticle "click" conjugates. Nano Lett 10(4):1477–1480

48. Zhang C, Macfarlane RJ, Young KL, Choi CHJ, Hao L, Auyeung E, Liu G, Zhou X, Mirkin CA (2013) A general approach to DNA-programmable atom equivalents. Nat Mater 12 (8):741–746
49. Mitchell GP, Mirkin CA, Letsinger RL (1999) Programmed assembly of DNA functionalized quantum dots. J Am Chem Soc 121(35):8122–8123
50. Young KL, Scott AW, Hao L, Mirkin SE, Liu G, Mirkin CA (2012) Hollow spherical nucleic acids for intracellular gene regulation based upon biocompatible silica shells. Nano Lett 12 (7):3867–3871
51. Banga RJ, Chernyak N, Narayan SP, Nguyen ST, Mirkin CA (2014) Liposomal spherical nucleic acids. J Am Chem Soc 136(28):9866–9869
52. Calabrese CM, Merkel TJ, Briley WE, Randeria PS, Narayan SP, Rouge JL, Walker DA, Scott AW, Mirkin CA (2015) Biocompatible infinite-coordination-polymer-nanoparticle—nucleic-acid conjugates for antisense gene regulation Angew Chem Int Ed 54(2):476−480
53. Cutler JI, Zhang K, Zheng D, Auyeung E, Prigodich AE, Mirkin CA (2011) Polyvalent nucleic acid nanostructures. J Am Chem Soc 133(24):9254–9257
54. Morris W, Briley WE, Auyeung E, Cabezas MD, Mirkin CA (2014) Nucleic acid-metal organic framework (MOF) nanoparticle conjugates. J Am Chem Soc 136(20):7261–7264
55. Alemdaroglu FE, Alemdaroglu NC, Langguth P, Herrmann A (2008) DNA block copolymer micelles—a combinatorial tool for cancer nanotechnology. Adv Mater 20(5):899–902
56. Li Z, Zhang Y, Fullhart P, Mirkin CA (2004) Reversible and chemically programmable micelle assembly with DNA block-copolymer amphiphiles. Nano Lett 4(6):1055–1058
57. Rouge JL, Hao L, Wu XA, Briley WE, Mirkin CA (2014) Spherical nucleic acids as a divergent platform for synthesizing RNA-nanoparticle conjugates through enzymatic ligation. ACS Nano 8(9):8837–8843
58. Alexis F, Pridgen E, Molnar LK, Farokhzad OC (2008) Factors affecting the clearance and biodistribution of polymeric nanoparticles. Mol Pharm 5(4):505–515
59. Kommareddy S, Amiji M (2007) Biodistribution and pharmacokinetic analysis of long-circulating thiolated gelatin nanoparticles following systemic administration in breast cancer-bearing mice. J Pharm Sci 96(2):397–407
60. Zhang K, Hao L, Hurst SJ, Mirkin CA (2012) Antibody-linked spherical nucleic acids for cellular targeting. J Am Chem Soc 134(40):16488–16491
61. Nanba D, Toki F, Barrandon Y, Higashiyama S (2013) Recent advances in the epidermal growth factor receptor/ligand system biology on skin homeostasis and keratinocyte stem cell regulation. J Dermatol Sci 72(2):81–86
62. Krex D, Klink B, Hartmann C, von Deimling A, Pietsch T, Simon M, Sabel M, Steinbach JP, Heese O, Reifenberger G, Weller M, Schackert G, Network ftGG (2007) Long-term survival with glioblastoma multiforme. Brain 130(10):2596–2606
63. Pardridge WM (2012) Drug transport across the blood-brain barrier. J Cereb Blood Flow Metab 32(11):1959–1972
64. Goti D, Hrzenjak A, Levak-Frank S, Frank S, Van Der Westhuyzen DR, Malle E, Sattler W (2001) Scavenger receptor class B, type I is expressed in porcine brain capillary endothelial cells and contributes to selective uptake of HDL-associated vitamin E. J Neurochem 76 (2):498–508
65. Mackic JB, Stins M, McComb JG, Calero M, Ghiso J, Kim KS, Yan SD, Stern D, Schmidt AM, Frangione B, Zlokovic BV (1998) Human blood-brain barrier receptors for Alzheimer's amyloid-beta 1–40. Asymmetrical binding, endocytosis, and transcytosis at the apical side of brain microvascular endothelial cell monolayer. J Clin Invest 102(4):734–743
66. Stegh AH, Brennan C, Mahoney JA, Forloney KL, Jenq HT, Luciano JP, Protopopov A, Chin L, DePinho RA (2010) Glioma oncoprotein Bcl2L12 inhibits the p53 tumor suppressor. Genes Dev 24(19):2194–2204
67. Stegh AH, Chin L, Louis DN, DePinho RA (2008) What drives intense apoptosis resistance and propensity for necrosis in glioblastoma? A role for Bcl2L12 as a multifunctional cell death regulator. Cell Cycle 7(18):2833–2839

68. Stegh AH, DePinho RA (2011) Beyond effector caspase inhibition Bcl2L12 neutralizes p53 signaling in glioblastoma. Cell Cycle 10(1):33–38
69. Stegh AH, Kesari S, Mahoney JE, Jenq HT, Forloney KL, Protopopov A, Louis DN, Chin L, DePinho RA (2008) Bcl2L12-mediated inhibition of effector caspase-3 and caspase-7 via distinct mechanisms in glioblastoma. Proc Natl Acad Sci USA 105(31):10703–10708
70. Stegh AH, Kim H, Bachoo RM, Forloney KL, Zhang J, Schulze H, Park K, Hannon GJ, Yuan J, Louis DN, DePinho RA, Chin L (2007) Bcl2L12 inhibits post-mitochondrial apoptosis signaling in glioblastoma. Genes Dev 21(1):98–111
71. Boveri M, Berezowski V, Price A, Slupek S, Lenfant A-M, Benaud C, Hartung T, Cecchelli R, Prieto P, Dehouck M-P (2005) Induction of blood-brain barrier properties in cultured brain capillary endothelial cells: comparison between primary glial cells and C6 cell line. Glia 51 (3):187–198
72. Cecchelli R, Dehouck B, Descamps L, Fenart L, Buée-Scherrer V, Duhem C, Lundquist S, Rentfel M, Torpier G, Dehouck MP (1999) In vitro model for evaluating drug transport across the blood–brain barrier. Adv Drug Deliv Rev 36(2–3):165–178
73. Culot M, Lundquist S, Vanuxeem D, Nion S, Landry C, Delplace Y, Dehouck M-P, Berezowski V, Fenart L, Cecchelli R (2008) An in vitro blood-brain barrier model for high throughput (HTS) toxicological screening. Toxicol In Vitro 22(3):799–811
74. Petros RA, DeSimone JM (2010) Strategies in the design of nanoparticles for therapeutic applications. Nat Rev Drug Discov 9(8):615–627
75. Huse JT, Holland EC (2009) Yin and yang: cancer-implicated miRNAs that have it both ways. Cell Cycle 8(22):3611–3612
76. Iorio MV, Croce CM (2009) MicroRNAs in cancer: small molecules with a huge impact. J Clin Oncol 27(34):5848–5856
77. Iorio MV, Croce CM (2012) Causes and Consequences of MicroRNA dysregulation. Cancer J 18(3):215–222 210.1097/PPO.1090b1013e318250c318001
78. Kouri FM, Hurley LA, Day ES, Hua Y, Merkel TJ, Queisser MA, Peng C-Y, Ritner C, Hao L, Daniel WL, Zhang H, Sznajder JI, Chin L, Giljohann DA, Kessler JA, Peter ME, Mirkin CA, Stegh AH (2015) miR-182 integrates apoptosis, growth and differentiation programs in glioblastoma Genes and Development, in press
79. Cancer Genome Atlas Research N (2008) Comprehensive genomic characterization defines human glioblastoma genes and core pathways. Nature 455(7216):1061–1068
80. Geusens B, Sanders N, Prow T, Van Gele M, Lambert J (2009) Cutaneous short-interfering RNA therapy. Expert Opin Drug Deliv 6(12):1333–1349
81. Leachman SA, Hickerson RP, Schwartz ME, Bullough EE, Hutcherson SL, Boucher KM, Hansen CD, Eliason MJ, Srivatsa GS, Kornbrust DJ, Smith FJD, McLean WHI, Milstone LM, Kaspar RL (2009) First-in-human mutation-targeted siRNA Phase Ib trial of an inherited skin disorder. Mol Ther 18(2):442–446
82. Proksch E, Brandner JM, Jensen J-M (2008) The skin: an indispensable barrier. Exp Dematol 17(12):1063–1072
83. Roberts PJ, Der CJ (2007) Targeting the Raf-MEK-ERK mitogen-activated protein kinase cascade for the treatment of cancer. Oncogene 26(22):3291–3310
84. Zhu H, Acquaviva J, Ramachandran P, Boskovitz A, Woolfenden S, Pfannl R, Bronson RT, Chen JW, Weissleder R, Housman DE, Charest A (2009) Oncogenic EGFR signaling cooperates with loss of tumor suppressor gene functions in gliomagenesis. Proc Natl Acad Sci USA 106(8):2712–2716
85. Dickens S, Van den Berge S, Hendrickx B, Verdonck K, Luttun A, Vranckx JJ (2010) Nonviral transfection strategies for keratinocytes, fibroblasts, and endothelial progenitor cells for ex vivo gene transfer to skin wounds. Tissue Eng Part C Methods 16(6):1601–1608
86. Jamieson ER, Lippard SJ (1999) Structure, recognition, and processing of cisplatin—DNA adducts. Chem Rev 99(9):2467–2498

87. Rosenberg B, Vancamp L, Trosko JE, Mansour VH (1969) Platinum compounds: a new class of potent antitumour agents. Nature 222(5191):385–386
88. Lorusso D, Petrelli F, Coinu A, Raspagliesi F, Barni S (2014) A systematic review comparing cisplatin and carboplatin plus paclitaxel-based chemotherapy for recurrent or metastatic cervical cancer. Gynecol Oncol 133(1):117–123
89. Dhar S, Daniel WL, Giljohann DA, Mirkin CA, Lippard SJ (2009) Polyvalent oligonucleotide gold nanoparticle conjugates as delivery vehicles for platinum(IV) warheads. J Am Chem Soc 131(41):14652–14653
90. Wang D, Lippard SJ (2005) Cellular processing of platinum anticancer drugs. Nat Rev Drug Discov 4(4):307–320
91. Zhang X-Q, Xu X, Lam R, Giljohann D, Ho D, Mirkin CA (2011) Strategy for increasing drug solubility and efficacy through covalent attachment to polyvalent DNA–nanoparticle conjugates. ACS Nano 5(9):6962–6970
92. Dubois J (2006) Recent progress in the development of docetaxel and paclitaxel analogues. Expert Opin Ther Pat 16(11):1481–1496
93. Marupudi NI, Han JE, Li KW, Renard VM, Tyler BM, Brem H (2007) Paclitaxel: a review of adverse toxicities and novel delivery strategies. Expert Opin Drug Saf 6(5):609–621
94. Panchagnula R (1998) Pharmaceutical aspects of paclitaxel. Int J Pharm 172(1–2):1–15
95. Skwarczynski M, Hayashi Y, Kiso Y (2006) Paclitaxel prodrugs: toward smarter delivery of anticancer agents. J Med Chem 49(25):7253–7269
96. Gavrieli Y, Sherman Y, Ben-Sasson SA (1992) Identification of programmed cell death in situ via specific labeling of nuclear DNA fragmentation. J Cell Biol 119(3):493–501
97. Baselga J, Swain SM (2009) Novel anticancer targets: revisiting ERBB2 and discovering ERBB3. Nat Rev Cancer 9(7):463–475
98. Hynes NE, Lane HA (2005) ERBB receptors and cancer: the complexity of targeted inhibitors. Nat Rev Cancer 5(5):341–354
99. Hendriks BS, Opresko LK, Wiley HS, Lauffenburger D (2003) Quantitative analysis of HER2-mediated effects on HER2 and epidermal growth factor receptor endocytosis: distribution of homo- and heterodimers depends on relative HER2 levels. J Biol Chem 278 (26):23343–23351
100. Song Y, Xu X, MacRenaris KW, Zhang X-Q, Mirkin CA, Meade TJ (2009) Multimodal gadolinium-enriched DNA–gold nanoparticle conjugates for cellular imaging. Angew Chem Int Ed 121(48):9307–9311
101. Aime S, Cabella C, Colombatto S, Geninatti Crich S, Gianolio E, Maggioni F (2002) Insights into the use of paramagnetic Gd(III) complexes in MR-molecular imaging investigations. J Magn Reson Imaging 16(4):394–406
102. Bloembergen N (1956) Spin relaxation processes in a two-proton system. Phys Rev 104 (6):1542–1547
103. Bloembergen N (1957) Proton relaxation times in paramagnetic solutions. Chem Phys 27 (2):572–573
104. Bloembergen N, Morgan LO (1961) Proton relaxation times in paramagnetic solutions. effects of electron spin relaxation. Chem Phys 34(3):842–850
105. Solomon I (1955) Relaxation processes in a system of two Spins. Phys Rev 99(2):559–565
106. Solomon I, Bloembergen N (1956) Nuclear magnetic interactions in the HF molecule. Chem Phys 25(2):261–266
107. Merbach AE, Toth E (eds) (2001) The chemistry of contrast agents in medical magnetic resonance imaging. Wiley, New York
108. Zheng J, Zhu G, Li Y, Li C, You M, Chen T, Song E, Yang R, Tan W (2013) A spherical nucleic acid platform based on self-assembled DNA biopolymer for high-performance cancer therapy. ACS Nano 7(8):6545–6554

109. Girvan AC, Teng Y, Casson LK, Thomas SD, Juliger S, Ball MW, Klein JB, Pierce WM Jr, Barve SS, Bates PJ (2006) AGRO100 inhibits activation of nuclear factor-kappaB (NF-kappaB) by forming a complex with NF-kappaB essential modulator (NEMO) and nucleolin. Mol Cancer Ther 5(7):1790–1799
110. Hwang DW, Ko HY, Lee JH, Kang H, Ryu SH, Song IC, Lee DS, Kim S (2010) A nucleolin-targeted multimodal nanoparticle imaging probe for tracking cancer cells using an aptamer. J Nucl Med 51(1):98–105
111. Wang K, You M, Chen Y, Han D, Zhu Z, Huang J, Williams K, Yang CJ, Tan W (2011) Self-assembly of a bifunctional DNA carrier for drug delivery. Angew Chem Int Ed 50 (27):6098–6101

Theranostic Magnetic Nanostructures (MNS) for Cancer

Vikas Nandwana, Mrinmoy De, Shihyao Chu, Manish Jaiswal, Matt Rotz, Thomas J. Meade and Vinayak P. Dravid

Abstract

Despite the complexities of cancer, remarkable diagnostic and therapeutic advances have been made during the past decade, which include improved genetic, molecular, and nanoscale understanding of the disease. Physical science and engineering, and nanotechnology in particular, have contributed to these developments through out-of-the-box ideas and initiatives from perspectives that are far removed from classical biological and medicinal aspects of cancer. Nanostructures, in particular, are being effectively utilized in sensing/diagnostics of cancer while nanoscale carriers are able to deliver therapeutic cargo for timed and controlled release at localized tumor sites. Magnetic nanostructures (MNS) have especially attracted considerable attention of researchers to address cancer diagnostics and therapy. A significant part of the promise of MNS lies in their potential for "theranostic" applications, wherein diagnostics makes use of the enhanced localized contrast in magnetic resonance imaging (MRI) while therapy leverages the ability of MNS to

V. Nandwana · V.P. Dravid (✉)
Department of Materials Science and Engineering, Northwestern University, Evanston, USA
e-mail: v-dravid@northwestern.edu

M. De
Department of Chemistry, Indian Institute of Science, Bengaluru, India

S. Chu
Sandia National Laboratory, Albuquerque, USA

M. Jaiswal
Department of Biomedical Engineering, Texas A&M University, College Station, USA

M. Rotz · T.J. Meade
Departments of Chemistry, Molecular BioSciences, Neurobiology, Biomedical Engineering, and Radiology, Northwestern University, Evanston, USA

C.A. Mirkin et al. (eds.), *Nanotechnology-Based Precision Tools for the Detection and Treatment of Cancer*, Cancer Treatment and Research 166,
DOI 10.1007/978-3-319-16555-4_3

heat under external radio frequency (RF) field for thermal therapy or use of thermal activation for release of therapy cargo. In this chapter, we report some of the key developments in recent years in regard to MNS as potential theranostic carriers. We describe that the r_2 relaxivity of MNS can be maximized by allowing water (proton) diffusion in the vicinity of MNS by polyethylene glycol (PEG) anchoring, which also facilitates excellent fluidic stability in various media and extended in vivo circulation while maintaining high r_2 values needed for T_2-weighted MRI contrast. Further, the specific absorption rate (SAR) required for thermal activation of MNS can be tailored by controlling composition and size of MNS. Together, emerging MNS show considerable promise to realize theranostic potential. We discuss that properly functionalized MNS can be designed to provide remarkable in vivo stability and accompanying pharmacokinetics exhibit organ localization that can be tailored for specific applications. In this context, even iron-based MNS show extended circulation as well as diverse organ accumulation beyond liver, which otherwise renders MNS potentially toxic to liver function. We believe that MNS, including those based on iron oxides, have entered a renaissance era where intelligent synthesis, functionalization, stabilization, and targeting provide ample evidence for applications in localized cancer theranostics.

Keywords
Magnetic nanostructures · Theranostics · Thermal activation · MR imaging · T_2 contrast agents

Contents

1 Introduction

Magnetic nanostructures (MNS) have emerged as promising functional probes for simultaneous diagnostics and therapeutics (theranostic) applications. The diagnostics potential of MNS arises from their role in enhancing the contrast in magnetic resonance imaging (MRI). The therapeutic prospects of MNS stem from thermal activation under external applied radio frequency (RF) field and/or localized release of therapeutic cargo, either through diffusive processes or triggered by thermal activation. Both of these attributes of MNS are related to their unique size-dependent physical properties as well as compatibility of size to typical biomolecules (Fig. 1) [1–7]. The characteristics of MNS are typically measured by saturation magnetization (M_s), remanent magnetization (M_r), and coercivity (H_c). Saturation magnetization is the maximum magnetization value of MNS under an applied magnetic field while remanent magnetization is the magnetization after removing that field. Coercivity is the strength of the applied magnetic field that is necessary to reverse the remanent magnetization back to zero [8]. Superparamagnetism (SPM) occurs when particles are small enough for thermal fluctuations to cause random flipping of magnetic moments, resulting in no remanent magnetization and coercivity in the absence of an applied magnetic field, akin to paramagnetism. However, under external applied magnetic field, the nanostructures can be magnetized and manipulated for transport and thermal activation. The superparamagnetic nanostructures are essential for biomedical studies where no remanent magnetization is critical in preventing their coagulation and sustaining a long period of circulation in the body. Additional properties of MNS required for biomedicine include greater magnetic susceptibility and high saturation magnetization that results in faster and stronger response, even at low external applied magnetic field, and also enhances thermal activation. The magnetic properties of MNS are dictated

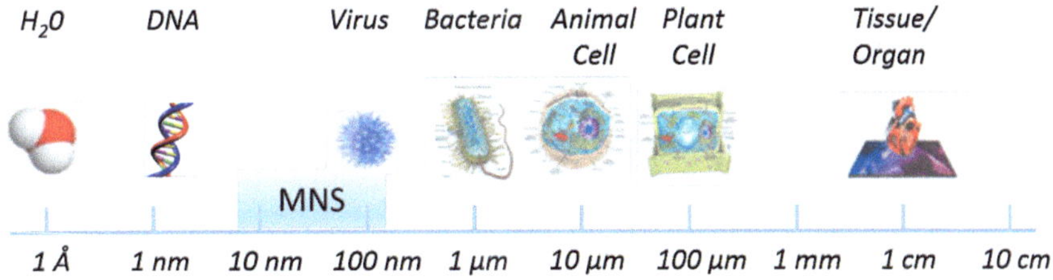

Fig. 1 Size scale of MNS as compared to biomolecules. MNS can be adapted to include biomolecules, drugs, or targeting and imaging molecules to form targeted MNS theranostic agents

by the composition, size, shape, and surface properties [9]. Hence, tuning control of these physical properties is essential for success of MNS for in vivo platform.

Typically, MNS comprise a magnetic core and biocompatible coating and/or surface functionalization that allows integration of targeting agents and bio/chemotherapeutics (Fig. 2). Targeting agents have been coupled with MNS for both diagnostic imaging and therapy of specific tumors [5, 10–12]. The diagnostic applications of MNS stem from the MRI [13]. MRI offers clinicians the ability to noninvasively obtain anatomic and metabolic/functional information with high spatial and temporal resolution [13–15]. The technique is based on the response of water proton spin in the presence of an applied magnetic field when triggered with a RF pulse. When external magnetic field is applied, protons align in one direction. Application of the RF pulse perturbs the alignment and the protons relax to the original state via two independent relaxation processes: longitudinal (T_1) and transverse (T_2) relaxation that are used to generate the MR images. The difference in water concentration and local environment between organs and tissues results in intrinsic contrast in MR images.

The spatial resolution as well as the sensitivity (S/N) of the MR images can be enhanced with the use of contrast agents. Paramagnetic molecular complexes, such as Gd(III) chelates, are used as T_1 contrast agents that increase signal intensity, i.e., higher r_1 relaxation, and appear bright in T_1-weighted images [16]. T_1 contrast agents are not covered in this chapter, however, a number of recent reviews have appeared [15–20]. MNS are used as T_2 contrast agents that decrease the signal

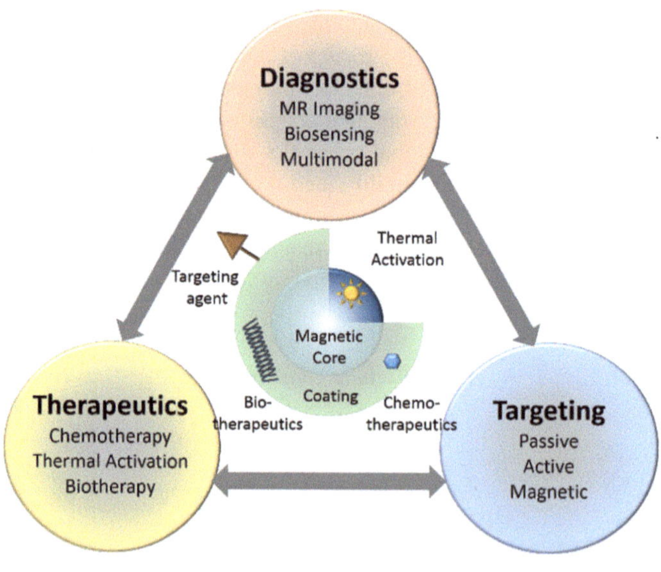

Fig. 2 Functional architecture of MNS and theranostic modalities. MNS are comprised of thermally active magnetic core and biocompatible coating and/or functionalization that allows integration of targeting agents and bio/chemotherapeutics

intensity, i.e., higher r_2 relaxation, and appear dark in T_2-weighted images. When water molecules (protons, more specifically) diffuse into the periphery of the induced dipole moment by MNS, the T_2 relaxation time of the protons is shortened, which enhances the negative contrast that helps in differentiating between pathogenic targets and normal tissues in T_2 weighted MRI images (Fig. 3).

Several MNS-based T_2 contrast agents (e.g., Feridex and Resovist) have been clinically approved [21, 22]. The MRI contrast enhancement effect is measured by the relaxation rate R_2 (s^{-1}) and the relaxivity coefficient r_2, a slope of R_2 against MNS concentration. The R_2 relaxation rate of MNS is defined as

$$R_2 = \frac{1}{T_2} = \frac{256\pi^2\gamma^2}{405} M_s^2 V \frac{r^2}{D(1 + \frac{L}{r})} \quad (1)$$

where T_2 is transverse relaxation time, γ is proton gyromagnetic ratio, M_s is saturation magnetization, V is volume of MNS, D is diffusion coefficient of water molecules, r is radius of MNS core, and L is thickness of MNS surface coating. [23]. The higher relaxivity corresponds to a better contrast effect. Based on Eq. (1), MNS should have high magnetization (M_s), large volume (V), and thin surface coating (small L) for better contrast effect.

MNS can generate heat under external RF field that make them very useful in cancer therapeutics. Under an external RF field (typically a few hundred kHz), superparamagnetic MNS switch their magnetization direction along the field directions, back and forth. The frictions caused by the physical rotation of the MNS (Brownian relaxation) and the magnetization reversal within the MNS (Neel relaxation) lead to the loss of magnetic energy and the generation of thermal energy [24]. The capability of generating heat at any targeted areas can be used for directly killing cancer cells via the thermal therapy and/or as an actuator for bio/chemo

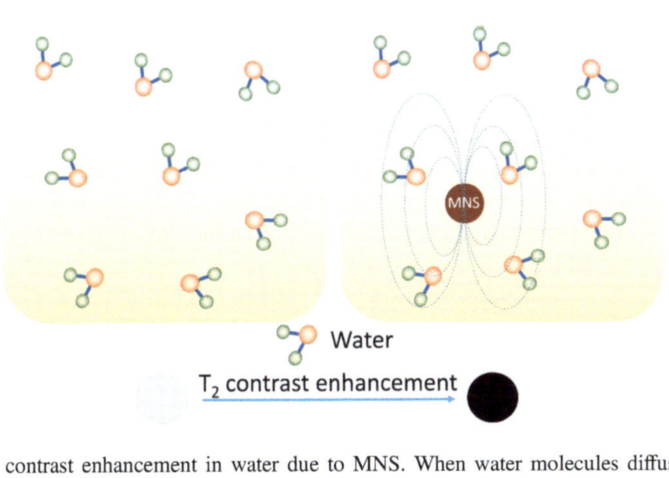

Fig. 3 T_2 contrast enhancement in water due to MNS. When water molecules diffuse into the periphery of the induced dipole moment by MNS, the T_2 relaxation time of the water protons is shortened which enhances the negative contrast

therapy cargo release. The thermal activation of MNS is measured by the specific absorption rate (SAR) that is measured as the initial temperature rise of the MNS solution per unit volume or mass.

$$SAR = \pi\mu_0\chi_0 H_0^2 f \frac{2\pi f \tau}{1 + (2\pi f \tau)^2} \tag{2}$$

where μ_0 is vacuum permeability, χ_0 is equilibrium susceptibility, H_0 is RF field amplitude, and f is frequency of the external RF field. The higher *SAR* is crucial for clinical use since that would require a smaller amount of MNS to be injected into the patient. According to Eq. (2), SAR highly depends on various parameters such as the size, size distribution, shape, chemical composition and surface modification, and saturation magnetization of the particles [25]. In addition, it is clear that SAR values depend on the frequency f and the field amplitude H_0 of the applied field. However, in order to apply hyperthermia safely to patients and avoid any detrimental effect on healthy tissues due to electromagnetic radiation exposure, the $H_0 f$ factor should not exceed a threshold that was experimentally estimated to equal 5×10^9 A m^{-1} s^{-1} [26]. Therefore, MNS with an exceptional SAR value that can generate heat under $H_0 f$ limit is highly desirable.

2 Synthesis and Characteristics of MNS: Prospects for Theranostics

For successful theranostic applications, MNS should be monodispersed and have uniform composition because the magnetic properties of MNS depend on the size, shape, and composition. It is clear from Eqs. (1) and (2) that the particles should possess high saturation magnetization and magnetic susceptibility, and be stable to a range of pH and salt concentrations. A key parameter for the magnetization of MNS is size. In a bulk magnetic material, all of the magnetic spins are aligned parallel to the applied magnetic field. However, in the nanoscale regime, a magnetically disordered spin-glass-like surface layer is formed. As the nanoparticle size decreases, such surface spin-canting effect becomes more pronounced and causes a drop in the saturation magnetization. While high saturation can be achieved with larger size particles, avoiding the surface-canting effect [27], the particle size should be under the superparamagnetic limit, which is typically less than ~20–30 nm for the majority of MNS. Further, the particles should have a coating or surface functional moieties that improve dispersion, biocompatibility, and provides a surface that can be functionalized. Strict attention to these parameters is essential during the design, synthesis, and formulation of MNS in order to be useful for in vivo applications.

MNS can be fabricated by either top-down (mechanical attrition) or bottom-up (chemical synthesis) approaches [9]. Since magnetic properties change with size and composition of MNS, chemical routes are preferred since they can synthesize

MNS with uniform composition and size. The chemical methods include co-precipitation, microemulsion, thermal decomposition and/or reduction, hydrothermal synthesis, and polyol synthesis. Two excellent reviews describing MNS fabrication methods have recently appeared and we will provide only a brief summary here [28, 29].

The most common synthetic strategy involves aqueous precipitation of iron salts with in situ, or post-synthesis addition of surfactant [30]. This strategy has notable limitations yielding monodispersity [31]. The microemulsion method does produce MNS of narrower size distribution compared to aqueous precipitation, but suffers from low yields [32]. Recently, the thermal decomposition/reduction method has gained considerable attention since this technique offers fine control over the final particle size, shape, and crystal structure compared to other methods and is scalable [33, 34]. Monodispersed MNS are formed due to the reaction conditions that yield a quick nucleation step followed by slower growth phase. However, the reaction occurs in organic solvent containing hydrophobic stabilizers, which requires additional surface modifications to the MNS to impart aqueous stability.

Here, we discuss different types of MNS synthesized using chemical methods and their magnetic properties. Table 1 summarizes different MNS core materials, their magnetic properties, and r_2 relaxivity.

2.1 Ferrite MNS

Fe_3O_4 MNS are extensively used in biomedicine because of their biocompatibility and ease of synthesis [47, 48]. The magnetic moment of superparamagnetic Fe_3O_4 MNS is dependent on the size with smaller particles producing lower magnetic moments [49]. Hence, the size of Fe_3O_4 MNS can be tuned by changing the reaction conditions such as reflux temperature, reflux time, and heating rate. The magnetic moment of Fe_3O_4 MNS was tuned from 25 to 43, 80, and 102 emu/(g Fe) with change in the size from 4, 6, 9, and 12 nm, respectively (resulting in r_2 values of 78, 106, 130, and 218 $mM^{-1}\ s^{-1}$) [27]. Further, the magnetic moment of Fe_3O_4 MNS can be modified by doping transition divalent metal ions (Co^{2+}, Ni^{2+}, Mn^{2+}, and Zn^{2+}). By adding different metal precursors during Fe_3O_4 synthesis, monodisperse MFe_2O_4 MNS were synthesized [38]. The metal ferrite nanostructures have an inverse spinel crystal structure composed of face-centered cubic packed lattice of oxygen atoms with octahedral sites (O_h) occupied by Fe^{3+} and M^{2+} ions and tetrahedral sites (T_d) occupied by Fe^{3+} ions (Fig. 4a). The magnetic spins of the ions at the O_h and T_h sites align opposite to each other. Hence the spins of Fe^{3+} ions at O_h and T_h cancel each other and net magnetization of MFe_2O_4 MNS is decided by magnetic moment of M^{2+} ions (Fig. 4b). The magnetization of $NiFe_2O_4$, $CoFe_2O_4$, Fe_3O_4, and $MnFe_2O_4$ were found to be 85, 99, 101, 110 (emu/g metal ions), respectively, depending upon the magnetic moments of the M^{2+} ions. This resulted in r_2 values of 152, 172, 218, and 358 $mM^{-1}\ s^{-1}$ (Fig. 4c) [38].

Table 1 Summary of MNS with core diameter, surface coating, magnetic properties, and r_2 relaxivity

MNS core material	Core diameter (nm)	Surface coating	Magnetic moment (emu/g)	B_0 (T)	r_2 ($mM^{-1}\ s^{-1}$)	References
Fe_3O_4 (Resovist)	4	Carboxy-Dextran	N/A	1.5	186	[35]
Fe_3O_4 (Feridex)	5	Dextran	45	1.5	120	[36]
Dy-SiO_2-$(Fe_3O_4)_n$	9	DMSA	N/A	9.4	397	[37]
Fe_3O_4	4–12	DMSA	25-101	1.5	78–218	[38]
Fe_3O_4	12	Nitrodopa-PEG600	N/A	1.5	396	Unpublished
Fe_3O_4	14	DSPE-mPEG1000	N/A	0.47	385	[39]
Fe_3O_4	58	DSPE-mPEG2000	132	1.5	324	[40]
$MnFe_2O_4$	6–12	DMSA	68–110	1.5	208–358	[38]
$CoFe_2O_4$	12	DMSA	99	1.5	172	[38]
$NiFe_2O_4$	12	DMSA	85	1.5	152	[38]
$Zn_{0.34}Fe_{0.66}OFe_2O_3$	5	DSPE-PEG	54.1	0.55	34.7	[41]
$Zn_{0.4}Fe_{0.6}Fe_2O_4$	15	DMSA	161	4.5	687	[42]
$Zn_{0.4}Mn_{0.6}Fe_2O_4$	15	DMSA	175	4.5	860	[42]
$Fe_{40}Co_{60}$	7	Phospholipid-PEG	215	1.5	644	[43]
Fe/Fe_3O_4	15	OAm-PEG	164	3	220	[44]
Fe/Fe_3O_4	16	DMSA	139	1.5	312	[45]
Fe/$MnFe_2O_4$	16	DMSA	149	0.47	356	[46]

r_2 transverse relaxivity; B_0 magnetic field strength; *DSPE-PEG* 1,2-Distearoyl-sn-glycero-3-phosphoethanolamine-*N*-[methoxy(polyethylene glycol)]; *DMSA* 2,3-dimercaptosuccinic acid; *OAm-PEG* oleylamine-α,ω-bis(2-carboxyethyl)poly(ethylene glycol)

Interestingly, doping of nonmagnetic Zn in Fe_3O_4 and $MnFe_2O_4$ MNS resulted in $(Zn_xFe_{1-x})Fe_2O_4$ and $(Zn_xMn_{1-x})Fe_2O_4$ MNS, respectively, that exhibit extremely high net magnetic moment and r_2 relaxivity [42]. The magnetization of $(Zn_xMn_{1-x})Fe_2O_4$ was dependent on the Zn doping level and were found to be 125, 140, 154, 166, 175, and 137 emu/g metal ions for $x = 0$, 0.1, 0.2, 0.3, 0.4, and 0.8, respectively, resulting in the r_2 values of 422, 516, 637, 754, 860, and 388 $mM^{-1}\ s^{-1}$, respectively (Fig. 5). The r_2 value for $Zn_{0.4}Mn_{0.6}Fe_2O_4$ MNS is the maximum r_2 value of reported to date for MNS [42], eight times higher than r_2 of Feridex [36]. $(Zn_xFe_{1-x})Fe_2O_4$ MNS exhibited a similar trend, but the magnetization and r_2 values were slightly lower than $(Zn_xMn_{1-x})Fe_2O_4$ MNS (Fig. 5) [42].

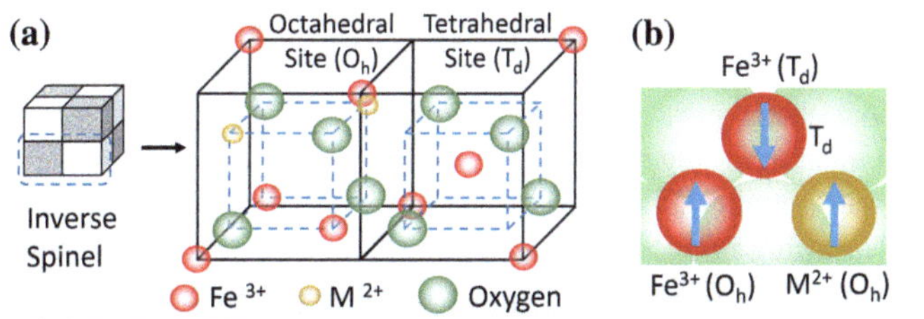

(c)	$MnFe_2O_4$	Fe_3O_4	$CoFe_2O_4$	$NiFe_2O_4$
Magnetic Moment	5 μ_B	4 μ_B	3 μ_B	2 μ_B
Saturation Magnetization (emu/g)	110	101	99	85
Relaxivity ($mM^{-1}s^{-1}$)	358	218	172	152

Fig. 4 *M*Fe_2O_4 (where *M* = Mn, Fe, Co, Ni) MNS with inverse spinel structure and its magnetic spin alignments. The mass magnetization values and r_2 relaxivity values of MFe_2O_4 MNS are proportional to the magnetic moments of the divalent ions (M^{2+}) [38]

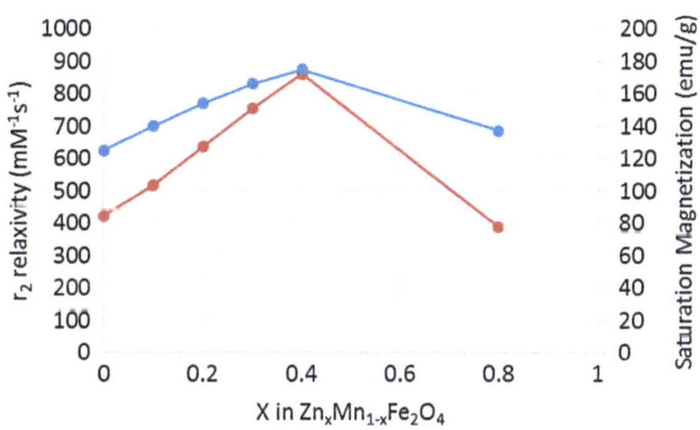

Fig. 5 Saturation magnetization and r_2 relaxivity (at 4.5 T) of $(Zn_xMn_{1-x})Fe_2O_4$ MNS at different Zn^{2+} doping levels. The $(Zn_xMn_{1-x})Fe_2O_4$ MNS showed significantly high r_2 relaxivities compared to conventional iron oxide MNS [42]

2.2 Metallic MNS

MNS based on transition metals of Fe, Co, Ni, and their alloys have higher magnetic moments than their oxide counterparts [44, 50–53]. Saturation magnetization of bulk FeCo (240 emu/g) and Fe (218 emu/g) is particularly high compared to bulk Fe_3O_4 (90 emu/g). Using the same mass of metallic MNS would then produce a far greater impact than their oxide counterparts, improving the T_2 contrast enhancement and therapeutic efficacy of drug delivery. However, the metallic MNS carries their own set of disadvantages like chemical instability, leaching of the noniron elements, and toxicity that renders them questionable for in vivo applications [9, 53, 54]. In addition, these pure metal nanoparticles are also ferromagnetic at room temperature, rather than superparamagnetic. This means that once they are magnetized, they will remain that way regardless of whether an external magnetic field is withdrawn, resulting in an aggregation. A number of reports have been published demonstrating coatings that prevent aggregation and ensure chemical stability of metallic MNS. Options under consideration include inert metals, such as Au and Ag, peptide capping ligands, ferrites, graphite, and silica [43, 44, 52, 55–57]. For example, after graphitic shell coating, FeCo MNS were stable up to 1 month and showed very high magnetic moment (215 emu/g) and r_2 relaxivity (644 mM^{-1} s^{-1}), which is far superior to conventional ferrite MNS [43, 52]. Crystalline Fe_3O_4 shell was also used to protect metallic Fe MNS that resulted in M_s and r_2 of 164 emu/g and 220 mM^{-1} s^{-1}, respectively [44, 56]. Co MNS were coated with Au shell to provide an inert, biocompatible and stable shell with a well-known surface chemistry. Though the bulk saturation magnetization of Co is ~160 emu/g, the measured value was found ~100 emu/g after Au coating, which was still higher than Fe_3O_4 MNS (75–80 emu/g) [55].

2.3 Multifunctional MNS

Hybrid MNS with two or more different functional units, such as Au–Fe_3O_4, FePt–CdS, and Fe_2O_3–carbon nanotube can be synthesized through seed mediated growth. In such a heterogeneous nanostructure, each unit exhibits its unique magnetic, optical, or electronic properties [10, 58–60]. Au–Fe_3O_4 nanostructures were prepared that preserved the optical property of Au (plasmonic absorption at ~530 nm) as well as the magnetic property of Fe_3O_4 MNS (M_s = 80 emu/g) [61]. This approach was extended to prepare semiconductor–metal alloy [62], semiconductor–metal oxide [63], and carbon nanotube–metal oxide complex [64]. MNS have been coupled with a wide range of fluorophores for multimodal imaging applications [65–67]. Lastly, T_1/T_2 MRI agent were prepared by conjugating Gd (III) based chelating agent with MNS [68].

3 Coating and Functionalization of MNS

In order to apply MNS in vitro and subsequently in vivo, the surface needs to be functionalized so that it; (i) protects against agglomeration; (ii) provides biocompatibility and chemical handles for the conjugation of drugs and targeting ligands; (iii) limits nonspecific cell interactions; and (iv) enhances MNS pharmacokinetics [69]. A diverse group of organic and inorganic coatings has been investigated including DMSA [27], PEG [70, 71], dextran [72], chitosan [73], liposomes [74], gold [75], and silica [76]. MNS coating can be achieved via a number of approaches, including in situ coating, post-synthesis adsorption, and post-synthesis end grafting [15]. Here, we discuss some of the most common coatings, their methods of attachment, and examples in cancer targeting.

3.1 Poly(Ethylene Glycol) (PEG)

PEG is a neutral and amphiphilic polymer that has been used clinically as excipients in FDA approved pharmaceutical formulations [77]. PEG coating of MNS improves their dispersion in biological media and increases blood circulation time since they are not readily recognized by the reticuloendothelial system (RES) [78]. Lutz et al. demonstrated in situ coating of PEG onto Fe_3O_4 MNS under aqueous conditions [79] while PEG grafting was achieved by single-point chemical anchoring through different functional groups including silanes [70], phosphate derivatives [80] and dopamine [81]. Peng and Sun reported ligand exchange with bifunctional PEG with dopamine [56]. Most recently, nitrodopamine has been proposed as an ultrastable chemical anchor for MNS [82]. We have developed MNS with high buffer stability by coating Fe_3O_4 with bifunctional PEG conjugated with nitrodopamine and carboxylate terminal groups [83]. The nirodopamine was covalently attached to the Fe_3O_4 surface by one end and the carboxylate group at the other end was kept open to functionalize with targeting ligands or therapeutic agents.

3.2 Dextran

Dextran is a branched polysaccharide comprised of glucose subunits and is widely used for MNS coatings because of its biocompatibility and polar interactions (chelation and hydrogen bonding). Addition of dextran during synthesis of Fe_3O_4 via the co-precipitation method resulted in dextran coated Fe_3O_4 MNS [84]. Subsequent iterations of this method produced clinically approved ferumoxtran-10 (AMI-277) and ferumoxides (AMI-25) [85–90]. These two structures have cores of ~5 nm, but differ significantly in dextran coating thickness (20–40 nm vs. 80–150 nm) which results in varying blood circulation times (24 h for ferumoxtran and 2 h for ferumoxides) [35]. Since the dextran molecules are adhered nonspecifically through hydroxyl interactions with the iron oxide core, there is always a possibility of desorption [91]. In order to prevent this, the dextran polymers were chemically cross-linked on MNS surface [92]. Using this strategy, clinically approved ferumoxytol [93] and ferucarbotran [22] have been synthesized.

3.3 Silica

Silica coating on MNS is popular because of the ease of synthesis and aqueous stability. By hydrolyzing silica precursors in basic solution, a uniform and thickness-controllable silica coating on MNS was obtained [76, 94, 95]. The silica shell have been used as a carrier for anticancer drugs (e.g., paclitaxel) and fluorescent molecules (e.g., fluorescein isothiocyanate (FITC)) [96]. Imparting additional functionalities to the silica coating have enabled targeting and labeling functionality. By addition of 3-aminopropyl-triethoxysilane (APS) to the silica precursors, we have coated silica shells with primary amine groups with controlled thickness [97]. Similarly, by reacting APS with isothiocyanate functionalized fluorescent dyes, Lu et al. were able to develop fluorescent MNS, a multimodal diagnostic agent [94]. Currently, silica coated MNS is available as ferumoxil (AMI-121), an orally ingested T_2 contrast agent for delineation of the intestinal loops from adjacent tissues and organs [21].

4 Pharmacokinetics and Biodistribution

The two most important factors that determine MNS pharmacokinetics are their surface characteristics and hydrodynamic size [98, 99]. Interplay of these properties with the reticuloendothelial system (RES) clearance determines plasma lifetime (blood circulation time). In RES clearance, circulating opsonin proteins adsorb to MNS surface (opsonization) that are recognized and removed from the bloodstream by tissue macrophages (Fig. 6). It has been shown that MNS with a hydrodynamic diameter of 10–100 nm are pharmacokinetically optimal for in vivo applications [100]. MNS smaller than 10 nm are subject to tissue extravasation and renal clearance, whereas those larger than 100 nm are quickly opsonized and eliminated from the circulation via the RES [31]. Decuzzi et al. produced models suggesting that within this range, smaller size nanostructures have longer blood circulation time [101].

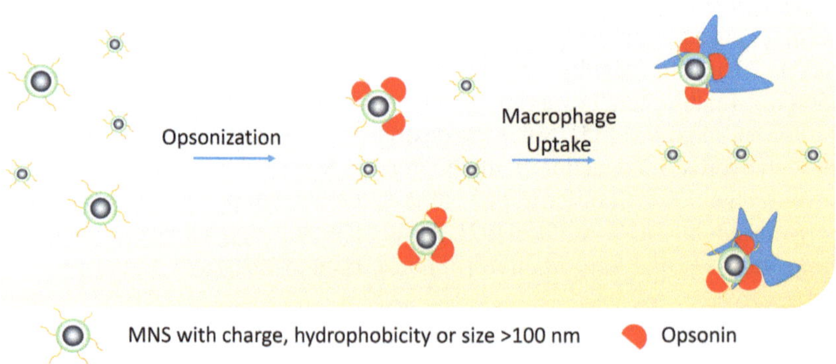

Fig. 6 Schematic illustration of RES clearance of MNS. MNS larger than 100 nm are absorbed by circulating opsonin proteins that are recognized by macrophages and removed from the bloodstream

With respect to the surface properties, charge can affect pharmacokinetics of MNS by enhancing their interactions with the plasma proteins and nontargeted cells, resulting in short blood circulation time [102]. In addition, hydrophobic groups on the surface of MNS induce the agglomeration of the MNS upon injection, leading to rapid removal by the RES. The rate of clearance, however, can be reduced by modification of MNS surfaces with coatings that resist RES interactions. As mentioned, surface modification with molecules such as the hydrophilic PEG has been a hallmark solution to many pharmacokinetic problems, including MNS [103, 104]. PEG chains linked to MNS reduce opsonization and macrophage uptake processes through steric repulsion, prolonging their circulation times [78, 105].

MNS biodistribution and cell uptake is significantly influenced by their physicochemical properties [99, 106, 107]. For example, it has been reported that MNS smaller than 150 nm accumulates in the bone marrow, heart, kidney, and stomach [108] while MNS larger than 150 nm are found in the liver and spleen [109]. Villanueva et al. showed that the charge and nature of surface functionalizing molecules on MNS affected their uptake of cancer cells [110]. They found that cells had effective uptake of positively charged aminodextran-MNS, minimal uptake of neutral charged dextran coated MNS, and low uptake of negatively charged DMSA-coated MNS [110]. Chouly et al. have found that negatively charged MNS gets opsonized quicker than neutral MNS and had greater liver uptake [99].

5 Targeting of MNS to Localized Cancer Tumors

The targeting of MNS for selected tumor tissues is critical in both diagnostic imaging and therapeutics [5, 10, 11]. Since nonspecific cell binding can place healthy tissue at risk, MNS have been engineered to target tumor tissues through passive and active targeting approaches.

5.1 Passive Targeting

Passive targeting uses the predetermined physicochemical properties of MNS to specifically migrate to selected tissues. The most common example of passive targeting is the enhanced permeability and retention (EPR) effect where MNS smaller than 200 nm can accumulate in many tumor tissues passively in solid tumors [111]. The compromised vasculature of a solid tumor facilitates passive MNS extravasation from the circulation into the tumor interstitium (Fig. 7) [112]. By contrast, endothelial cells of normal tissue vessels are closely packed and present a barrier for MNS penetration. However, passive targeting is limited to specific tumors since success of EPR effect depends on a number of factors such as lymphatic drainage rate, degree of capillary disorder, and blood flow which varies in different tumor types [113, 114].

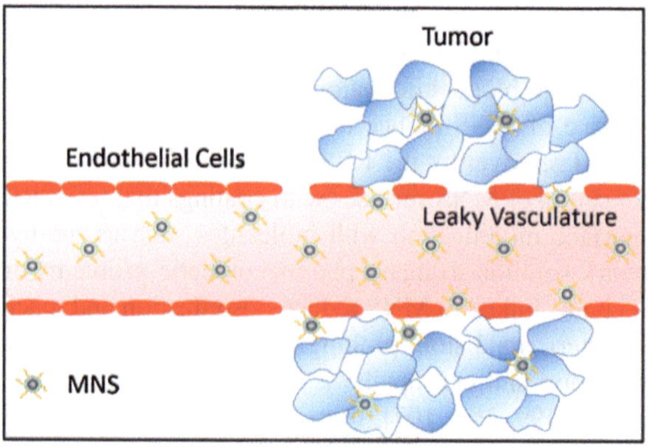

Fig. 7 Passive targeting of MNS via enhanced permeability and retention (EPR) effect. The compromised vasculature of a solid tumor facilitates extravasation of MNS of size less than 200 nm from the circulation into the tumor interstitium, while endothelial cells are closely packed and present a barrier for MNS penetration

5.2 Active Targeting with Targeting Agents

Because passive targeting is available for only certain types of tumors and does not necessarily guarantee internalization of MNS by targeted cells, MNS can be modified with tumor-selective agents to employ active targeting [115, 116]. These agents are complementary to unique receptors that are overexpressed or present on tumor cells. A variety of targeting agents have been used for MNS, depending on the specific target, and these are reviewed elsewhere [5, 10]. Some of the studies include: small organic molecules [115, 117, 118], peptides [119], proteins [120], and antibodies [27]. The density and molecular organization of these ligands significantly influence MNS binding to target cells due to the multivalency phenomenon [121]. Some of the targeting agents can be used to facilitate MNS internalization into cells, primarily via endocytosis [5]. However, synthesis of these targeting agents is expensive and involves complicated chemistry. Therefore, the process of scaling up the synthesis is challenging and may be a hurdle for clinical applications.

5.3 Active Targeting with External Magnetic Field

Accumulation of MNS can be realized by applying external magnetic field on the target site, a unique feature for MNS. Magnetic targeting has been studied for a number of tumor models [122, 123]. This technique was successfully implemented in a clinical trial to deliver the chemotherapeutic, doxorubicin, to hepatocarcinoma cells [124]. David et al. have explored magnetic targeting to brain tumors with PEI functionalized MNS [102, 122]. While successful, the efficacy of magnetic targeting is limited to target tissue that is close to the body's surface, since the magnetic field strength decreases with the distance from the magnetic source.

6 MNS for Diagnostic Imaging of Cancer

The diagnostic imaging applications of MNS have been realized as T_2 contrast agents in MR imaging over the past 20 years [13–15, 28, 125]. Owing to their significant deposition in liver, several MNS-based clinically approved T_2 contrast agents (Feridex I.V.®, Resovist®, and Gastromark®) have been used for liver imaging of humans [21, 22]. In order to extend MR visibility to image tumor at other parts, targeting agents have been coupled to MNS [85]. For example, Artemov et al. used Fe_3O_4 MNS conjugated with biotinylated Her-2/neu antibody Herceptin to generate strong T_2 contrast in breast cancer cell lines (AU-565, MCF-7, and MDA-MB-231) overexpressed with tyrosine kinase Her-2/neu receptors [126]. Contrast observed in MR images was found to be proportional to the expression level of kinase her-2/neu receptors for the given cell lines. Gao et al. successfully performed targeted MR imaging of human colon carcinoma xenograft tumors in mice by conjugating a cancer-targeting antibody (rch 24 mAb), to an 11 nm Fe_3O_4 MNS [127]. T_2 and T_2*-weighted MR images acquired before and after injection showed that the tumor site turned dark as early as 10 min after the injection of the rch 24 mAb conjugates and became darker and bigger until 24 h. In contrast, the Fe_3O_4 MNS without rch 24 mAb showed nearly no variation after injection [127]. Sun et al. reported c(RGDyK) peptide-coated Fe_3O_4 MNS and demonstrated their in vivo tumor-specific targeting capability [128]. When administrated intravenously in a mice bearing U87MG tumors, the c(RGDyK) peptide-coated Fe_3O_4 MNS accumulated preferentially in the integrin $\alpha v\beta 3$-rich tumor area resulting in a significant drop in the tumor MR signal intensity [128].

Other than target-specific molecular imaging, MNS-based contrast agents have been used in MRI for cell-based therapy since cells must be tracked in vivo to optimize cell therapy [129–134]. MRI-based immune cell tracking using MNS has been applied to many types of preclinical studies, such as tumor targeting of cytotoxic T cells and natural killer cells [135, 136], organ-specific targeting of autoimmune T cells [137], and neural stem cells [138]. De Vries et al. have shown that in vivo magnetic resonance tracking of MNS labeled dendritic cells is feasible in humans in conjunction with detailed anatomical information in melanoma patients. In contrast to scintigraphic imaging, MRI allowed assessment of the accuracy of dendritic cell delivery and of inter- and intranodal cell migration patterns [139–143]. Recently, Bulte et al. evaluated the long-term clinical tracking of MNS labeled stem cells after intracerebroventricular transplantation in an 18-month-old patient with global cerebral ischemia [144]. Twenty-four hours post-transplantation, MRI was able to detect hypointense cells in the occipital horn of the lateral ventricle. The signal gradually decreased over 4 months and became undetectable at 33 months.

To improve T_2 contrast for advanced MR imaging, researchers have been studying various parameters that affect the r_2 relaxivity of MNS. In the following sections, we discuss the parameters that have improved T_2 contrast of MNS for cancer diagnostic imaging. They have been divided into four categories; (i) size, shape, and composition control; (ii) nanoassembly of MNS; (iii) coating of MNS; and (iv) MNS with metal core.

6.1 Size, Shape, and Composition Control

According to Eq. 1, the R_2 of MNS is proportional to saturation magnetization and volume. Since the saturation magnetization of an MNS is proportional to its size due to surface spin-canting effects, the r_2 value of MNS can be increased by increasing the size of MNS. However, for biological imaging applications, the hydrodynamic size of MNS should be below 100 nm in order to have longer circulation times and to avoid nonspecific uptake [145]. Cheon et al. investigated the size effect where Fe_3O_4 MNS of diameter 4, 6, 9, and 12 nm resulted in the r_2 relaxivity of 78, 106, 130, and 218 $mM^{-1}\ s^{-1}$, respectively (Fig. 8). The MR contrast changed from light gray to black or from red to blue in color-coded images (Fig. 8) [27]. The 9 nm Fe_3O_4 MNS conjugated with Herceptin was used to image a breast cancer cell line, SK-BR-3, which possesses overexpressed HER2/neu cancer markers. Herceptin was selected due to its specific binding properties against a HER2/neu receptor. In the T_2-weighted MR images, treatment of 9 nm Fe_3O_4 MNS-Herceptin probe conjugates to the SK-BR03 breast cancer cell lines resulted in the significant negative contrast of the MR images compared to nontreated cell lines [27]. Chen et al. studied the size effect of polyvinylpyrrolidone (PVP)-coated Fe_3O_4 MNS (core size 8, 23, 37, and 65) on MRI of hepatic lesions in vivo [146]. PVP-Fe_3O_4 MNS with core size 37 and 65 nm showed higher r_2 relaxivity (239 and 248 $mM^{-1}\ s^{-1}$, respectively) compared to other sizes. When administered in nude mice bearing orthotopic Huh7 liver cancer, PVP-Fe_3O_4 MNS with 37 nm core size showed higher contrast change compared to Feridex in T_2 and T_2* weighted MR images [146].

The change in shape of the MNS has been used to increase r_2 relaxivity. We reported size and shape effects of $CoFe_2O_4$ MNS on their r_2 relaxivity. Spherical $CoFe_2O_4$ MNS of various sizes were synthesized via seed mediated growth method, while faceted irregular $CoFe_2O_4$ MNS were synthesized via the same method but in the presence of a magnetic field [147]. While the r_2 relaxivity coefficient of $CoFe_2O_4$ MNS increased with an increase in size for spherical particles, faceted $CoFe_2O_4$ MNS showed higher r_2 relaxivity than spherical $CoFe_2O_4$ MNS [147]. Recently, Gao et al. reported octapod shape Fe_3O_4 MNS that exhibited highest relaxivity of 679.3+/30 $mM^{-1}\ s^{-1}$ for Fe_3O_4 and demonstrated in vivo imaging and tumor detection [148].

As discussed earlier, the magnetization (and hence r_2) of MNS can be influenced by doping with magnetically susceptible elements. The effect of metal doping on

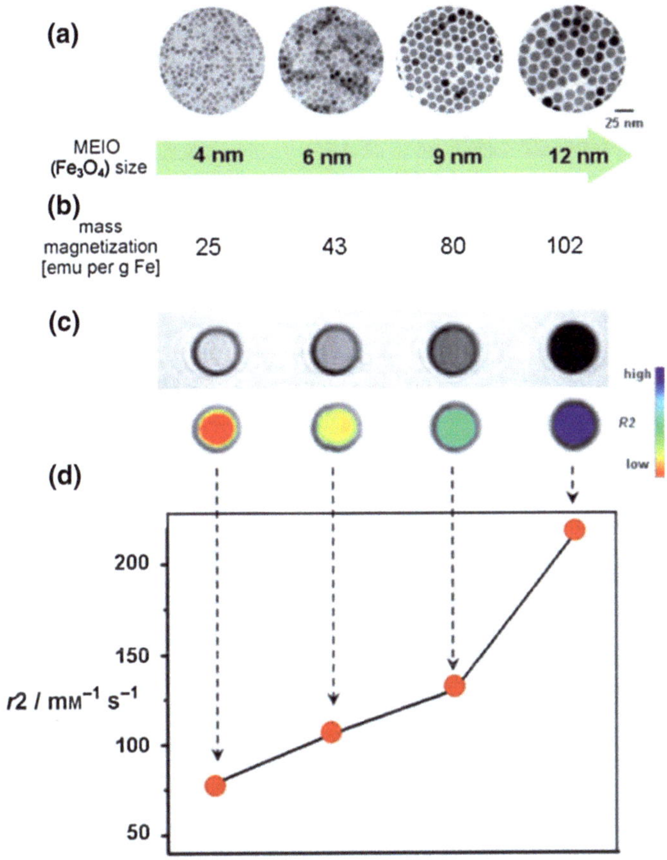

Fig. 8 Size effects of Fe_3O_4 MNS on r_2 relaxivity. **a** TEM images, **b** saturation magnetization values, **c** T_2-weighted MR images (*top black* and *white*, *bottom* color), and **d** the r_2 relaxivity values of 4, 6, 9, and 12 nm sized Fe_3O_4 MNS. The r_2 relaxivity value increased with size of Fe_3O_4 MNS which resulted in the T_2 contrast change from *light gray* to *black* in T_2 weighted MR images or from *red* to *blue* in the corresponding color-coded images. Reprinted with permission from [27]. Copyright 2005 American Chemical Society

the r_2 relaxivity was investigated for MFe_2O_4 in which Fe^{2+} ions were replaced by other transition metal dopants (Mn^{2+}, Zn^{2+}, Ni^{2+}, and Co^{2+}). The r_2 of 12 nm $MnFe_2O_4$ MNS was observed 358 mM^{-1} s^{-1} compared to 218 and 62 mM^{-1} s^{-1} from 12 nm Fe_3O_4 and dextran coated cross-linked 4 nm Fe_3O_4 (CLIO) MNS, respectively (Fig. 9) [38]. This increased MR contrast enhancement was tested to detect breast and ovarian cancer tumor in mice after Herceptin conjugation and intravenous injection (Fig. 9) [38]. $MnFe_2O_4$-Herceptin conjugates produced higher contrast than CLIO-Herceptin conjugates at the tumor site after 2 h (Fig. 9 a–d). Quantitatively, r_2 increase up to 34 % was observed for $MnFe_2O_4$-Herceptin

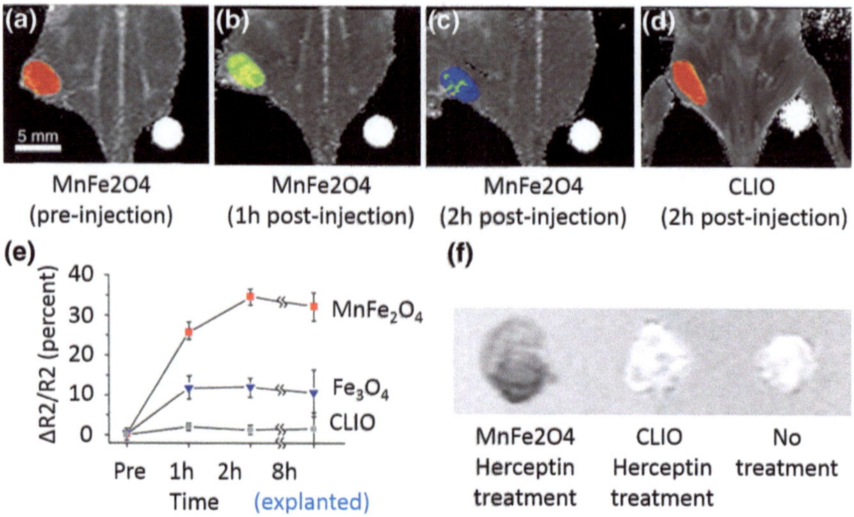

Fig. 9 In vivo MR detection of cancer in a mouse implanted with the cancer cell line NIH3T6.7 using 12 nm $MnFe_2O_4$, 12 nm Fe_3O_4 and dextran coated cross-linked 4 nm Fe_3O_4 (CLIO) MNS. T_2-weighted MR images of the mouse (**a**) before injection, (**b**) after 1 h injection, and (**c**) after 2 h injection of $MnFe_2O_4$ in comparison to (**d**) after 2 h injection of CLIO. $MnFe_2O_4$-Herceptin conjugates produced higher contrast than CLIO-Herceptin conjugates at the tumor site after 2 h. **e** Plot of R_2 change versus time. Increase in R_2 up to 34 % was observed for $MnFe_2O_4$-Herceptin conjugates in comparison to 5 and 13 % for CLIO-Herceptin conjugate (*dots*) and 12 nm Fe_3O_4-Herceptin conjugates, respectively. **f** Change in R_2 values was confirmed in the ex vivo MR images of explanted tumors (8 h). Reprinted with permission from [38]. Copyright 2007 Nature Publishing Group

conjugates in comparison to 5 and 13 % for CLIO-Herceptin conjugates and 12 nm Fe_3O_4-Herceptin conjugates, respectively (Fig. 9 e,f).

6.2 Nanoassembly of the MNS

Nanoassembly of MNS made up of numerous MNS have shown to increase r_2 relaxivity since an individual MNS in the nanoassembly is more efficient at dephasing the spins of the surrounding water protons [149]. For example, Feridex exhibits much higher r_2 relaxivity than monocrystalline iron oxide MNS because Feridex consists of several iron oxide nanoparticles embedded in a dextran shell [36]. Advances in chemical synthesis have enabled the preparation of nanoassemblies of tunable size and shape. Fe_3O_4 MNS have been assembled on dye-doped mesoporous silica nanoparticles (Fig. 10a) [37, 150]. The r_2 value of Fe_3O_4 coated silica nanoparticles was 2.8 times higher than that of well-dispersed Fe_3O_4 nanoparticles (Fig. 10b) [150]. After subcutaneous injection into each dorsal shoulder of a nude mouse, MCF-7 cells labeled with Fe_3O_4-MNS were clearly detected as a dark volume of subcutaneous tumor in the T_2-weighted MR image, while the

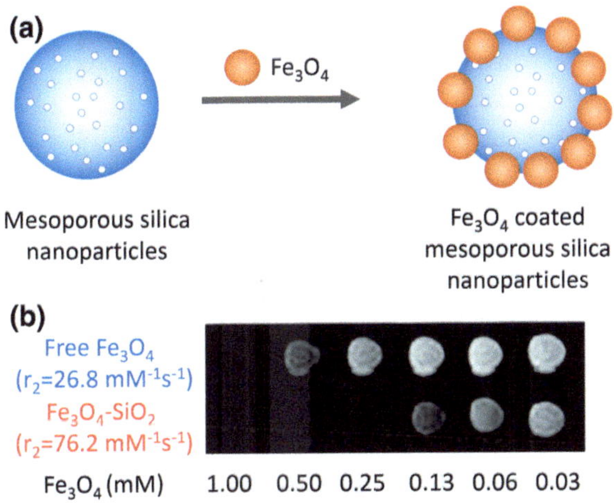

Fig. 10 MR signal enhancement by assembly of Fe_3O_4 on SiO_2 nanoparticles. **a** Schematic illustration of the synthetic procedure for Fe_3O_4 decorated mesoporous silica nanoparticles. **b** Relaxivity values and T_2 weighted MR image of Fe_3O_4 decorated SiO_2 nanoparticles (Fe_3O_4-MSN) and free Fe_3O_4 nanoparticles. The r_2 relaxivity of Fe_3O_4 decorated SiO_2 nanoparticles was increased by 2.8 times as compared to free Fe_3O_4 nanoparticles, hence darker signal was observed in T_2 weighted MR image at the same concentration of Fe. Reprinted with permission from Ref. [150]. Copyright 2010 American Chemical Society

unlabeled MCF-7 cells did not show any MR contrast enhancement [150]. We have demonstrated controlled assembly of 6 nm amine functionalized Fe_3O_4 MNS [151]. The size of aggregates was $\sim$40 nm and the r_2 value was 315 $mM^{-1}\ s^{-1}$, which was significantly higher than r_2 of monodisperse 6 nm Fe_3O_4 ($\sim$100 $mM^{-1}\ s^{-1}$) [38].

6.3 Coating of MNS

Typically, the role of the MNS coating is to provide stability, biocompatibility, and enhanced blood circulation times. In addition, the coating on MNS affect the r_2 relaxivity since it can increase the residence time of the surrounding water molecules by forming hydrogen bonds [152]. For example, polyethylene glycol (PEG) is one of the most common coatings that is used to make MNS stable in aqueous/biological media. Each ethylene glycol subunit in the PEG associates with two or three water molecules which slows water diffusion, resulting in a high r_2 relaxivity [152].

We have developed MNS with Fe_3O_4 core and PEG coating that shows excellent buffer stability [83]. It was found that the PEG coating not only provides stability but the thickness of PEG coating also affects the r_2 relaxivity of Fe_3O_4 MNS. The thickness was varied by changing the molecular weight of PEG that resulted in different r_2 relaxivities. The r_2 relaxivity of 12 nm Fe_3O_4 MNS was found at 160, 194, 277, and 396 $mM^{-1}\ s^{-1}$ when the molecular weight of PEG was 200, 400, 500,

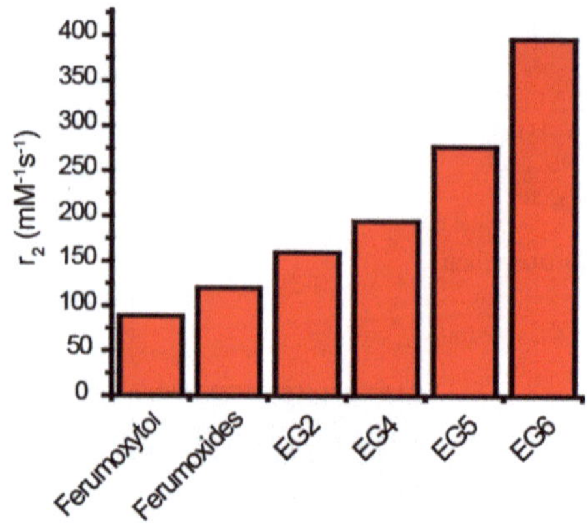

Fig. 11 r_2 relaxivity values of 12 nm nitrodopamine-PEG functionalized Fe_3O_4 MNS with molecular weight of PEG 200 (EG2), 400 (EG4), 500 (EG5), and 600 Da (EG6) in comparison to Ferumoxytol and Ferumoxides (unpublished). It was found that the PEG coating not only provides stability but the thickness of PEG coating also affects the r_2 relaxivity of Fe_3O_4 MNS. The highest r_2 value of 396 $mM^{-1}\ s^{-1}$ with PEG 600 (EG6) was almost four times that of Feridex (Ferumoxide)

and 600 Da (Fig. 11). Bao et al. reported different r_2 relaxivities for phospholipid (DSPE)-PEG-coated Fe_3O_4 MNS by changing their core size (5 and 14 nm) and PEG length (mol. wt. 550, 750, 1000, 2000, and 5000). The highest r_2 value of 385 $mM^{-1}\ s^{-1}$ was obtained for 14 nm Fe_3O_4 MNS coated with DSPE-PEG1000 [39]. In vivo tumor imaging was done using Fe_3O_4 MNS (5 and 14 nm) coated with DSPE-PEG1000 and conjugated with antibodies against mouse VEGFR-1. Tumors were induced by implanting human U87 glioblastoma cells subcutaneously in the nude mice. Tail vein injection of the 14 nm Fe_3O_4 MNS with r_2 relaxivity ~385 $mM^{-1}\ s^{-1}$ produced more T_2 contrast enhancement of the tumor tissues in comparison to 5 nm Fe_3O_4 whose relaxivity was 130 $mM^{-1}\ s^{-1}$.

Silica coating on MNS is another common method to provide aqueous stability. It has been observed that r_2 relaxivity of MNS decreases with increment in silica shell thickness. Pinho et al. observed a systematic decrease in the r_2 relaxivity, from 228 to 23 $mM^{-1}\ s^{-1}$, after coating silica shell up to 20 nm on a 10 nm γ-Fe_2O_3 MNS [95]. The authors attributed this effect to two regions of silica shell coating, an inner water-impermeable layer and an outer water-permeable layer. The outer layer provided colloidal stability while the impermeable layer reduced the interaction between the MNS and water protons significantly, resulting in a drop in r_2 relaxivity.

6.4 MNS with Metal Core

Although MNS with ferrite cores are promising MRI contrast agents, their saturation magnetization can be further improved since a portion of their magnetic spins cancels each other. MNS with metallic core such as Fe, Co, FeCo, FePt, and CoPt exhibit higher saturation magnetization since all their magnetic spins align in one direction and there is no canceled spin. Saturation magnetization of bulk FeCo and Fe is 240 and 218 emu/g, respectively compared to 90 emu/g of bulk Fe_3O_4 MNS. However, each metallic MNS has a specific limitation as MRI contrast agents. For example, FePt and CoPt are chemically stable but potentially toxic due to the possibility of leaching of Pt [153]. Fe, Co, and FeCo nanoparticles are biocompatible but are chemically unstable in air and prone to oxidation. To stabilize MNS with metallic core and maintain their magnetic properties, a number of coatings have been applied. MNS with Fe core and ferrite shell were reported that exhibited long-term stability in air [44–46]. Sun et al. reported bcc Fe/Fe_3O_4 core/shell MNS. Due to the high magnetization of Fe core, the r_2 relaxivity of Fe/Fe_3O_4 MNS (220 mM^{-1} s^{-1}) was found ~10 times higher than r_2 of Fe_3O_4 MNS (24 mM^{-1} s^{-1}) of the same size and coating [44, 56] and 2 times higher than the typical iron oxide NP contrast agent Feridex (110 mM^{-1} s^{-1}). However, no in vitro or in vivo studies were reported. Weissleder et al. further improved the r_2 relaxivity up to 430 mM^{-1} s^{-1} by coating $MnFe_2O_4$ shell on Fe nanoparticles. The authors intravenously injected Fe/$MnFe_2O_4$ MNS, Fe_3O_4 MNS of the same size and cross-linked iron oxide (CLIO). Taken at 3 h postinjection, the images verified that Fe/$MnFe_2O_4$ MNS resulted in the most significant darkening compared to CLIO and Fe_3O_4 MNS [46]. Dai et al. developed FeCo nanoparticles embedded in graphitic carbon (GC) shell that were discrete, chemically functionalized, and water soluble as desired for biological applications. Due to the high saturation magnetization, FeCo/GC MNS exhibited very high r_2 values (644 mM^{-1} s^{-1}). During in vivo intravascular MR imaging of the blood pool in the rabbit, mesenchymal stem cells labeled with FeCo/GC MNS showed significantly higher T_2 negative contrast enhancement compared to the ones labeled with Feridex [43, 52].

7 Thermally Activated MNS for Cancer Therapeutics

With the capability to generate thermal energy at targeted areas, MNS can be used in cancer therapeutics [5, 12, 154]. Compared with photodynamic therapy agents such as gold and graphene, MNS are advantageous for targets that reside deep inside the biological system without penetration depth problem. In addition, the fact that magnetic field causes no adverse effect on biological tissues serves as a distinctive benefit for noninvasive, in vivo applications.

The use of MNS in cancer therapeutic has been divided into three categories; (i) magnetic hyperthermia where MNS kill tumor cells via increase in tissue temperature; (ii) chemotherapy where MNS deliver a drug and trigger release at the tumor site; (iii) biotherapy where MNS are highly effective carrier platforms for bioactive molecules such as siRNA, oligos, and genes and facilitate transport of biomolecules in plasma membrane penetration necessary for cell internalization.

7.1 Magnetic Hyperthermia: MNS as Heat Generators

Targeted MNS can accumulate at the tumor site and increase the tissue temperature under an external RF field. Since the cancer tissues have higher heat sensitivity than normal tissues, thermal activation of MNS can be used to selectively kill tumor cells in the range of 41–47 °C [69]. As mentioned earlier, as MNS with high SAR have high efficacy for killing cancer cells, a variety of next-generation MNS with high SAR have been developed [155–157]. $(Zn_{0.4}Mn_{0.6})Fe_2O_4$ MNS have shown high SAR value of 432 W/g which is ~4 times higher than SAR of Feridex (115 W/g). The high SAR of $(Zn_{0.4}Mn_{0.6})Fe_2O_4$ MNS resulted in 84.4 % death of HeLa cancer cells in comparison to 13.5 % from Feridex (Fig. 12) [42]. Similarly, 82 % of Hela cells died when treated with $CoFe_2O_4$ MNS with SAR value of 238 W/g at 168 kHz [158]. Cheon et al. reported a significant increase in the SAR values of MNS by synthesizing core/shell MNS with hard ferromagnetic core and soft ferromagnetic shell [159]. Due to the exchange coupling between core and shell of MNS, the core/shell MNS showed SAR up to 3034 W/g which are an order of magnitude larger than SAR of conventional ferrite MNS (~300 W/g). Due to the high SAR values, the therapeutic efficacy of these MNS under RF was found superior to a common anticancer drug (doxorubicin). The same amounts (75 μg) of MNS and doxorubicin were injected into the tumor of a nude mice xenografted with cancer cells (U87MG). The tumor was clearly eliminated in 18 days in the group treated with the core/shell MNS, while in the doxorubicin-treated group tumor growth slowed initially, but then regrew after 18 days [159]. Recently, the same group reported Gd (III) texaphyrins (GdTx) conjugated $ZnFe_2O_4$ MNS with SAR of 471 W/g designed

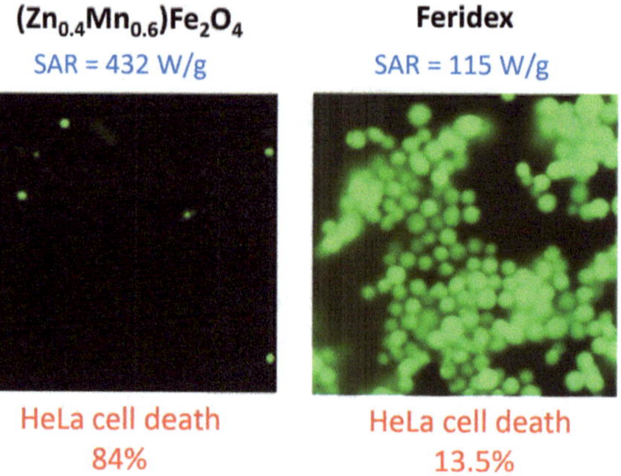

Fig. 12 SAR values and percentage of HeLa cells killed after treatment with $(Zn_{0.4}Mn_{0.6})Fe_2O_4$ MNS or Feridex in AMF. Fluorescence microscopy images of HeLa cells treated with $(Zn_{0.4}Mn_{0.6})Fe_2O_4$ nanoparticles (or Feridex) and stained with calcein show live cells as green fluorescence. $(Zn_{0.4}Mn_{0.6})Fe_2O_4$ MNS have shown SAR value of 432 W/g, ~4 times higher than SAR of Feridex (115 W/g) which resulted in 84.4 % death of HeLa cancer cells in comparison to 13.5 % from Feridex. Reprinted with permission from Ref. [42] Copyright © 2009 WILEY-VCH Verlag GmbH & Co. KGaA, Weinheim

for hyperthermic treatment for apoptosis. This system is a double-effector MNS that generates heat as well as reactive oxygen species (ROS) which remarkably increased the degree of apoptotic cell death. Xenograft tumors in mice treated with the double-effector MNS were eliminated within 8 days whereas the tumor in untreated mice increased eightfold in 8 days [160].

7.2 Biotherapy: MNS as Carrier for Gene Therapeutics

Gene therapy is a technique that uses DNA and antisense RNA (siRNA) to treat and prevent disease via gene expression and gene silencing of defective genes [161, 162]. The coupling of nucleic acids with MNS improves the plasma pharmacokinetics and plasma membrane penetration of nucleic acids necessary for internalization into cells [162]. MNS designed for gene therapy have been coated with cationic polymers such as polyethylenimine (PEI), polyamidoamine, or chitosan in order to conjugate with negatively charged nucleic acids. While the cationic MNS have shown great success in vitro, their applicability in vivo has been limited because of toxicity and instability in biological media [163]. Zhang et al. coated MNS with a copolymer of PEI, PEG, and chitosan (NP-CP-PEI) [164]. The chitosan and PEG suppressed the PEI toxicity while PEG also provided the stability. NP-CP-PEI demonstrated an innocuous toxic profile and a high level of expression of the delivered plasmid DNA in a C6 xenograft mouse model, while MNS coated with only PEI or chitosan showed high toxicity or low gene transfection efficiency, respectively [164]. The attachment of the targeting ligand, chlorotoxin (CTX), to NP-CP-PEI enhanced the gene transfection efficiency. Histology analysis and confocal microscopy of the C6 xenograft tumor sections showed more cells expressing GFP in tumors treated with the NP-CP-PEI attached with CTX compared to NP-CP-PEI without CTX [165]. One alternative to the cationic coatings was offered by conjugating siRNA to MNS by covalent bonding. Medarova et al. developed a dual purpose probe for the simultaneous noninvasive imaging and delivery of siRNAs to tumors. This probe consisted of MNS labeled with Cy5.5 dye and conjugated to a synthetic siRNA duplex targeting a gene of interest. With use of model (green fluorescent protein, GFP) and therapeutic (surviving) genes, the authors demonstrated that the targeting and delivery of the probe could be monitored in vivo by MRI and optical imaging. In addition, they were able to follow the silencing process by optical imaging and to correlate it with histological data [86, 87].

7.3 Chemotherapy: MNS as Drug Carrier/Release Trigger for Chemotherapeutics

Chemotherapy focuses on the treatment of disease through delivery of small molecule drug formulations [88]. Most of the drugs do not have cell-targeting capabilities which results in undesirable side effects when internalized by healthy cells. The success of MNS in diagnostic imaging has generated considerable interest in their use as drug delivery vehicles. MNS coatings provide anchor points to which drug molecules can be coupled. Integrating the drugs into MNS improves their targeting abilities, limits their side effects, and allows increment of the drug dosage at the diseased tissue [5, 10]. Currently, several drugs have been combined with MNS for cancer chemotherapy, including paclitaxel (PTX), doxorubicin (DOX), and methotrexate (MTX) [89, 90]. The therapeutic moieties can be covalently bonded to MNS with cleavable linkages, encapsulated in the hydrophobic coating on MNS, or physically absorbed on the surface of MNS.

An ideal drug delivery vehicle should have efficient drug loading and controllable drug release. In applications where the drug (such as MTX) has an affinity for the target cell, it can be advantageous to graft the drug to the surface of the MNS. Kohler et al. demonstrated covalent attachment of MTX to the surface of a PEG-coated MNS via a cleavable amide linkage [89]. However, the drug loading capacity via direct conjugation was found low due to the small number of functional groups on the surface of MNS. Hollow MNS have been used to increase chemotherapeutic efficacy due to higher drug loading capacity [166]. Sun et al. utilized porous hollow Fe_3O_4 MNS with 5 times higher cisplatin loading compared to solid Fe_3O_4 MNS [167]. Once coupled with Herceptin to the surface, the cisplatin-loaded hollow NPs targeted breast cancer SK-BR-3 cells with IC_{50} reaching 2.9 μM, much lower than 6.8 μM needed for free cisplatin. Labheshwar et al. coated a PEO-PPO diblock copolymer (Pluronic F127) on oleic acid coated MNS, where a hydrophobic region in the oleic acid/PPO layer provided drug loading of 8.2 and 9.5 % for DOX and PTX, respectively [168]. They found that MNS loaded with both DOX and PTX in a 1:1 ratio demonstrated highly synergistic antiproliferative activity in MCF-7 breast cancer cells relative to MNS loaded with only DOX or PTX.

The thermal energy from MNS has been used as an external trigger for controlled drug release. Thomas et al. loaded mesoporous silica nanoparticles with DOX and 15 nm $(Zn_{0.4}Fe_{0.6})Fe_2O_4$ MNS and capped the pores with cucurbit [6] uril that functioned as a heat labile molecular valve [169]. There was no drug release at the room temperature since the pores of the mesoporous silica remain capped, but under an external RF field, decapping occurred due to thermal activation of MNS, releasing most of the drug. In vitro, this controlled drug release killed 7-times more breast cancer cells. MNS coated with thermally responsive agents (e.g., hydrogels, thermosensitive polymers, lipids) have been explored where temperature works as a trigger for drug release [74, 170]. Recently, we reported poly(*N*-isopropylacrylamide) encapsulated Fe_3O_4 MNS in which DOX was loaded into the hydrogel via absorption. Drug release in presence of RF field was found 2 times higher than in absence of RF field due to thermal activation of MNS. In vitro localized drug delivery studies of the DOX loaded hydrogel-MNS composite with HeLa cell lines resulted in more than 80 % cell death under external RF field compared to 40 % cell death without RF field (Fig. 13) [83].

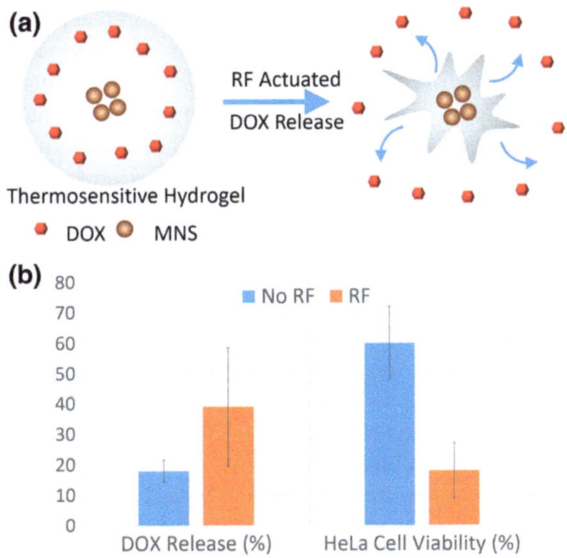

Fig. 13 **a** Schematic illustration of drug (DOX) release from thermoresponsive hydrogel-MNS composite. **b** Percent DOX release and cell viability of HeLa cell lines treated with the hydrogel-MNS composite with and without external RF field [83]

8 Summary and Outlook

Magnetic nanostructures (MNS) truly represent a prototypical nanotechnology platform in the sense that their properties and phenomena are unquestionably size dependent in the nanoscale regime. Nominal ferrimagnetic behavior of MNS changes to superparamagnetism below ~10–15 nm size scale, which is essential for colloidal stability of MNS. The perturbations of protons in vicinity of the MNS provide the essential characteristics for contrast in MRI, while enthalpic contributions from external RF field generate localized thermal energy for therapeutic purpose. Thus, the combined *theranostic* attributes of MNS arise from diagnostic imaging and thermal therapy combination.

Over and beyond the technical and scientific aspects of theranostic administration of cancer, MNS also embody the other important attribute of nanotechnology in terms of complementarity, integration, and synergy of nominally disparate fields and subjects. For just MNS alone, these subjects and technical themes include: physics of magnetism, chemistry of synthesis, materials science of structure–property relationship, surface science of functionalization, biomedical engineering in MR imaging protocols and RF activation parameters, and the core biology and medical themes of cancer targeting, diagnostics imaging, and therapy. As a result, this has brought together scientists, engineers, and clinical practitioners from diverse backgrounds for more than a decade to advance biomedical sensing, diagnostics, and therapeutics.

As demonstrated by the examples highlighted in this chapter, remarkable advances have been made in the recent decade to harness the size, composition, and size-dependent properties of MNS for cancer diagnostics, diagnostic imaging, and localized therapy. MNS continue to exhibit realistic potential to address diagnostic imaging by MRI and localized therapy via thermal activation and/or timed therapeutic cargo release. In particular, it has been shown that appropriate choice of composition, size, and surface functionalization has the potential for synergistic combination of diagnostics MR imaging and thermally activated therapy.

Despite some promising results obtained so far, including in vivo animal studies, there are specific challenges for effective use of MNS in humans; the final objective for any cancer theranostic platform. Regulatory approval for use in humans will require further and extensive safety and toxicology studies. The composition, surface properties, drug loading, biodistribution, and pharmacokinetics are the diverse factors that may affect the toxicity of the MNS in a direct or indirect manner and need to be understood thoroughly. Limitations also exist for targeting efficiency, the lack of homogeneous MNS penetration, and inadequate delivery of therapeutics into the tumor volume.

Research is continuing in this regard; including development of new magnetic core materials with higher relaxivity and thermal activation properties, along with design of new coating materials to improve the pharmacokinetics, biodistribution, and biocompatibility. Success with MRI and progress over the past few years offer considerable prospects for eventual diagnostic and therapeutic translation of MNS technology. Indeed, several diagnostic clinical trials using MNS have been initiated over the past few years. The increasing trend toward in vivo studies in animals and subsequent escalation to clinical trials are expected to help translate MNS from the laboratory to the clinic.

References

1. Kim J, Piao Y, Hyeon T (2009) Multifunctional nanostructured materials for multimodal imaging, and simultaneous imaging and therapy. Chem Soc Rev 38(2):372–390
2. Sun C, Lee JSH, Zhang MQ (2008) Magnetic nanoparticles in MR imaging and drug delivery. Adv Drug Deliv Rev 60(11):1252–1265
3. Ho D, Sun XL, Sun SH (2011) Monodisperse magnetic nanoparticles for theranostic applications. Acc Chem Res 44(10):875–882
4. Yoo D et al (2011) Theranostic magnetic nanoparticles. Acc Chem Res 44(10):863–874
5. Veiseh O, Gunn JW, Zhang MQ (2010) Design and fabrication of magnetic nanoparticles for targeted drug delivery and imaging. Adv Drug Deliv Rev 62(3):284–304
6. Xu CJ, Sun SH (2013) New forms of superparamagnetic nanoparticles for biomedical applications. Adv Drug Deliv Rev 65(5):732–743
7. Cole AJ, Yang VC, David AE (2011) Cancer theranostics: the rise of targeted magnetic nanoparticles. Trends Biotechnol 29(7):323–332
8. Livingston JD (1981) A review of coercivity mechanisms. J Appl Phys 52(3):2544–2548
9. Reddy LH et al (2012) Magnetic nanoparticles: design and characterization, toxicity and biocompatibility. Pharmaceutical and biomedical applications. Chem Rev 112 (11):5818–5878
10. McCarthy JR, Weissleder R (2008) Multifunctional magnetic nanoparticles for targeted imaging and therapy. Adv Drug Deliv Rev 60(11):1241–1251
11. Misra RDK (2008) Magnetic nanoparticle carrier for targeted drug delivery: perspective, outlook and design. Mater Sci Technol 24(9):1011–1019

12. Kievit FM, Zhang MQ (2011) Surface engineering of iron oxide nanoparticles for targeted cancer therapy. Acc Chem Res 44(10):853–862
13. Jun YW, Lee JH, Cheon J (2008) Chemical design of nanoparticle probes for high-performance magnetic resonance imaging. Angew Chem Int Ed 47(28):5122–5135
14. Lee N, Hyeon T (2012) Designed synthesis of uniformly sized iron oxide nanoparticles for efficient magnetic resonance imaging contrast agents. Chem Soc Rev 41(7):2575–2589
15. De M et al (2011) Hybrid magnetic nanostructures (MNS) for magnetic resonance imaging applications. Adv Drug Deliv Rev 63(14–15):1282–1299
16. Caravan P et al (1999) Gadolinium(III) chelates as MRI contrast agents: structure, dynamics, and applications. Chem Rev 99(9):2293–2352
17. Caravan P (2006) Strategies for increasing the sensitivity of gadolinium based MRI contrast agents. Chem Soc Rev 35(6):512–523
18. Frullano L, Meade TJ (2007) Multimodal MRI contrast agents. J Biol Inorg Chem 12 (7):939–949
19. Na HB, Hyeon T (2009) Nanostructured T1 MRI contrast agents. J Mater Chem 19 (35):6267–6273
20. Na HB, Song IC, Hyeon T (2009) Inorganic nanoparticles for MRI contrast agents. Adv Mater 21(21):2133–2148
21. Hahn PF et al (1990) 1st clinical-trial of a new superparamagnetic iron-oxide for use as an oral gastrointestinal contrast agent in MR imaging. Radiology 175(3):695–700
22. Reimer P, Balzer T (2003) Ferucarbotran (Resovist): a new clinically approved RES-specific contrast agent for contrast-enhanced MRI of the liver: properties, clinical development, and applications. Eur Radiol 13(6):1266–1276
23. Koenig SH, Kellar KE (1995) Theory of $1/T_1$ and $1/T_2$ NMRD profiles of solutions of magnetic nanoparticles. Magn Reson Med 34(2):227–233
24. Jordan A et al (1999) Magnetic fluid hyperthermia (MFH): cancer treatment with AC magnetic field induced excitation of biocompatible superparamagnetic nanoparticles. J Magn Magn Mater 201:413–419
25. Fortin JP et al (2007) Size-sorted anionic iron oxide nanomagnets as colloidal mediators for magnetic hyperthermia. J Am Chem Soc 129(9):2628–2635
26. Hergt R, Dutz S (2007) Magnetic particle hyperthermia-biophysical limitations of a visionary tumour therapy. J Magn Magn Mater 311(1):187–192
27. Jun YW et al (2005) Nanoscale size effect of magnetic nanocrystals and their utilization for cancer diagnosis via magnetic resonance imaging. J Am Chem Soc 127(16):5732–5733
28. Laurent S et al (2008) Magnetic iron oxide nanoparticles: synthesis, stabilization, vectorization, physicochemical characterizations, and biological applications. Chem Rev 108(6):2064–2110
29. Roca AG et al (2009) Progress in the preparation of magnetic nanoparticles for applications in biomedicine. J Phys D Appl Phys 42(22):224002
30. Massart R (1981) Preparation of aqueous magnetic liquids in alkaline and acidic media. IEEE Trans Magn 17(2):1247–1248
31. Gupta AK, Gupta M (2005) Synthesis and surface engineering of iron oxide nanoparticles for biomedical applications. Biomaterials 26(18):3995–4021
32. Loo AL et al (2008) Synthesis of magnetic nanoparticles in bicontinuous microemulsions. Effect of surfactant concentration. J Mater Sci 43(10):3649–3654
33. Murray CB, Kagan CR, Bawendi MG (2000) Synthesis and characterization of monodisperse nanocrystals and close-packed nanocrystal assemblies. Annu Rev Mater Sci 30:545–610
34. Park J et al (2004) Ultra-large-scale syntheses of monodisperse nanocrystals. Nat Mater 3 (12):891–895
35. Wang YXJ, Hussain SM, Krestin GP (2001) Superparamagnetic iron oxide contrast agents: physicochemical characteristics and applications in MR imaging. Eur Radiol 11 (11):2319–2331

36. Jung CW, Jacobs P (1995) Physical and chemical-properties of superparamagnetic iron-oxide MR contrast agents—ferumoxides, ferumoxtran, ferumoxsil. Magn Reson Imaging 13 (5):661–674
37. Lee JH et al (2006) Dual-mode nanoparticle probes for high-performance magnetic resonance and fluorescence imaging of neuroblastoma. Angew Chem Int Ed 45 (48):8160–8162
38. Lee JH et al (2007) Artificially engineered magnetic nanoparticles for ultra-sensitive molecular imaging. Nat Med 13(1):95–99
39. Tong S et al (2010) Coating optimization of superparamagnetic iron oxide nanoparticles for high T-2 relaxivity. Nano Lett 10(11):4607–4613
40. Lee N et al (2011) Magnetosome-like ferrimagnetic iron oxide nanocubes for highly sensitive MRI of single cells and transplanted pancreatic islets. Proc Natl Acad Sci USA 108 (7):2662–2667
41. Barcena C et al (2008) Zinc ferrite nanoparticles as MRI contrast agents. Chem Commun 19:2224–2226
42. Jang JT et al (2009) Critical enhancements of MRI contrast and hyperthermic effects by dopant-controlled magnetic nanoparticles. Angew Chem Int Ed 48(7):1234–1238
43. Seo WS et al (2006) FeCo/graphitic-shell nanocrystals as advanced magnetic-resonance-imaging and near-infrared agents. Nat Mater 5(12):971–976
44. Lacroix LM et al (2011) Stable single-crystalline body centered cubic Fe nanoparticles. Nano Lett 11(4):1641–1645
45. Lee H, Yoon TJ, Weissleder R (2009) Ultrasensitive detection of bacteria using core-shell nanoparticles and an NMR-filter system. Angew Chem Int Ed 48(31):5657–5660
46. Yoon TJ et al (2011) Highly magnetic core-shell nanoparticles with a unique magnetization mechanism. Angew Chem Int Ed 50(20):4663–4666
47. Hu FQ et al (2010) High-performance nanostructured MR contrast probes. Nanoscale 2 (10):1884–1891
48. Hu FQ et al (2010) Highly dispersible, superparamagnetic magnetite nanoflowers for magnetic resonance imaging. Chem Commun 46(1):73–75
49. Morales MP et al (1999) Surface and internal spin canting in gamma-Fe2O3 nanoparticles. Chem Mater 11(11):3058–3064
50. Puntes VF, Krishnan KM, Alivisatos AP (2001) Colloidal nanocrystal shape and size control: the case of cobalt. Science 291(5511):2115–2117
51. Cordente N et al (2001) Synthesis and magnetic properties of nickel nanorods. Nano Lett 1 (10):565–568
52. Desvaux C et al (2005) Multimillimetre-large superlattices of air-stable iron-cobalt nanoparticles. Nat Mater 4(10):750–753
53. Schutz-Sikma EA et al (2011) Probing the chemical stability of mixed ferrites: implications for magnetic resonance contrast agent design. Chem Mater 23(10):2657–2664
54. Colombo M et al (2012) Biological applications of magnetic nanoparticles. Chem Soc Rev 41(11):4306–4334
55. Bao YP, Krishnan KM (2005) Preparation of functionalized and gold-coated cobalt nanocrystals for biomedical applications. J Magn Magn Mater 293(1):15–19
56. Peng S et al (2006) Synthesis and stabilization of monodisperse Fe nanoparticles. J Am Chem Soc 128(33):10676–10677
57. Ni XM et al (2010) Silica-coated iron nanoparticles: shape-controlled synthesis, magnetism and microwave absorption properties. Mater Chem Phys 120(1):206–212
58. Cozzoli PD, Pellegrino T, Manna L (2006) Synthesis, properties and perspectives of hybrid nanocrystal structures. Chem Soc Rev 35(11):1195–1208
59. Donega CD (2011) Synthesis and properties of colloidal heteronanocrystals. Chem Soc Rev 40(3):1512–1546
60. Gao JH, Gu HW, Xu B (2009) Multifunctional magnetic nanoparticles: design, synthesis, and biomedical applications. Acc Chem Res 42(8):1097–1107

61. Xu CJ, Wang BD, Sun SH (2009) Dumbbell-like Au-Fe_3O_4 nanoparticles for target-specific platin delivery. J Am Chem Soc 131(12):4216–4217
62. Gu HW et al (2004) Facile one-pot synthesis of bifunctional heterodimers of nanoparticles: a conjugate of quantum dot and magnetic nanoparticles. J Am Chem Soc 126(18):5664–5665
63. Kwon KW, Shim M (2005) gamma-Fe_2O_3/II-VI sulfide nanocrystal heterojunctions. J Am Chem Soc 127(29):10269–10275
64. Choi JH et al (2007) Multimodal biomedical imaging with asymmetric single-walled carbon nanotube/iron oxide nanoparticle complexes. Nano Lett 7(4):861–867
65. Gao JH et al (2007) Fluorescent magnetic nanocrystals by sequential addition of reagents in a one-pot reaction: a simple preparation for multifunctional nanostructures. J Am Chem Soc 129(39):11928–11935
66. Pittet MJ et al (2006) Labeling of immune cells for in vivo imaging using magnetofluorescent nanoparticles. Nat Protoc 1(1):73–79
67. Wang DS et al (2004) Superparamagnetic Fe_2O_3 Beads-CdSe/ZnS quantum dots core-shell nanocomposite particles for cell separation. Nano Lett 4(3):409–413
68. Choi JS et al (2010) Self-confirming "AND" logic nanoparticles for fault free MRI. J Am Chem Soc 132(32):11015–11017
69. Gupta AK et al (2007) Recent advances on surface engineering of magnetic iron oxide nanoparticles and their biomedical applications. Nanomedicine 2(1):23–39
70. Kohler N, Fryxell GE, Zhang MQ (2004) A bifunctional poly(ethylene glycol) silane immobilized on metallic oxide-based nanoparticles for conjugation with cell targeting agents. J Am Chem Soc 126(23):7206–7211
71. Xie J et al (2007) Controlled PEGylation of monodisperse Fe_3O_4 nanoparticles for reduced non-specific uptake by macrophage cells. Adv Mater 19(20):3163–3166
72. Mornet S, Portier J, Duguet E (2005) A method for synthesis and functionalization of ultrasmall superparamagnetic covalent carriers based on maghemite and dextran. J Magn Magn Mater 293(1):127–134
73. Kim DH et al (2009) Targeting to carcinoma cells with chitosan- and starch-coated magnetic nanoparticles for magnetic hyperthermia. J Biomed Mater Res Part A 88A(1):1–11
74. Pradhan P et al (2010) Targeted temperature sensitive magnetic liposomes for thermo-chemotherapy. J Controlled Release 142(1):108–121
75. Wang LY et al (2008) Core@shell nanomaterials: gold-coated magnetic oxide nanoparticles. J Mater Chem 18(23):2629–2635
76. Ma DL et al (2007) Superparamagnetic Fe(x)Oy@SiO_2 core-shell nanostructures: controlled synthesis and magnetic characterization. J Phys Chem C 111(5):1999–2007
77. Fuertges F, Abuchowski A (1990) The clinical efficacy of poly(ethylene glycol)-modified proteins. J Controlled Release 11(1–3):139–148
78. Harris JM, Chess RB (2003) Effect of pegylation on pharmaceuticals. Nat Rev Drug Discov 2(3):214–221
79. Lutz JF et al (2006) One-pot synthesis of PEGylated ultrasmall iron-oxide nanoparticles and their in vivo evaluation as magnetic resonance imaging contrast agents. Biomacromolecules 7(11):3132–3138
80. Kim SW et al (2005) Phosphine oxide polymer for water-soluble nanoparticles. J Am Chem Soc 127(13):4556–4557
81. Xu CJ et al (2004) Dopamine as a robust anchor to immobilize functional molecules on the iron oxide shell of magnetic nanoparticles. J Am Chem Soc 126(32):9938–9939
82. Amstad E et al (2009) Ultrastable iron oxide nanoparticle colloidal suspensions using dispersants with catechol-derived anchor groups. Nano Lett 9(12):4042–4048
83. Jaiswal MK et al (2014) Thermoresponsive magnetic hydrogels as theranostic nanoconstructs. ACS Appl Mater Interfaces 6(9):6237–6247
84. Molday RS, Mackenzie D (1982) Immunospecific ferromagnetic iron-dextran reagents for the labeling and magnetic separation of cells. J Immunol Methods 52(3):353–367

85. Weissleder R, Pittet MJ (2008) Imaging in the era of molecular oncology. Nature 452 (7187):580–589
86. Medarova Z et al (2007) In vivo imaging of siRNA delivery and silencing in tumors. Nat Med 13(3):372–377
87. Kumar M et al (2010) Image-guided breast tumor therapy using a small interfering RNA nanodrug. Cancer Res 70(19):7553–7561
88. DeVita VT, Chu E (2008) A history of cancer chemotherapy. Cancer Res 68(21):8643–8653
89. Kohler N et al (2006) Methotrexate-immobilized poly(ethylene glycol) magnetic nanoparticles for MR imaging and drug delivery. Small 2(6):785–792
90. Sun C et al (2008) In vivo MRI detection of gliomas by chlorotoxin-conjugated superparamagnetic nanoprobes. Small 4(3):372–379
91. Bautista MC et al (2005) Surface characterisation of dextran-coated iron oxide nanoparticles prepared by laser pyrolysis and coprecipitation. J Magn Magn Mater 293(1):20–27
92. Wunderbaldinger P, Josephson L, Weissleder R (2002) Crosslinked iron oxides (CLIO): a new platform for the development of targeted MR contrast agents. Acad Radiol 9:S304–S306
93. Li W et al (2005) First-pass contrast-enhanced magnetic resonance angiography in humans using ferumoxytol, a novel ultrasmall superparamagnetic iron oxide (USPIO)-based blood pool agent. J Magn Reson Imaging 21(1):46–52
94. Lu Y et al (2002) Modifying the surface properties of superparamagnetic iron oxide nanoparticles through a sol-gel approach. Nano Lett 2(3):183–186
95. Pinho SLC et al (2010) Fine tuning of the relaxometry of gamma-Fe_2O_3@SiO_2 nanoparticles by tweaking the silica coating thickness. ACS Nano 4(9):5339–5349
96. Liong M et al (2008) Multifunctional inorganic nanoparticles for imaging, targeting, and drug delivery. ACS Nano 2(5):889–896
97. Aslam M, Li S, Dravid VP (2007) Controlled synthesis and stability of Co@SiO_2 aqueous colloids. J Am Ceram Soc 90(3):950–956
98. Yoo JW, Chambers E, Mitragotri S (2010) Factors that control the circulation time of nanoparticles in blood: challenges, solutions and future prospects. Curr Pharm Des 16 (21):2298–2307
99. Chouly C et al (1996) Development of superparamagnetic nanoparticles for MRI: effect of particle size, charge and surface nature on biodistribution. J Microencapsul 13(3):245–255
100. Gupta AK, Wells S (2004) Surface-modified superparamagnetic nanoparticles for drug delivery: preparation, characterization, and cytotoxicity studies. IEEE Trans Nanobiosci 3 (1):66–73
101. Decuzzi P et al (2006) The effective dispersion of nanovectors within the tumor microvasculature. Ann Biomed Eng 34(4):633–641
102. Chertok B, David AE, Yang VC (2010) Polyethyleneimine-modified iron oxide nanoparticles for brain tumor drug delivery using magnetic targeting and intra-carotid administration. Biomaterials 31(24):6317–6324
103. Sun CR et al (2010) PEG-mediated synthesis of highly dispersive multifunctional superparamagnetic nanoparticles: their physicochemical properties and function in vivo. ACS Nano 4(4):2402–2410
104. Larsen EKU et al (2009) Size-dependent accumulation of PEGylated silane-coated magnetic iron oxide nanoparticles in murine tumors. ACS Nano 3(7):1947–1951
105. Yallapu MM et al (2010) PEG-functionalized magnetic nanoparticles for drug delivery and magnetic resonance imaging applications. Pharm Res 27(11):2283–2295
106. Longmire M, Choyke PL, Kobayashi H (2008) Clearance properties of nano-sized particles and molecules as imaging agents: considerations and caveats. Nanomedicine 3(5):703–717
107. Gratton SEA et al (2008) The effect of particle design on cellular internalization pathways. Proc Natl Acad Sci USA 105(33):11613–11618
108. Moghimi SM (1995) Exploiting bone-marrow microvascular structure for drug-delivery and future therapies. Adv Drug Deliv Rev 17(1):61–73

109. Moghimi SM (1995) Mechanisms of splenic clearance of blood-cells and particles—towards development of new splenotropic agents. Adv Drug Deliv Rev 17(1):103–115
110. Villanueva A et al (2009) The influence of surface functionalization on the enhanced internalization of magnetic nanoparticles in cancer cells. Nanotechnology 20(11):115103
111. Maeda H et al (2000) Tumor vascular permeability and the EPR effect in macromolecular therapeutics: a review. J Controlled Release 65(1–2):271–284
112. Jain RK (1999) Transport of molecules, particles, and cells in solid tumors. Annu Rev Biomed Eng 1:241–263
113. Fang J, Nakamura H, Maeda H (2011) The EPR effect: unique features of tumor blood vessels for drug delivery, factors involved, and limitations and augmentation of the effect. Adv Drug Deliv Rev 63(3):136–151
114. Prabhakar U et al (2013) Challenges and key considerations of the enhanced permeability and retention effect for nanomedicine drug delivery in oncology. Cancer Res 73 (8):2412–2417
115. Zhang Y, Kohler N, Zhang MQ (2002) Surface modification of superparamagnetic magnetite nanoparticles and their intracellular uptake. Biomaterials 23(7):1553–1561
116. Sinha R et al (2006) Nanotechnology in cancer therapeutics: bioconjugated nanoparticles for drug delivery. Mol Cancer Ther 5(8):1909–1917
117. Kohler N et al (2005) Methotrexate-modified superparamagnetic nanoparticles and their intracellular uptake into human cancer cells. Langmuir 21(19):8858–8864
118. Sun C, Sze R, Zhang MQ (2006) Folic acid-PEG conjugated superparamagnetic nanoparticles for targeted cellular uptake and detection by MRI. J Biomed Mater Res Part A 78A(3):550–557
119. Montet X, Weissleder R, Josephson L (2006) Imaging pancreatic cancer with a peptide-nanoparticle conjugate targeted to normal pancreas. Bioconjug Chem 17(4):905–911
120. Gunn J et al (2008) A multimodal targeting nanoparticle for selectively labeling T cells. Small 4(6):712–715
121. Hong S et al (2007) The binding avidity of a nanoparticle-based multivalent targeted drug delivery platform. Chem Biol 14(1):107–115
122. Chertok B et al (2007) Glioma selectivity of magnetically targeted nanoparticles: a role of abnormal tumor hydrodynamics. J Controlled Release 122(3):315–323
123. Dobson J (2006) Magnetic nanoparticles for drug delivery. Drug Dev Res 67(1):55–60
124. Wilson MW et al (2004) Hepatocellular carcinoma: regional therapy with a magnetic targeted carrier bound to doxorubicin in a dual MR imaging/conventional angiography suite—initial experience with four patients. Radiology 230(1):287–293
125. Bulte JWM, Kraitchman DL (2004) Iron oxide MR contrast agents for molecular and cellular imaging. NMR Biomed 17(7):484–499
126. Artemov D et al (2003) MR molecular imaging of the Her-2/neu receptor in breast cancer cells using targeted iron oxide nanoparticles. Magn Reson Med 49(3):403–408
127. Hu FQ et al (2006) Preparation of biocompatible magnetite nanocrystals for in vivo magnetic resonance detection of cancer. Adv Mater 18(19):2553–2556
128. Xie J et al (2008) Ultrasmall c(RGDyK)-coated Fe(3)O(4) nanoparticles and their specific targeting to integrin alpha(v)beta(3)-rich tumor cells. J Am Chem Soc 130(24):7542
129. Bulte JWM et al (2001) Magnetodendrimers allow endosomal magnetic labeling and in vivo tracking of stem cells. Nat Biotechnol 19(12):1141–1147
130. Frank JA et al (2003) Clinically applicable labeling of mammalian and stem cells by combining; superparamagnetic iron oxides and transfection agents. Radiology 228 (2):480–487
131. Bulte JWM (2009) In vivo MRI cell tracking: clinical studies. Am J Roentgenol 193 (2):314–325
132. Berman SMC, Walczak P, Bulte JWM (2011) Tracking stem cells using magnetic nanoparticles. Wiley Interdisc Rev Nanomed Nanobiotechnol 3(4):343–355

133. Ahrens ET, Bulte JWM (2013) Tracking immune cells in vivo using magnetic resonance imaging. Nat Rev Immunol 13(10):755–763
134. Srivastava AK, Bulte JWM (2014) Seeing stem cells at work in vivo. Stem Cell Rev Rep 10 (1):127–144
135. Kircher MF et al (2003) In vivo high resolution three-dimensional imaging of antigen-specific cytotoxic T-lymphocyte trafficking to tumors. Cancer Res 63(20):6838–6846
136. Daldrup-Link HE et al (2005) In vivo tracking of genetically engineered, anti-HER2/neu directed natural killer cells to HER2/neu positive mammary tumors with magnetic resonance imaging. Eur Radiol 15(1):4–13
137. Anderson SA et al (2004) Magnetic resonance imaging of labeled T-Cells in a mouse model of multiple sclerosis. Ann Neurol 55(5):654–659
138. Berman SMC et al (2013) Cell motility of neural stem cells is reduced after SPIO-labeling, which is mitigated after exocytosis. Magn Reson Med 69(1):255–262
139. de Vries IJM et al (2005) Magnetic resonance tracking of dendritic cells in melanoma patients for monitoring of cellular therapy. Nat Biotechnol 23(11):1407–1413
140. Verdijk P et al (2007) Sensitivity of magnetic resonance imaging of dendritic cells for in vivo tracking of cellular cancer vaccines. Int J Cancer 120(5):978–984
141. Verdijk P et al (2009) Limited amounts of dendritic cells migrate into the T-Cell area of lymph nodes but have high immune activating potential in melanoma patients. Clin Cancer Res 15(7):2531–2540
142. Schuurhuis DH et al (2009) In situ expression of tumor antigens by messenger RNA-electroporated dendritic cells in lymph nodes of melanoma patients. Cancer Res 69 (7):2927–2934
143. Cruz LJ et al (2011) multimodal imaging of nanovaccine carriers targeted to human dendritic cells. Mol Pharm 8(2):520–531
144. Janowski M et al (2014) Long-term MRI cell tracking after intraventricular delivery in a patient with global cerebral ischemia and prospects for magnetic navigation of stem cells within the CSF. PLoS ONE 9(6):e97631
145. Moghimi SM, Hunter AC, Murray JC (2001) Long-circulating and target-specific nanoparticles: theory to practice. Pharmacol Rev 53(2):283–318
146. Huang J et al (2010) Effects of nanoparticle size on cellular uptake and liver MRI with polyvinylpyrrolidone-coated iron oxide nanoparticles. ACS Nano 4(12):7151–7160
147. Joshi HM et al (2009) Effects of shape and size of cobalt ferrite nanostructures on their MRI contrast and thermal activation. J Phys Chem C 113(41):17761–17767
148. Zhao ZH et al (2013) Octapod iron oxide nanoparticles as high-performance T-2 contrast agents for magnetic resonance imaging. Nat Commun 4:2266
149. Berret JF et al (2006) Controlled clustering of superparamagnetic nanoparticles using block copolymers: design of new contrast agents for magnetic resonance imaging. J Am Chem Soc 128(5):1755–1761
150. Lee JE et al (2010) Uniform mesoporous dye-doped silica nanoparticles decorated with multiple magnetite nanocrystals for simultaneous enhanced magnetic resonance imaging, fluorescence imaging, and drug delivery. J Am Chem Soc 132(2):552–557
151. Barick KC et al (2009) Nanoscale assembly of amine-functionalized colloidal iron oxide. J Magn Magn Mater 321(10):1529–1532
152. LaConte LEW et al (2007) Coating thickness of magnetic iron oxide nanoparticles affects R-2 relaxivity. J Magn Reson Imaging 26(6):1634–1641
153. Gao JH et al (2008) Multifunctional yolk-shell nanoparticles: a potential MRI contrast and anticancer agent. J Am Chem Soc 130(35):11828–11833
154. Mahmoudi M et al (2011) Superparamagnetic iron oxide nanoparticles (SPIONs): development, surface modification and applications in chemotherapy. Adv Drug Deliv Rev 63(1–2):24–46
155. Pollert E et al (2009) Search of new core materials for magnetic fluid hyperthermia: preliminary chemical and physical issues. Prog Solid State Chem 37(1):1–14

156. Lacroix LM et al (2009) Magnetic hyperthermia in single-domain monodisperse FeCo nanoparticles: evidences for Stoner-Wohlfarth behavior and large losses. J Appl Phys 105 (2):023911
157. Mehdaoui B et al (2010) Large specific absorption rates in the magnetic hyperthermia properties of metallic iron nanocubes. J Magn Magn Mater 322(19):L49–L52
158. Franchini MC et al (2010) Bovine serum albumin-based magnetic nanocarrier for MRI diagnosis and hyperthermic therapy: a potential theranostic approach against cancer. Small 6 (3):366–370
159. Lee J-H et al (2011) Exchange-coupled magnetic nanoparticles for efficient heat induction. Nat Nano 6(7):418–422
160. Yoo D et al (2012) Double-effector nanoparticles: a synergistic approach to apoptotic hyperthermia. Angew Chem Int Ed 51(50):12482–12485
161. Mykhaylyk O et al (2008) siRNA delivery by magnetofection. Curr Opin Mol Ther 10 (5):493–505
162. McBain SC et al (2008) Magnetic nanoparticles as gene delivery agents: enhanced transfection in the presence of oscillating magnet arrays. Nanotechnology 19(40):405102
163. Petri-Fink A et al (2008) Effect of cell media on polymer coated superparamagnetic iron oxide nanoparticles (SPIONs): colloidal stability, cytotoxicity, and cellular uptake studies. Eur J Pharm Biopharm 68(1):129–137
164. Kievit FM et al (2009) PEI-PEG-Chitosan-copolymer-coated iron oxide nanoparticles for safe gene delivery: synthesis, complexation, and transfection. Adv Funct Mater 19 (14):2244–2251
165. Kievit FM et al (2010) Chlorotoxin labeled magnetic nanovectors for targeted gene delivery to glioma. ACS Nano 4(8):4587–4594
166. Shin JM et al (2009) Hollow manganese oxide nanoparticles as multifunctional agents for magnetic resonance imaging and drug delivery. Angew Chem Int Ed 48(2):321–324
167. Cheng K et al (2009) Porous hollow Fe_3O_4 nanoparticles for targeted delivery and controlled release of cisplatin. J Am Chem Soc 131(30):10637–10644
168. Jain TK et al (2008) Magnetic nanoparticles with dual functional properties: drug delivery and magnetic resonance imaging. Biomaterials 29(29):4012–4021
169. Thomas CR et al (2010) Noninvasive remote-controlled release of drug molecules in vitro using magnetic actuation of mechanized nanoparticles. J Am Chem Soc 132 (31):10623–10625
170. Purushotham S, Ramanujan RV (2010) Thermoresponsive magnetic composite nanomaterials for multimodal cancer therapy. Acta Biomater 6(2):502–510

Nanodiamond-Based Chemotherapy and Imaging

Dean Ho

Abstract

The advent of cancer nanomedicine has forged new pathways for the enhanced imaging and treatment of a broad range of cancers using new classes of materials. Among the many platforms being developed for drug delivery and imaging, nanodiamonds (NDs) possess several important attributes that may be beneficial toward improving the efficacy and safety of cancer nanomedicine applications. These include the uniquely faceted surfaces of the ND particles that result in electrostatic properties that mediate enhanced interactions with water and loaded therapeutic compounds, scalable processing and synthesis parameters, versatility as platform carriers, and a spectrum of other characteristics. In addition, comprehensive in vitro and in vivo studies have demonstrated that NDs are well tolerated. This chapter will examine several recent studies that have

D. Ho (✉)
Division of Oral Biology and Medicine, UCLA School of Dentistry,
10833 Le Conte Avenue, Room B3-068A, Los Angeles, CA 90095, USA
e-mail: dean.ho@ucla.edu

D. Ho
Department of Bioengineering, UCLA School of Engineering and Applied Science,
Los Angeles, USA

D. Ho
The Jane and Jerry Weintraub Center for Reconstructive Biotechnology, UCLA,
Los Angeles, USA

D. Ho
Jonsson Comprehensive Cancer Center, UCLA School of Medicine, Los Angeles, USA

D. Ho
California NanoSystems Institute, Los Angeles, USA

C.A. Mirkin et al. (eds.), *Nanotechnology-Based Precision Tools for the Detection and Treatment of Cancer*, Cancer Treatment and Research 166,
DOI 10.1007/978-3-319-16555-4_4

harnessed the ND agent as a foundation for both systemic and localized drug delivery, as well as the marked improvements in magnetic resonance imaging efficiency that has been observed following ND-contrast agent conjugation. In addition, insight into the important steps toward bringing the ND translational pathway to the clinic will be discussed.

Keywords
Nanodiamond · Nanomedicine · Drug delivery · Imaging · Targeting

Abbreviations

CMC	Chemistry, Manufacturing and Controls
CED	Convection-Enhanced Delivery
Cryo-TEM	Cryogenic Transmission Electron Microscopy
DNR	Daunorubicin
Dox	Doxorubicin
DLS	Dynamic Light Scattering
ELISA	Enzyme-Linked Immunosorbent Assay
EGF	Epidermal Growth Factor
EGFR	Epidermal Growth Factor Receptor
Epi	Epirubicin
FDA	Food and Drug Administration
FTIR	Fourier Transform Infrared Spectroscopy
Gd(III)	Gadolinium(III)
GLP	Good Laboratory Practice
GMP	Good Manufacturing Practice
HPHT	High Pressure High Temperature
LDH	Lactate Dehydrogenase
MRI	Magnetic Resonance Imaging
ND	Nanodiamond
NDX	Nanodiamond-Doxorubicin
NDLP	Nanodiamond-Lipid Hybrid Particles
PET	Positron Emission Tomography
qRT-PCR	Quantitative Real Time Polymerase Chain Reaction
TNBC	Triple Negative Breast Cancer

Contents

1 Introduction

A broad range of nanomaterials has been previously explored as drug delivery and imaging platforms for cancer nanomedicine. These have included polymer, carbon, metallic, and other agents that have demonstrated promise as potential translational approaches to improve the outcomes of treatment and diagnosis [1–17]. Nanodiamonds (NDs) can be classified into different domains, each with their unique characteristics that can be harnessed for studies that range from fundamental to translational in the area of cancer nanomedicine. For example, High Pressure High Temperature (HPHT) NDs can be embedded with nitrogen vacancies that result in remarkable photostability and brightness with simultaneously retained biocompatibility. As such, HPHT NDs have been widely explored as cellular imaging agents, and have been utilized in tracking ND fate in C. Elegans and stem cell localization, among other important studies [18–21]. Another class of ND is the Detonation ND that possesses diameters of approximately 5 nm per particle, and a truncated octahedral architecture. The facets of detonation NDs result in electrostatic properties that are conducive to water binding as well as potent drug interactions (Fig. 1).

Furthermore, the versatile chemical properties on the detonation ND surfaces enable either the conjugation or reversible adsorption of a broad collection of therapeutic and imaging compounds, depending on the intended application or indication being addressed [22–25]. Given these interesting properties and findings, NDs have emerged as promising platforms for drug delivery and imaging, particularly for applications in cancer, because they have improved the efficacy *and* safety of treatment and diagnosis of different types of cancer over conventional clinical standards via ND-drug synthesis methods that are reliable and scalable. These complexes were capable of mediating order of magnitude improvements in per-gadolinium (Per-GD(III)) relaxivity which are among the highest ever reported values compared to all clinical and nanoparticle agents [26]. In addition, ND-doxorubicin (NDX) agents are able to bypass drug-resistant breast and liver tumors to demonstrate marked improvements in tumor reduction capabilities with no apparent myelosuppression compared to the administration of doxorubicin (Dox) alone, which is a clinical standard [27].

While NDs come in different shapes and sizes, and both HPHT and detonation NDs have demonstrated significant promise in the basic and translational development domains, this chapter will focus on detonation NDs given that they have

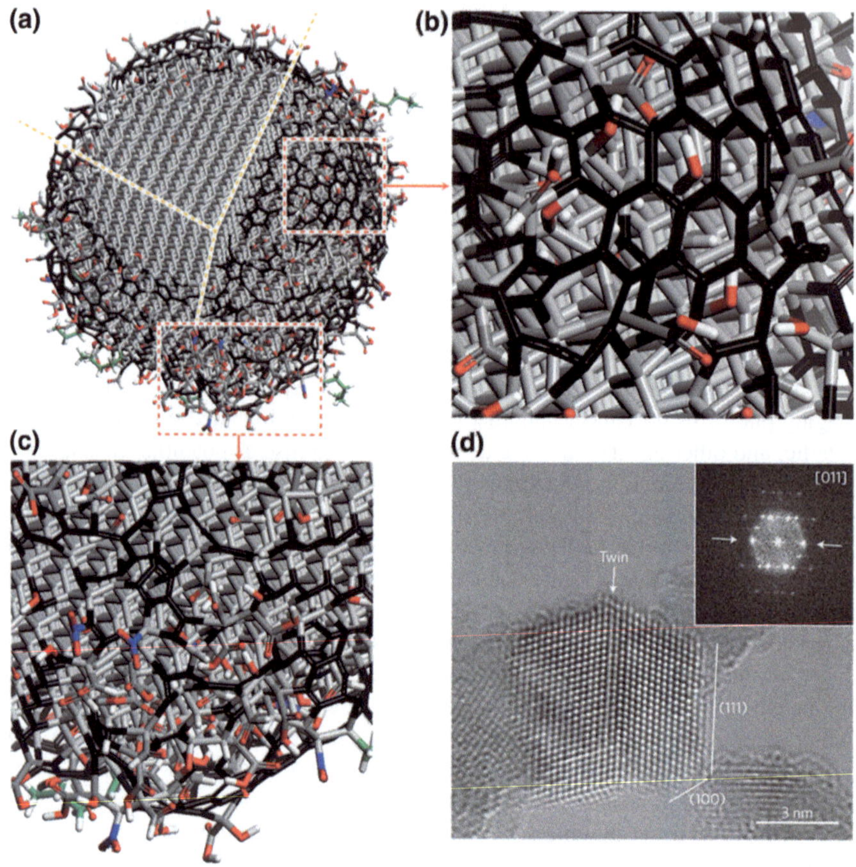

Fig. 1 **a–d** Nanodiamond surfaces possess versatile chemical–physical properties that can mediate potent drug binding and marked improvements in MRI contrast efficiency levels. Reprinted with permission from Nature Publishing Group

been extensively studied in preclinical models and proposed as platforms for magnetic resonance imaging (MRI). Current achievements as well as challenges toward clinical implementation will be addressed.

2 Nanodiamond Drug Delivery

Among the early demonstrations of the promise of NDs toward applications in biology and medicine was their use as chemotherapeutic delivery vehicles in vitro with Dox as the payload as well as other compounds [28–33]. Early studies demonstrated that while the Dox molecules were bound to the ND surfaces, their activity was attenuated such that the cytotoxic nature of Dox could potentially be reduced while adsorbed to the surface. The interactions between the ND surface and various classes of therapeutic compounds have been studied using various modeling/simulation methods to provide further insight into the unique chemical–physical properties of the ND facets [34, 35]. This was important in that the systemic toxicity of Dox can lead to significant side effects including myelosuppression, cardiotoxicity, superinfections, as well as mortality.

Therefore, this initial study paved the way for additional work that resulted in the preclinical validation of NDX safety and efficacy in liver and breast tumor models, preclinical anthracycline targeting toward TNBC, and localized glioblastoma therapy. Collectively, these findings have addressed particularly hard-to-treat cancers, demonstrating the promise of NDs as broadly applicable drug delivery platforms.

2.1 Nanodiamond Particles as Systemic Chemotherapeutic Delivery Agents for Drug-Resistant Breast and Liver Tumor Treatment

Drug resistance is a pervasive problem in cancer therapy that results in a vast majority of tumor treatment failure in metastatic cancer cases. Of all of the ND-drug agents that have been synthesized, ND-anthracycline agents are highly scalable and they potently bind the drug compounds that result in marked improvements in efficacy and safety. The ability to rapidly synthesize nanoparticle-tagged drugs that are too large to be effluxed from a tumor mass, coupled with an ability to prevent early systemic release and systemic toxicity and off-target effects would represent a very promising nanomedicine-based route to overcome drug resistance. A recent study demonstrated that NDX is among the most promising platforms for this indication due to the remarkably scalable nature of its synthesis protocol. NDX particles were systemically administered in mice via tail vein injection [27]. Subsequent toxicity measurements of serum IL-6 and serum ALT even at high ND doses indicated no apparent toxicity, demonstrating that the NDs were well tolerated. The impact of NDX injections on myelosuppression was evaluated. NDX administration resulted in no apparent myelosuppression, while the administration of Dox alone resulted in marked reductions in white blood cell count below threshold values. Myelosuppression is the dose-limiting side effect of chemotherapy and can cause major side effects such as superinfections and patient mortality.

With regard to efficacy, unmodified Dox administration at 100 μg equivalence resulted in virtually no efficacy. NDX administration at the same dosage equivalence resulted in a marked decrease in tumor size. When the drug dosage equivalence was doubled, unmodified Dox administration resulted in accelerated animal mortality. When the lethal dosage was delivered via ND, all of the animals survived, and the tumor treatment efficacy was further enhanced, resulting in the smallest tumors observed in the study (Fig. 2). This demonstrated that the ND platform was capable of mediating major improvements to chemotherapeutic tolerance [27].

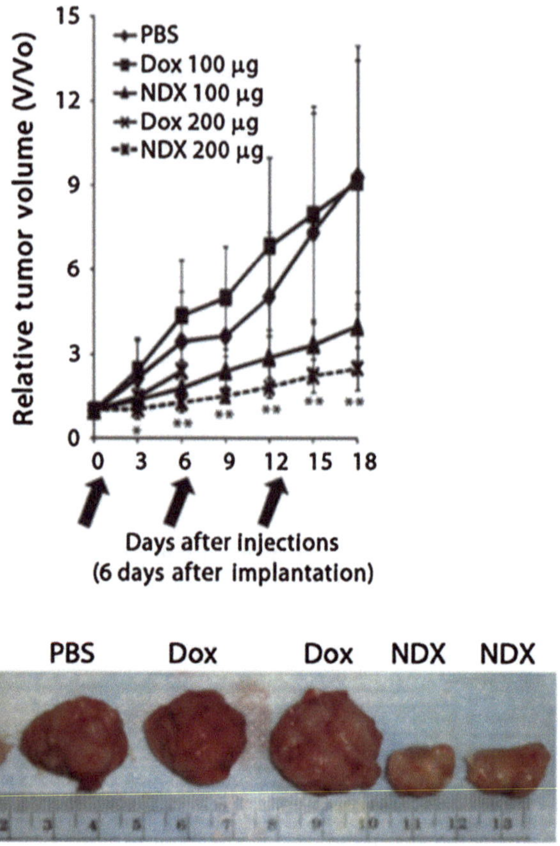

Fig. 2 *Top* NDX administration at 100 µg equivalence delays tumor growth compared to Dox alone. NDX at a doubled equivalent dose results in even further improved efficacy and drug tolerance. *Bottom* Unmodified Dox administration results in little to no efficacy given the drug resistant nature of the breast cancer studied. NDX administration results in markedly reduced tumor sizes. Reprinted with permission from the American Association for the Advancement of Science

This initial demonstration of preclinical NDX administration demonstrated that there was no premature drug elution even with a therapeutic that is bound via physisorption. This was confirmed by the fact that premature drug release would have resulted in both apparent myelosuppression due to the circulation of free drug, as well as no therapeutic efficacy since the Dox would be separate from the ND and effluxed. The absence of myelosuppression and marked improvements in efficacy confirmed that potent interaction of NDs with Dox served as an important mediator of both significantly enhanced efficacy and safety, even in a passively targeted system.

2.2 Liposome-Encapsulated Nanodiamonds for Targeted Triple-Negative Breast Cancer Treatment

The initial validation of NDX systemic administration as an effective strategy for drug-resistant breast and liver tumor therapy served as a promising passively targeted approach. The improved drug tolerance and therapeutic efficacy observed even with the lethal Dox dosage served as a foundation for the continued development of NDX toward clinical use. A recent study has synthesized a variation of the NDX complex in that the Dox compound was replaced with Epi [36]. The disease model in this study was the MDA-MB-231 triple-negative breast cancer (TNBC) that has been shown to overexpress the epidermal growth factor receptor (EGFR) [36]. Therefore, the ND-Epi complexes were encapsulated within liposomes that were surface functionalized with the EGFR antibody (Fig. 3). This architecture was employed in order to maximize drug loading on the ND surface due to the fact that the NDX study was able to confirm that the ND-anthracycline interaction was capable of enhancing intratumoral drug retention to prevent efflux and improve efficacy, circulatory half-life by a factor of 10, and eliminate myelosuppression, among other benefits. These ND-Lipid hybrid particles (NDLPs) thus served as an actively targeted variation of the ND-anthracycline complexes that were previously scalably synthesized [36].

NDLP synthesis was accomplished using a rehydration process of lipid thin films that were comprised of cholesterol and biotinylated lipids and ND solution [36]. By harnessing biotin–streptavidin chemistry, the anti-EGFR antibodies were conjugated to the liposome surface. Characterization of NDLP synthesis was done using a broad range of methodologies including zeta potential analysis and dynamic light scattering (DLS) in order to assess particle size, Fourier Transform Infrared Spectroscopy (FTIR), flow cytometry in order to quantify successful NDLP synthesis and to differentiate the NDLPs from free NDs and lipids, and cryogenic transmission electron microscopy (cryo-TEM) in order to image NDLP formation. Inductively coupled plasma atomic emission spectroscopy (ICP-AES) and enzyme-linked immunosorbent assay (ELISA) were used to assess lipid presence and confirm anti-EGFR antibody conjugation, respectively [36].

Comprehensive in vitro assessment of anti-EGFR functionalized NDLP targeting efficacy was performed. Epidermal Growth Factor (EGF) flooding studies resulted in gradually decreasing targeting efficacy as a result of increasing the EGF concentration which was a demonstration of successful anti-EGFR antibody conjugation and targeting abilities. With regard to preclinical efficacy, targeted (anti-EGFR antibody functionalized) and untargeted NDLP complexes, as well as phosphate buffered saline (PBS) and unmodified Epi were utilized as test conditions [36]. Unmodified Epi administration resulted in accelerated animal mortality that was similarly observed with unmodified Dox administration. Untargeted NDLP administration immediately improved drug tolerance as early mortality was no longer observed. Furthermore, tumor growth was clearly delayed. Interestingly, the

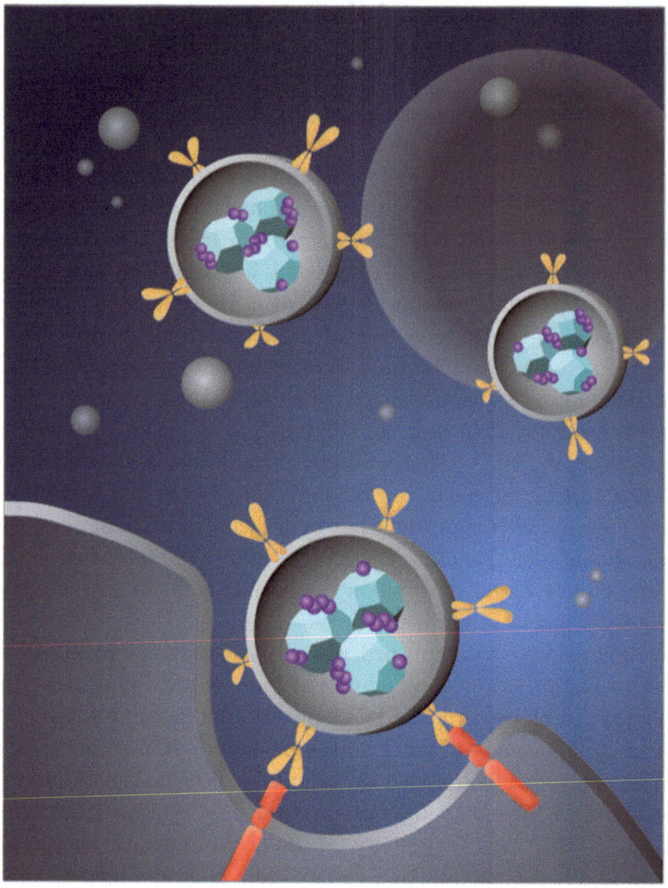

Fig. 3 A schematic demonstrates the synthesis of liposome-encapsulated nanodiamonds that are functionalized with epirubicin. The conjugation of antiepidermal growth factor receptor antibodies on the particle-enabled TNBC targeting. *Image* courtesy of Laura Moore

administration of the anti-EGFR functionalized NDLPs resulted in tumor regression to the point where they were no longer visible, which is a promising finding for actively targeted ND-chemotherapeutic compounds (Fig. 4).

With regard to ND biocompatibility, this study provided among the most comprehensive assessments of ND serum and urine toxicity. A broad range of markers was analyzed including serum ALT, ALK and AST, hematocrit, blood urea nitrogen (BUN), bilirubin, white blood cell differential, hemoglobin, and red blood cell counts, among other readouts. In all cases, ND and NDLP administration resulted in no apparent toxicity that further substantiated the promise of the ND platform for drug delivery.

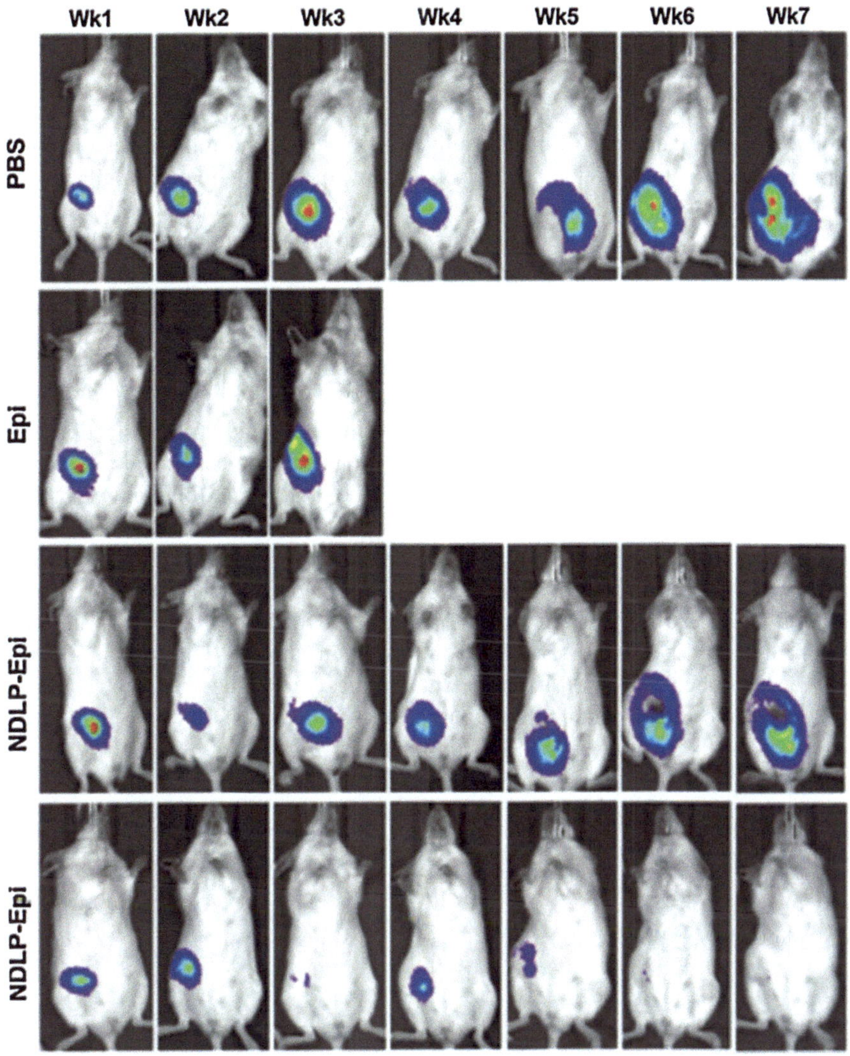

Fig. 4 Liposome-encapsulated nanodiamond complexes (NDLP-Epi) mediated tumor regression, while the administration of drug alone resulted in accelerated animal mortality. This study demonstrated that NDLP-Epi complexes were capable of improving drug tolerance in addition to efficacy. Reprinted with permission from Wiley-VCH

2.3 Nanodiamond-Based Glioblastoma Therapy

A key barrier to glioblastoma therapy is the need for the therapeutic to traverse the blood–brain barrier (BBB) [37]. In addition, off-target toxicity is of a major concern because nonspecific cell death in the central nervous system (CNS) results in irreversible damage since CNS neurons do not regenerate. Therefore, localizing drug release to the specific tumor-containing regions, when relevant, is of particular importance.

While the ability for NDX to cross the BBB should be further investigated, a recent study that utilized convection-enhanced delivery (CED), the stereotactic application of a catheter to directly deliver NDX into brain tumors demonstrated improved tumor localization and clear improvements to treatment efficacy in multiple tumor models [38]. Specifically, the C6 rodent model and U251MG, an aggressive human model, were both investigated in a rat model. Initial in vitro studies demonstrated that NDX retention within both cell lines were improved compared to free drug administration. In addition Ki67 and caspase studies demonstrated enhanced cancer cell death mediated by NDX administration compared to free Dox administration. Preclinical retention studies showed that NDX administration in the right lobe of the brain resulted in the presence of Dox at the 72 h time point, while the administration of Dox alone resulted in rapid drug dissipation with no drug presence at 72 h. Importantly, toxicity studies were conducted to examine the presence of edema and swelling in the left lobe of the brain following right lobe administration. While PBS and ND administration in the right lobe resulted in healthy left lobe, unmodified Dox administration into the right lobe resulted in significant swelling in the left lobe, serving as an indicator that free Dox administration could cause major off-target toxicity, even with localized CED administration. It should be noted that NDX administration into the right lobe resulted in healthy left lobe. This was an important finding confirming that Dox could be confined to the tumor-containing regions of the brain via NDX which may markedly improve the tolerance of CED Dox injection (Fig. 5).

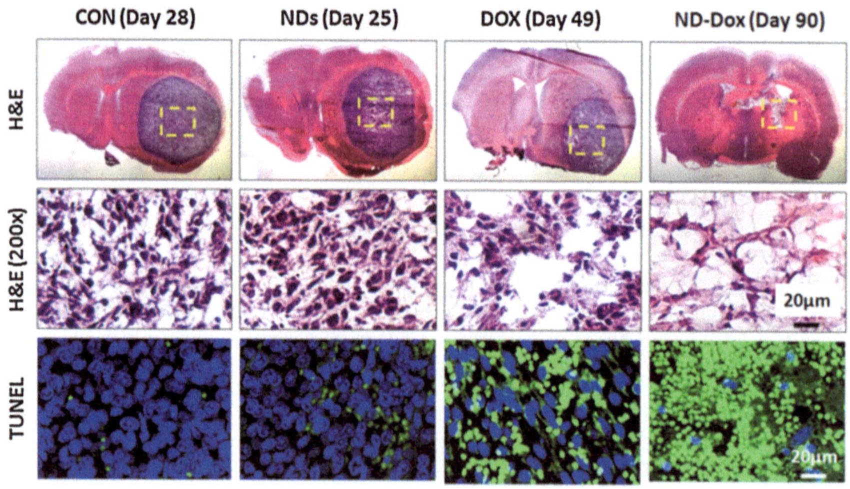

Fig. 5 Nanodiamond-Dox (NDX) administration via CED to treat glioblastoma in a rat model results in tumor regression compared to unmodified drug administration and controls. This was further confirmed via tissue staining and apoptosis assays. Reprinted with permission from Elsevier

Extensive preclinical studies pertaining to NDX efficacy and safety in both C6 and U251MG were conducted. In the U251MG model, tumors observed following PBS and ND administration resulted in animal mortality. Dox administration resulted in an observable reduction in tumor size. However, NDX administration resulted in a significant lengthening of animal survival and cancer cell apoptosis. In the C6 model, the control and Dox-only conditions resulted in largely the same outcomes as those observed in the U251 model. However, NDX administration resulted in tumor regression to the point where the luciferase signal emitted from the tumors were no longer observable and pronounced improvements in animal survival were observed. Hematoxylin and Eosin (H&E) staining showed that the tumors were virtually undetectable, and TUNEL staining revealed significant increases in apoptosis mediated by NDX.

CED is being studied as a promising direct drug delivery route via several clinical trials, and may serve as a strategy for localized ND administration. This study demonstrated that NDX-mediated glioblastoma therapy may limit drug dissipation to reduce off-target toxicity, while improving intratumoral retention to mediate tumor regression. Given the importance of preventing drug distribution to healthy tissue in the CNS, the integration of CED with NDX delivery warrants further investigation.

2.4 Localized Nanodiamond Drug Delivery

A majority of ND-based drug delivery studies have pertained to systemic administration, and the potent adsorption of drugs to the ND surface. NDs can also improve inherent material properties including mechanical robustness, among others, that can result in improved treatment efficacy and safety [20, 23, 29, 36, 39]. Prior studies have demonstrated the use of scalable parylene thin film devices embedded with NDX to reduce burst release while preserving drug activity. In this study, chemical vapor deposition was used to synthesize parylene thin film bases which were then coated with NDX. A semi-porous layer of parylene was subsequently deposited on the NDX and the devices were then removed from the deposition substrate to result in flexible, "plastic wrap"-like devices [40]. These microfilm devices may be applicable as localized release platforms to eliminate residual cancer cells following resection surgeries.

In addition to parylene/ND-based approaches, a recent study served as an example where ND incorporation into polymer matrices for an ophthalmology indication was capable of improving several properties simultaneously [41]. A diamond-embedded chitosan nanogels was used to trigger timolol release using lysozyme. Timolol is a widely used glaucoma drug, typically administered via eyedrops. However, due to problems with patient compliance and drug leakage, adherence to timolol administration schedules is often disrupted. In this work, the ND nanogels were embedded within contact lenses and drug release profiles were compared against drug-imprinted lenses and drug-soaked lenses. The soaked and imprinted lenses mediated burst release which resulted in virtually all of the timolol

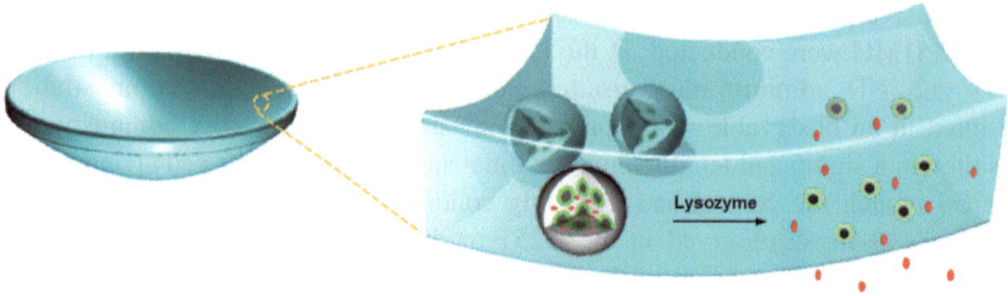

Fig. 6 Nanodiamond-embedded contact lens devices were able to simultaneously mediate lysozyme-triggered drug release, while also possessing improved mechanical properties and water content. Reprinted with permission from the American Chemical Society

being released within a few hours. The ND-based lenses were able to release the timolol following lysozyme exposure. This was an important finding in that this may allow for drug-loaded contact lenses to be stored in liquid with no early or complete drug release. In addition, lysozyme is an enzyme found in tears that allows for localized elution only after lens use in the eye. With regard to the contact lens properties, water content was unaffected, likely due to the unique ND-water interactions. This is important because water content can determine gas permeability of the lens which influences wear comfort. Furthermore, water content at adequate levels is required for proper lysozyme access to release timolol. The Young's Modulus of the contact lenses was increased as the formation of an ND-based composite is known to improve the mechanical robustness of the device (Fig. 6).

The aforementioned examples serve as cases where ND incorporation into polymeric matrices can be used to improve local drug delivery efficacy and safety. Lysozyme triggering serves as a viable model for future ND-embedded devices to potentially elute chemotherapeutic compounds following exposure to relevant stimuli. Finally, as the faceted ND surfaces are capable of coordinating water molecules, proper stimulus access into the device can be preserved.

3 Nanodiamond-Based Imaging

Nanomaterials are being widely explored as carriers for medical imaging agents (e.g., positron emission tomography (PET), MRI, etc.) because of their high surface area-to-volume ratios which can mediate high loading capacities [42]. However, it is important to note that carriers that can enable imaging agents to function more efficiently to reduce the dosages required may have unique advantages in the clinic due to commonly observed toxicity with these imaging probes. The use of NDs as imaging agents is gaining interest in the diagnostic communities because of recent work demonstrating per-Gd(III) relaxivity values that are among the highest ever reported compared to all clinical and nanoparticle-based agents as well as other studies demonstrating the photostable nature of ND fluorescence as well as ND-based targeted imaging (Fig. 7) [39, 43–53]. By conjugating Gd(III) to ND

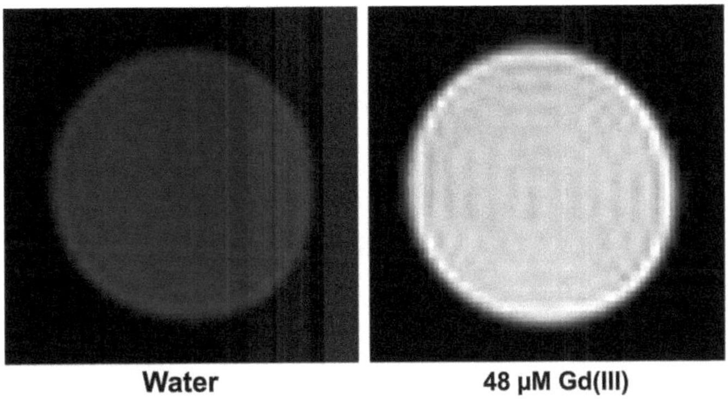

Fig. 7 Nanodiamond-Gadolinium(III) complexes mediated a 12-fold enhancement in per-gadolinium relaxivity. This is among the highest ever reported values compared to all clinical and nanoparticle-based contrast agents. Reprinted with permission from the American Chemical Society

surfaces, this study demonstrated a 12-fold increase in per-Gd(III) relaxivity that may enable one order of magnitude reduction in Gd(III) dosages needed for adequate contrast [26]. This would represent a major advance in medicine, and as such, NDs continue to receive attention as promising contrast imaging platforms.

4 Assessment of Nanodiamond Safety

As NDs continue to be evaluated as potential drug delivery and imaging agents for applications in cancer nanomedicine, thorough investigations into material safety are required using both in vitro and in vivo studies. Initial comparative studies between varying types of carbon nanomaterials, hemocompatibility studies, and comprehensive studies investigating inflammatory, apoptotic, and other mechanistic responses following ND administration have indicated that they are well tolerated [54–58]. While some recent studies have examined responses such as thromboembolism and embryonic stem cell genetic response, preclinical studies have indicated that NDs do not exhibit apparent toxicity [59, 60].

More recently, the most comprehensive assessment of ND biocompatibility was performed in that multiple ND subtypes. These included detonation NDs, HPHT NDs, amine-functionalized NDs, and anthracycline-functionalized NDs. They were incubated with multiple cell types and cell death, cell metabolism, apoptosis induction, gene expression, and drug cytotoxicity attenuation studies were conducted (Fig. 8) [14]. The cell lines that were interrogated included the HeLa and HepG2 cell lines given the prevalence of HeLa cells as platforms for new drug testing, and the fact that HepG2 liver cells can serve as readouts for toxicity against nanomaterial incubation.

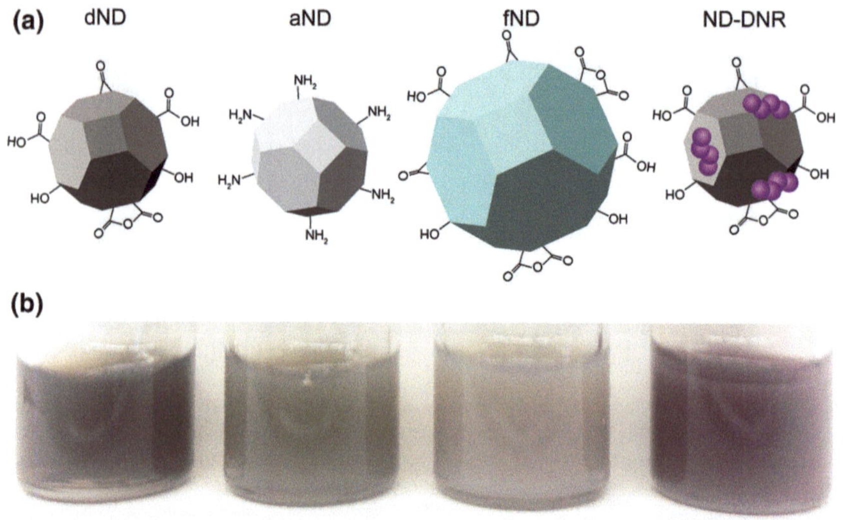

Fig. 8 **a** A schematic of the different nanodiamond subtypes examined in the biocompatibility study is shown. These include detonation nanodiamonds (dND), amine-functionalized nanodiamonds (aND), fluorescence nanodiamonds (fND), and nanodiamond–daunorubicin complexes (ND-DNR). **b** Dispersed solutions of all tested ND subtypes are shown. Reprinted with permission from the Royal Society of Chemistry

Biocompatibility assays conducted included the XTT cell proliferation assay to examine cell metabolism, lactate dehydrogenase (LDH) cytotoxicity assay, and caspase 3/7 apoptotic induction assay. Of note, even high ND dosages (250 μg/ml) did not result in any apparent toxicity across all ND subtypes in both cell lines. Transcriptional regulation was probed using quantitative real-time polymerase chain reaction (qRT-PCR). Genes studied in the HepG2 cell lines included *Bax* (pro-apoptotis), *c-Myc* (antiapoptosis, pro-proliferation), and *Ki-67* (pro-proliferation). In these studies, 25 μg/ml of NDs were incubated with the cells. *Bax* expression was not upregulated following exposure with any type of ND, which indicated that there was no apoptotic response to ND exposure. Downregulation of *c-Myc* was not observed following incubation with any subtype of ND, also demonstrating that the NDs do not promote apoptosis nor do they inhibit cellular proliferation. Assessing *Ki-67* levels also showed that the NDs do not mediate antiproliferative responses in HepG2 cells. In addition to probing gene expression levels for the same markers in HeLa cells, *COX-2* levels were examined to assess potential inflammatory responses following ND exposure. Following ND incubation, no apparent gene expression changes were observed, confirming that the NDs are well tolerated.

In addition to validating that the NDs do not elicit changes to cell metabolism and proliferation or mediate cellular apoptosis, additional studies examining the role of ND sequestering of anthracycline cytotoxicity were performed [14]. More specifically, the daunorubicin (DNR) chemotherapeutic was loaded onto the NDs and XTT, LDH, and Caspase assays were performed on both the HeLa and HepG2 cells to compare the impact of ND-DNR and free DNR administration. These studies showed that ND-DNR delivery resulted in reduced cell death, increased cellular proliferation, and reduced cellular apoptosis compared to free drug administration.

Preclinical studies have provided significant insight into the fate and toxicity of NDs as well as ND tolerance even at high doses [36, 61]. As previously mentioned, the NDLP study provided the most comprehensive preclinical demonstrations that NDs are well tolerated following systemic injections and both comprehensive blood, and urinalysis, did not reveal any apparent changes to key markers for toxicity. Future studies involving larger animal models (e.g., nonhuman primates, etc.) will serve as vital precursors to potential clinical validation.

5 Challenges and Steps Toward Translation

The significant advances demonstrated thus far by the collection of nanomaterials being engineered for applications in cancer nanomedicine may open new doors for changing the way that cancer is treated and diagnosed. Before the widespread implementation of cancer nanomedicine in the clinic is realized, several important steps remain to be addressed. These include areas such as engineering that nanomaterial synthesis and processing are compliant with chemistry, manufacturing and controls (CMC) standards, good laboratory practice (GLP), and good manufacturing practice (GMP) guidelines, among other important considerations. Furthermore, as the U.S. Food and Drug Administration (FDA) and other international agencies (e.g., European Medicines Agency, Pharmaceutical and Medical Devices Agency of Japan) continue to develop streamlined regulatory pathway guidances for the approvals of nanomedicine-based therapeutic and imaging drugs, more nanomedicine-based constructs may transition from the benchtop toward the clinic to ultimately improve patient—treatment outcomes.

6 Concluding Remarks

Nanodiamonds combine many properties that are favorable for both drug delivery and imaging into a united platform. These include a faceted truncated octahedral architecture with unique electrostatic properties that can mediate potent drug and water binding properties to improve both drug retention and efficacy as well as MRI contrast efficiency. Both passively and actively targeted drug delivery studies have resulted in delayed tumor growth as well as tumor regression. Significantly, ND-drug complexes have mediated tumor treatment with improved safety profiles when compared to clinical standards. From breast, liver and brain cancer therapy, to MR imaging applications, the future of ND development for cancer nanomedicine is certainly promising.

Acknowledgments D.H. gratefully acknowledges support from the National Science Foundation CAREER Award (CMMI-1350197), Center for Scalable and Integrated NanoManufacturing (DMI-0327077), CMMI-0856492, DMR-1343991, V Foundation for Cancer Research Scholars Award, Wallace H. Coulter Foundation Translational Research Award, National Cancer Institute grant U54CA151880 (The content is solely the responsibility of the authors and does not necessarily represent the official views of the National Cancer Institute or the National Institutes of Health), Society for Laboratory Automation and Screening Endowed Fellowship, and Beckman Coulter.

References

1. Levy M et al (2011) Long term in vivo biotransformation of iron oxide nanoparticles. Biomaterials 32:3988
2. Hao J et al (2011) Development and optimization of solid lipid nanoparticle formulation for ophthalmic delivery of chloramphenicol using a Box-Behnken design. Int J Nanomed 6:683
3. Xu X et al (2013) Enhancing tumor cell response to chemotherapy through nanoparticle-mediated codelivery of siRNA and cisplatin prodrug. In: Proceedings of the national academy of sciences, 28 Oct 2013
4. Peppas NA, Hilt JZ, Khademhosseini A, Langer R (2006) Hydrogels in biology and medicine: from molecular principles to bionanotechnology. Adv Mater 18:1345
5. Langer R, Peppas NA (2003) Advances in biomaterials, drug delivery, and bionanotechnology. AIChE J 49:2990
6. Hrkach J et al (2012) Preclinical development and clinical translation of a PSMA-targeted docetaxel nanoparticle with a differentiated pharmacological profile. Sci Transl Med 4:128ra39
7. Bennewitz MF et al (2011) Biocompatible and pH-sensitive PLGA encapsulated MnO nanocrystals for molecular and cellular MRI. ACS Nano 5:3438
8. Liu Z et al (2008) Drug delivery with carbon nanotubes for in vivo cancer treatment. Cancer Res 68:6652
9. Li Y et al (2010) A nonenzymatic cholesterol sensor constructed by using porous tubular silver nanoparticles. Biosens Bioelectron 25:2356
10. Kannan S et al (2012) Dendrimer-based postnatal therapy for neuroinflammation and cerebral palsy in a rabbit model. Sci Transl Med 4:130ra46
11. Jia LJ et al (2010) Preparation and characterization of silybin-loaded nanostructured lipid carriers. Drug Deliv 17:11
12. Dai W et al (2010) Preparation and characteristics of oridonin-loaded nanostructured lipid carriers as a controlled-release delivery system. J Microencapsul 27:234
13. Zhang X-Q et al (2011) Strategy for increasing drug solubility and efficacy through covalent attachment to polyvalent DNA–nanoparticle conjugates. ACS Nano 5:6962
14. Moore L et al (2014) Comprehensive interrogation of the cellular response to fluorescent, detonation and functionalized nanodiamonds. Nanoscale 6:11712
15. Loo C, Lowery A, Halas N, West J, Drezek R (2005) Immunotargeted nanoshells for integrated cancer imaging and therapy. Nano Lett 5:709
16. Gobin AM et al (2007) Near-infrared resonant nanoshells for combined optical imaging and photothermal cancer therapy. Nano Lett 7:1929
17. Napier ME, DeSimone JM (2007) Nanoparticle drug delivery platform. Polym Rev 47:321
18. Mochalin VN, Shenderova O, Ho D, Gogotsi Y (2012) The properties and applications of nanodiamonds. Nat Nano 7:11
19. Faklaris O et al (2009) Photoluminescent diamond nanoparticles for cell labeling: study of the uptake mechanism in mammalian cells. ACS Nano 3:3955
20. Mohan N, Chen C-S, Hsieh H-H, Wu Y-C, Chang H-C (2010) In vivo imaging and toxicity assessments of fluorescent nanodiamonds in caenorhabditis elegans. Nano Lett 10:3692

21. Wu T-J et al (2013) Tracking the engraftment and regenerative capabilities of transplanted lung stem cells using fluorescent nanodiamonds. Nat Nano 8:682
22. Mochalin VN, Gogotsi Y (2009) Wet chemistry route to hydrophobic blue fluorescent nanodiamond. J Am Chem Soc 131:4594
23. Liang Y, Ozawa M, Krueger A (2009) A general procedure to functionalize agglomerating nanoparticles demonstrated on nanodiamond. ACS Nano 3:2288
24. Barnard AS (2008) Self-assembly in nanodiamond agglutinates. J Mater Chem 18:4038
25. Huang H, Pierstorff E, Osawa E, Ho D (2008) Protein-mediated assembly of nanodiamond hydrogels into a biocompatible and biofunctional multilayer nanofilm. ACS Nano 2:203
26. Manus LM et al (2010) Gd(III)-nanodiamond conjugates for MRI contrast enhancement. Nano Lett 10:484
27. Chow EK et al (2011) Nanodiamond therapeutic delivery agents mediate enhanced chemoresistant tumor treatment. Sci Transl Med 3:73ra21
28. Huang H, Pierstorff E, Osawa E, Ho D (2007) Active nanodiamond hydrogels for chemotherapeutic delivery. Nano Lett 7:3305
29. Mochalin VN et al (2013) Adsorption of drugs on nanodiamond: toward development of a drug delivery platform. Mol Pharm 10:3728
30. Xiao J et al (2013) Nanodiamonds-mediated doxorubicin nuclear delivery to inhibit lung metastasis of breast cancer. Biomaterials 34:9648
31. Moore L, Gatica M, Kim H, Osawa E, Ho D (2013) Multi-protein delivery by nanodiamonds promotes bone formation. J Dent Res 92:976
32. Smith AH et al (2011) Triggered release of therapeutic antibodies from nanodiamond complexes. Nanoscale 3:2844
33. Zhao L et al (2014) Polyglycerol-coated nanodiamond as a macrophage-evading platform for selective drug delivery in cancer cells. Biomaterials 35:5393
34. Kim H et al (2012) Multiscale simulation as a framework for the enhanced design of nanodiamond-polyethylenimine-based gene delivery. J Phys Chem Lett 3:3791
35. Adnan A et al (2010) Atomistic simulation and measurement of pH dependent cancer therapeutic interactions with nanodiamond carrier. Mol Pharm 8:368
36. Moore L, Chow EK-H, Osawa E, Bishop JM, Ho D (2013) Diamond-lipid hybrids enhance chemotherapeutic tolerance and mediate tumor regression. Adv Mater 25:3532
37. Deeken JF, Loscher W (2007) The blood-brain barrier and cancer: transporters, treatment, and trojan horses. Clin Cancer Res 13:1663
38. Xi G et al (2014) Convection-enhanced delivery of nanodiamond drug delivery platforms for intracranial tumor treatment. Nanomed Nanotechnol Biol Med 10:381
39. Lin Y-C, Perevedentseva E, Tsai L-W, Wu K-T, Cheng C-L (2012) Nanodiamond for intracellular imaging in the microorganisms in vivo. J Biophotonics 5:838
40. Lam R et al (2008) Nanodiamond-embedded microfilm devices for localized chemotherapeutic elution. ACS Nano 2:2095
41. Kim H-J, Zhang K, Moore L, Ho D (2014) Diamond nanogel-embedded contact lenses mediate lysozyme-dependent therapeutic release. ACS Nano 8:2998
42. Bui T et al (2010) Novel Gd nanoparticles enhance vascular contrast for high-resolution magnetic resonance imaging. PLoS ONE 5:e13082
43. Yuen Yung H, Chia-Liang C, Huan-Cheng C (2010) Nanodiamonds for optical bioimaging. J Phys D: Appl Phys 43:374021
44. Hegyi A, Yablonovitch E (2013) Nanodiamond molecular imaging with enhanced contrast and expanded field of view. Biomedo 19:011015
45. Igarashi R et al (2012) Real-time background-free selective imaging of fluorescent nanodiamonds in vivo. Nano Lett 12:5726
46. McGuinness LP et al (2011) Quantum measurement and orientation tracking of fluorescent nanodiamonds inside living cells. Nat Nano 6:358
47. Barnard AS (2009) Diamond standard in diagnostics: nanodiamond biolabels make their mark. Analyst 134:1751

48. Hui YY et al (2014) Wide-field imaging and flow cytometric analysis of cancer cells in blood by fluorescent nanodiamond labeling and time gating. Sci Rep 4. doi:10.1038/srep05574
49. Perevedentseva E, Lin Y-C, Jani M, Cheng C-L (2013) Biomedical applications of nanodiamonds in imaging and therapy. Nanomedicine 8:2041
50. Heyer S et al (2014) Toward deep blue nano hope diamonds: heavily boron-doped diamond nanoparticles. ACS Nano 8:5757
51. Zurbuchen MA, Lake MP, Kohan SA, Leung B, Bouchard L-S (2013) Nanodiamond landmarks for subcellular multimodal optical and electron imaging. Sci. Rep 3:2668
52. Tisler J et al (2009) Fluorescence and spin properties of defects in single digit nanodiamonds. ACS Nano 3:1959
53. Liu K-K, Wang C-C, Cheng C-L, Chao J-I (2009) Endocytic carboxylated nanodiamond for the labeling and tracking of cell division and differentiation in cancer and stem cells. Biomaterials 30:4249
54. Shenderova O et al (2014) Carbon-dot-decorated nanodiamonds. Part Part Syst Charact 31:580
55. Vaijayanthimala V, Chang HC (2008) Functionalized fluorescent nanodiamonds for biomedical applications. Nanomedicine 4:47
56. Zhang Q et al (2011) Fluorescent PLLA-nanodiamond composites for bone tissue engineering. Biomaterials 32:87
57. Zhang X, Hu W, Li J, Tao L, Wei Y (2012) A comparative study of cellular uptake and cytotoxicity of multi-walled carbon nanotubes, graphene oxide, and nanodiamond. Toxicol Res 1:62
58. Li H-C et al (2013) The hemocompatibility of oxidized diamond nanocrystals for biomedical applications. Sci Rep 3:2668
59. Xing Y et al (2011) DNA damage in embryonic stem cells caused by nanodiamonds. ACS Nano 5:2376
60. Kumari S, Singh MK, Singh SK, Grácio JJA, Dash D (2013) Nanodiamonds activate blood platelets and induce thromboembolism. Nanomedicine 9:427
61. Chow EK-H, Ho D (2013) Cancer nanomedicine: from drug delivery to imaging. Sci Transl Med 5:216rv4

Theranostic Lipid Nanoparticles for Cancer Medicine

Danielle M. Charron, Juan Chen and Gang Zheng

Abstract

Disease heterogeneity within and between patients necessitates a patient-focused approach to cancer treatment. This exigency forms the basis for the medical practice termed personalized medicine. An emerging, important component of personalized medicine is theranostics. Theranostics describes the co-delivery of therapeutic and imaging agents in a single formulation. Co-delivery enables noninvasive, real-time visualization of drug fate, including drug pharmacokinetic and biodistribution profiles and intratumoral accumulation. These technological advances assist drug development and ultimately may translate to improved treatment planning at the bedside. Nanocarriers are advantageous for theranostics as their size and versatility enables integration of multiple functional components in a single platform. This chapter focuses on recent developments in advanced lipid theranostic nanomedicine from the perspective of the "all-in-one" or the "one-for-all" approach. The design paradigm of "all-in-one" is the most common approach for assembling theranostic lipid nanoparticles, where the advantages of theranostics are achieved by combining multiple components that each possesses a specific singular function for therapeutic activity or imaging contrast. We will review lipoprotein nanoparticles and liposomes as representatives of the "all-in-one" approach. Complementary to the "all-in-one" approach

D.M. Charron · G. Zheng
Institute of Biomaterials and Biomedical Engineering,
University of Toronto, Toronto, Canada

D.M. Charron · J. Chen · G. Zheng
Princess Margaret Cancer Center, University Health Network, Toronto, Canada

G. Zheng (✉)
Department of Medical Biophysics, University of Toronto, Toronto, Canada
e-mail: gang.zheng@uhnres.utoronto.ca

C.A. Mirkin et al. (eds.), *Nanotechnology-Based Precision Tools for the Detection and Treatment of Cancer*, Cancer Treatment and Research 166,
DOI 10.1007/978-3-319-16555-4_5

is the emerging paradigm of the "one-for-all" approach where nanoparticle components are intrinsically multifunctional. We will discuss the "one-for-all" approach using porphysomes as a representative. We will further discuss how the concept of "one-for-all" might overcome the regulatory hurdles facing theranostic lipid nanomedicine.

Keywords
Multifunctional nanoparticle · Cancer imaging · Cancer therapy · Lipoprotein · Liposome · Porphyrin

Contents

1 Introduction

Every individual is unique, so unsurprisingly the nature of disease is likewise diverse. Personalized medicine is a patient-focused approach to administering the appropriate therapy for the individual, to the correct location in the body, at the most beneficial time. Recent advances in scientific, technical, and medical research together bring the concept of personalized medicine to the clinic. Genetic and molecular profiling now offer detailed assessment of disease and metrics to predict response to therapy. Targeted and focal therapies provide the clinician with the means to combat the identified disease appropriately. Noninvasive imaging modalities such as magnetic resonance imaging (MRI), computed tomography (CT), ultrasound imaging, positron emission tomography (PET), single photon emission computed tomography (SPECT), fluorescence imaging, and photoacoustic imaging (PAI) are useful tools for understanding signaling pathways and fundamental biological processes at the cellular and molecular level, diagnosing disease, delineating tumors, monitoring physiological responses to therapy, and evaluating drug efficacy longitudinally. Together, these tools enable clinicians to address the heterogeneity of disease within and between patients. There are many comprehensive reviews of the different imaging modalities [1, 2]. A brief summary of their key characteristics is listed in Table 1.

Table 1 Comparison of clinical imaging modalities used in nanotheranostics

Modality	Signal	Contrast agent	Sensitivity [1] (mol label)	Resolution [1]	Tissue penetration	Risk
Primarily morphological/anatomical imaging modalities						
MRI	Magnetic fields	Gadolinium, iron oxide nanoparticles	10^{-9}–10^{-6}	50 μm	Unlimited	Harmless
CT	X-rays	Iodine	10^{-6}	50 μm	Unlimited	Radiation
US	Acoustic waves	Microbubbles	10^{-8}	50 μm	Unlimited	Harmless
Primarily molecular imaging modalities						
PET	Positrons from radionuclides	Positron-emitting radionuclides (e.g., ^{18}F, ^{64}Cu, ^{124}I)	10^{-15}	1–2 mm	Unlimited	Radiation
SPECT	γ-rays	Single photon-emitting radionuclides (e.g., ^{99m}Tc, ^{67}Ga, ^{111}In)	10^{-14}	1–2 mm	Unlimited	Radiation
Fluorescence imaging	Light (NIR)	Fluorescent probes	10^{-12}	1–2 mm	3 mm	Harmless
PAI	Acoustic waves	Probes that absorb light and emit heat	10^{-12}	50 μm	7 cm	Harmless

MR magnetic resonance; *CT* computed tomography; *US* ultrasound imaging; *PET* positron emission tomography; *SPECT* single photon emission computed tomography; *NIR* near-infrared; *PAI* photoacoustic imaging

As cancer research pushes toward the goal of personalized medicine, theranostics will play an increasingly important role [3]. Theranostics is the co-delivery of therapeutic and imaging agents in a single formulation [4, 5]. In this way the drug and contrast agent share the same pharmacokinetic and biodistribution profiles, and therefore imaging provides an accurate assessment of drug distribution. Regarding theranostics, drug delivery is not a black box having injected dose as input and clearance rate as output. Imaging functionality can quantify the intratumoral drug dose [6], specify the activation state of a therapeutically shielded drug [7], and indicate the intracellular uptake pathway [8]. Nanocarriers offer many advantages for drug delivery and theranostics compared to molecular systems. Encapsulation of molecular theranostic agents in nanocarriers protects them from serum degradation and rapid renal clearance, and enables the use of hydrophobic agents. It is widely recognized that nanoparticle encapsulation passively improves tumour targeting based on the enhanced permeability and retention (EPR) effect [9–11]. In addition, nanoparticle size and surface composition may be designed to avoid protein adsorption and removal by the reticuloendothelial system (RES), prolonging blood circulation, and increasing the probability of tumour uptake

[12–15]. The targets of many chemotherapeutics (e.g., doxorubicin (dox), siRNA, Photofrin®) are intracellular; therefore, drug efficacy is dependent on its ability to enter the target cell. Conjugation of molecular therapeutics to targeting agents such as antibodies enhances both intracellular uptake and selectivity for tumour cells over normal cells [16, 17]. However, to maintain tumour-targeting functionality the number of therapeutic agents that can be loaded on each targeting moiety is low, and labeling can interfere with therapeutic activity. In contrast, a large number of therapeutics can be encapsulated in a single targeted nanoparticle, significantly increasing the quantity of drug delivered during each internalization event. High loading capacity for imaging agents also improves imaging contrast, and is particularly advantageous for modalities with low sensitivity such as MRI and CT. Environmental factors can also be used to differentiate tumour cells from normal cells. These factors may be internal to the tumour microenvironment (e.g., tumour acidity), or external conditions locally applied to the tumour region (e.g., light exposure). Stimuli-responsive nanoparticles are designed to respond to these environmental factors and enhance payload release [18–21]. In theranostics, actively targeted or stimuli-responsive nanoparticles are called "smart delivery" or "activatable" systems [20]. During systemic circulation, the nanoparticles are therapeutically shielded and imaging functionally silenced to minimize off-target effects and background noise. In the tumour region, the drug and imaging agent are released or activated. With nanotechnology, it is possible to integrate multiple properties in a single platform, including passive targeting, target-triggered activatable release, multimodal imaging, and therapy capabilities. Recent advances in the technological capabilities of nanotheranostic systems impact drug development as well as treatment planning at the patient bedside (Fig. 1). For example, real-time imaging of drug accumulation in the tumour region may be used to optimize drug formulation during the research and development stage, as well as to predict patient response to an administered drug dose.

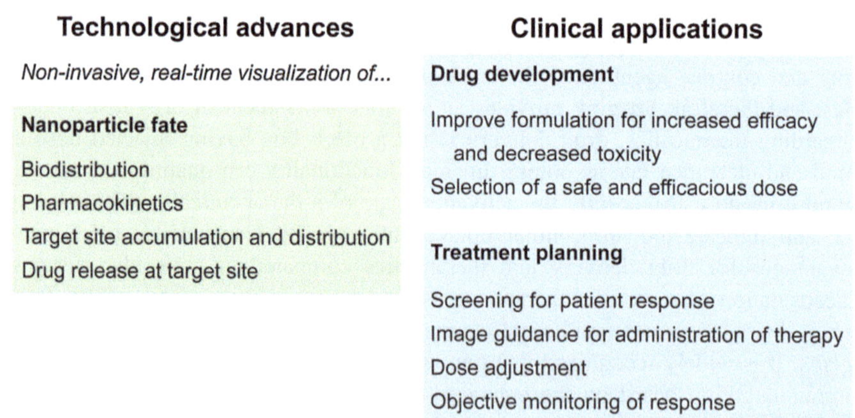

Fig. 1 Recent technological advances in nanotheranostics and their clinical applications for drug development and treatment planning

Various nanocarriers have been investigated for theranostics, including lipid nanoparticles, polymer conjugates, micelles, and dendrimers, as well as inorganic nanoparticles such as silica, gold, quantum dots, iron oxide, and carbon nanotubes [22–24]. Among them, lipid nanoparticles are in the most advanced stage of development and have shown favorable biocompatibility and biodegradation in comparison to inorganic nanoparticles [25]. Extensive knowledge has been obtained on the exploitation of these platforms for cancer therapy, particularly for their ability to enhance therapeutic efficacy over free drugs through improved encapsulation, prolonged circulation half-life, and sustained or triggered release [16, 26–28]. This chapter focuses on recent developments in advanced lipid theranostic nanomedicine from the perspective of the "all-in-one" or the "one-for-all" approach.

2 Engineering Theranostic Lipid Nanoparticles by Assembling Multiple Functional Components—The "All-in-One" Approach

The design paradigm of "all-in-one" is the most common approach for assembling multifunctional lipid nanoparticles, where the advantages of multifunctionality and theranostics are achieved by combining multiple components that each possesses a specific singular function, such as therapeutic activity, imaging contrast, targeting, clearance avoidance, and triggered drug release. Here, we will review lipoprotein nanoparticles and liposomes as representatives of the "all-in-one" approach.

2.1 Lipoprotein Theranostics

Lipoproteins are normally comprised of a hydrophobic lipid core and an outer shell of phospholipids and amphipathic apolipoproteins that confer water solubility and precise size control [29]. As natural cholesterol transporters, lipoproteins integrate many advantages for drug delivery [30–36]. For example:

1. As endogenous carriers, they can escape recognition as foreign entities by the human immune system and clearance by RES [37].
2. Their long blood circulation time provides favorable systemic drug delivery without the need for a polyethylene glycol (PEG) coating to improve steric stability.
3. Their natural structure enables stable systemic delivery of hydrophobic bioactive compounds [38]. This permits various drug loading approaches, including core loading, surface loading, and protein conjugation.
4. Apolipoprotein interaction with the lipid layer encourages a consistent and reproducible number of apolipoproteins per particle and a narrow size distribution, overcoming concerns of polydispersity in size and also providing ligand-targeted delivery [8].

Lipoproteins have been employed for delivery of many chemotherapeutics including dox, paclitaxel, siRNA, hypericin, valrubicin, curcumin, statin, etc. [39–48], and there are many comprehensive reviews available on their uses for drug delivery [35, 37, 38, 49–51]. This chapter focuses on the expansion of their development for theranostic applications.

2.1.1 Building Imaging Functionality for Lipoprotein Nanomedicine

Lipoproteins have a rich history as imaging contrast agents due to their ability to transport large loads of hydrophobic agents and ease of loading. Three main strategies have been explored to functionalize lipoproteins: (1) surface loading—noncovalent intercalation of the imaging agent within the surface of the lipoprotein; (2) covalent modification—conjugating the imaging agent onto the surface of the apolipoprotein or onto the phospholipid headgroup; (3) and core loading through reconstitution—encapsulating hydrophobic agents in the nanoparticle core (Fig. 2a) [35]. Table 2 provides an overview of the imaging modalities demonstrated to date. These modalities generally focus on tracking different aspects of lipoprotein fate. MRI, CT, and PET monitor lipoproteins during systemic circulation to determine their biodistribution and ability to accumulate in the tumour region. Fluorescence imaging is widely used to determine the cell delivery mechanism [30, 53–56]. Core-loaded fluorophores permit low background fluorescence imaging by means of a "smart delivery system" approach that takes advantage of the well-established phenomenon of aggregation-induced chromophore quenching [68]. The intracellular uptake mechanisms of low-density lipoprotein (LDL) and high-density lipoprotein (HDL) have been disclosed using fluorescence microscopy. LDL enters cells through the LDL receptor (LDLR)-mediated endocytosis pathway [69], while HDL interacts with scavenger receptor class B member 1 (SR-B1) to drive direct transport of the payload molecules into the cell cytoplasm without internalization of the entire particle [8].

Beyond natural lipoproteins, many advanced lipoprotein nanocarriers have been developed to overcome the significant challenges that limit the use of lipoproteins as general theranostic nanoplatforms. For example, the lipoprotein rerouting approach developed by Zheng et al. demonstrates a plausible approach to redirecting lipoprotein theranostics to other cancer associated biomarkers beyond their normal receptors that are only overexpressed in a limited number of cancers [31]. Furthermore, original lipoprotein nanocarriers were natural structures directly isolated from human donors, and faced issues of purity, quantity, length of processing time, as well as safety. New synthetic lipoproteins have been developed, including reconstituted lipoproteins that use isolated apolipoproteins (recombinant or naturally derived) to construct the nanoparticle [41], and lipoprotein-mimetic nanoparticles constrained by apolipoprotein-mimetic peptides [70]. These synthetic systems possess the known functions ascribed to native lipoproteins, including natural targeting, enzyme/pathway activation, intracellular uptake, and lipid transfer. More importantly, their controlled synthesis offers advantages of uniform size, zeta potential, and stable drug loading (core and surface loading). The formulation

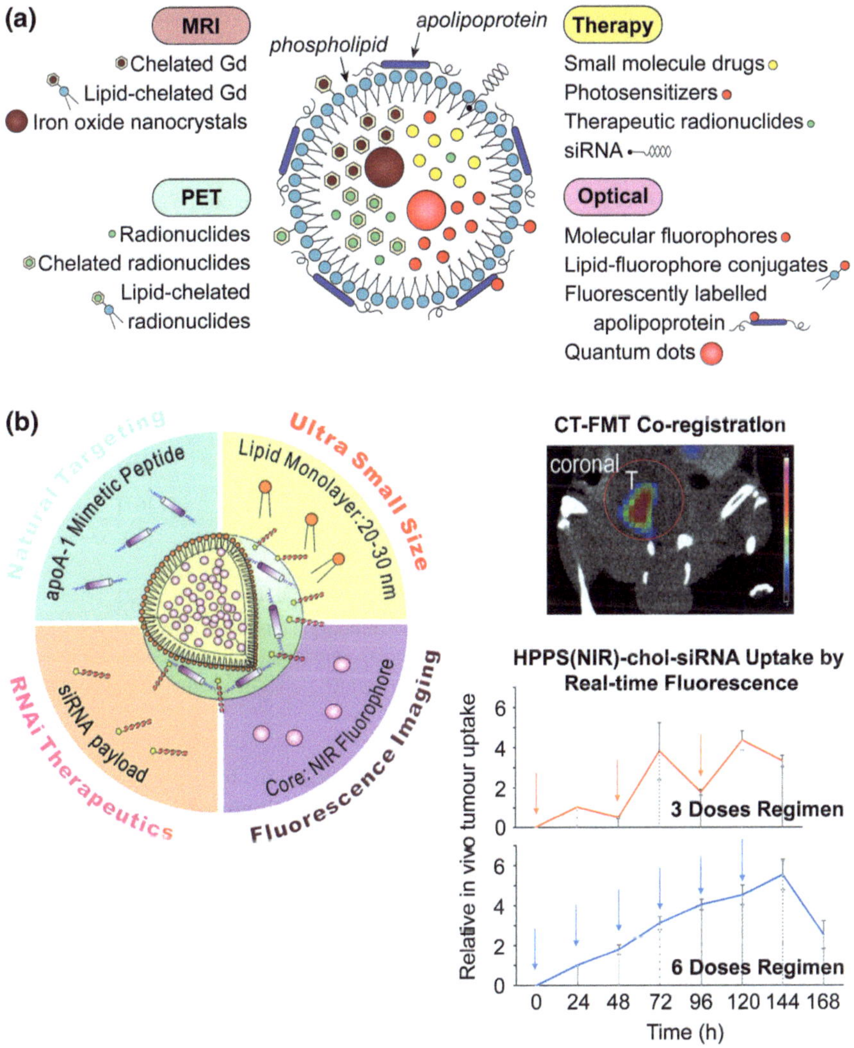

Fig. 2 **a** Schematic representation of the strategies available for incorporating imaging and therapeutic agents in lipoproteins. Hydrophobic molecules such as drugs, photosensitizers, and fluorophores, as well as inorganic nanoparticles (e.g., iron oxide, QD) are core encapsulated through reconstitution. Gadolinium and radionuclides are core encapsulated with and without assistance of chelation (e.g., by diethylene triamine pentaacetic acid (DTPA)). Imaging agents are incorporated into the nanoparticle shell through lipid conjugation or fluorescence labeling of the apolipoprotein. Cholesterol-conjugated siRNA is embedded within the lipid layer though noncovalent intercalation. **b** *Left* HDL-mimetic lipoprotein nanoparticle "HPPS" incorporating inherent targeting functionality, gene therapy, and fluorescence imaging in a compact spherical structure. *Right*, *top* CT-FMT co-registered images for visualizing siRNA delivery in a prostate tumour model; *bottom* siRNA delivery quantified longitudinally by FMT. Adapted with permission from Lin et al., copyright Wiley-VCH (2014) [52]

Table 2 Imaging modalities incorporated in lipoproteins for visualizing biodistribution, metabolism, fate, and cytosolic delivery

Modality	Label	Lipoprotein type	Purpose	References
MRI	Gd-DTPA	LDL	Biodistribution, metabolism	[59]
	Gd-DTPA-PE	LDL, HDL	Imaging arteriosclerotic plaques	[57, 60]
	SPION	LDL, HDL	Tumour accumulation, metabolism kinetics	[58, 61, 62]
CT	Poly-iodinated triglyceride	LDL	Tumour accumulation	[63]
	Au nanocrystals	LDL, HDL	Tumour accumulation	[57, 61, 64, 65]
PET/SPECT	^{99m}Tc	LDL	Biodistribution, tumour accumulation	[66]
	^{68}Ga-DTPA	LDL	Metabolism	[67]
	^{111}In-DTPA	LDL	Metabolism	[67]
Fluorescence imaging	Fluorescent probes	LDL, HDL	Delivery function	[30, 53–56]
	Quantum dots	HDL	Delivery function	[62]

MRI magnetic resonance imaging; *DTPA* diethylene triamine pentaacetic acid; *LDL* low-density lipoprotein; *HDL* high-density lipoprotein; *SPION* superparamagnetic iron oxide nanoparticles; *CT* computed tomography; *PET* positron emission tomography; *SPECT* single photon emission computed tomography

can be adjusted for the desired phospholipid composition, hydrophobic core moiety, payload, and protein or peptide type. Synthetic lipoproteins may therefore significantly simplify the scale-up process for manufacturing lipoproteins for human use, as well as accelerate the clinical translation of lipoprotein drug delivery.

The development of multifunctional and theranostic lipoproteins has been reliant on the advancement of methods for encapsulating imaging agents. Allijn et al. developed a sonication method for core-labeling native LDL particles with a range of diagnostically active nanocrystals or hydrophobic agents, allowing for detection of LDL delivery from the anatomical level (whole-body imaging by conventional and spectral CT and MRI), to the microscopic level (by fluorescence imaging), and down to the nanometer scale for subcellular localization (by transmission electron microscopy (TEM)) [57, 58]. Using additional well-developed reconstitution methods [53], a range of photosensitizers (e.g., phthalocyanine, naphthalocyanine, pyropheophorbide, bacteriochlorin-e6) have been stably incorporated within lipoprotein particles for effective targeted delivery of PDT agents [30, 34, 54, 71, 72]. Their intrinsic NIR fluorescence can be tracked through fluorescence imaging, thus providing a noninvasive tool to assess in vivo tumour accumulation and guide laser placement for effective PDT. Zheng et al. recently developed a synthetic lipoprotein termed HPPS (HDL phospholipid scaffold) [43, 70, 73], comprised of a lipid shell with a hydrophobic core, constrained by an amphipathic peptide which functionally mimics the apolipoprotein ApoA-1 to confer HDL-like structure and function [74].

HPPS is easily synthesized by sonication and modified for incorporation of functional payloads. Both the hydrophobic core and lipid shell can be replaced with hydrophobic [72] or amphipathic [75] therapeutic or imaging agents [70, 73], making it a versatile platform for theranostics.

Benefiting from their inherent cell uptake mechanism, lipoproteins are useful nanocarriers for treatments requiring cell delivery to be effective, such as gene therapy. SR-B1-mediated delivery is particularly advantageous as it does not involve endosomal entrapment typical of receptor-mediated endocytosis, thus minimizing degradation of siRNA [8, 37, 51, 52, 76, 77]. Cholesterol-conjugated siRNA (chol-siRNA) can be stably loaded into the HDL shell, protecting it from rapid renal clearance. The siRNA is selectively delivered to tumour cells that upregulate SR-B1, significantly improving cell uptake compared to free siRNA. Adding additional imaging functionality to track siRNA delivery would be useful for determining the optimal therapeutic regimen and predicting patient response. Therefore, a 30 nm fluorescent HDL-like nanoparticle was developed for siRNA delivery (HPPS(NIR)-chol-siRNA), with a NIR fluorescent core and siRNA payloads intercalated within the particle membrane (Fig. 2b). In vitro and in vivo fluorescence hyperspectral imaging demonstrated that the NIR core is a suitable surrogate for fluorescently labeling therapeutic siRNA [52]. Directly labeling siRNA itself is disadvantageous as labeling can affect therapeutic activity. Using this system and a CT-FMT (fluorescence molecular tomography) co-registration approach, the authors were able to evaluate the efficacy of two different treatment regimens in vivo: (1) administration of three doses every other day; compared to (2) administration of six smaller daily doses. Monitoring the real-time fluorescence uptake curve demonstrated that six daily doses gives a continuous increase in siRNA accumulation compared to three larger doses spaced further apart in time, thus guiding the treatment regimen to achieve efficacious RNAi therapy in an orthotopic PC3 tumour model.

2.2 Liposome Theranostics

Among all lipid nanoparticles, liposomes have demonstrated significant advances both as drug delivery and diagnostic tools, and represent the nanocarriers furthest along in clinical application [78–80]. Liposomes are spherical nanocarriers comprised of a self-assembled lipid bilayer and an aqueous core [81]. Liposomes have been extensively studied for drug delivery owing to their unique structure that allows for loading of hydrophilic therapeutic agents within the core, as well as hydrophobic agents within the bilayer. Recent reviews of the experimental development and clinical use of liposomes as drug delivery vehicles have focused on the emerging trend of multifunctionality (Fig. 3a) [16, 26–28]. Inclusion of functional components transforms conventional liposomes into stealth, ligand-targeted, and triggered-release liposomes. More recently, liposomes have been recognized as suitable nanocarriers for theranostics because they have ample room for both therapeutic and diagnostic agents. Inclusion of molecular and nanoparticle imaging

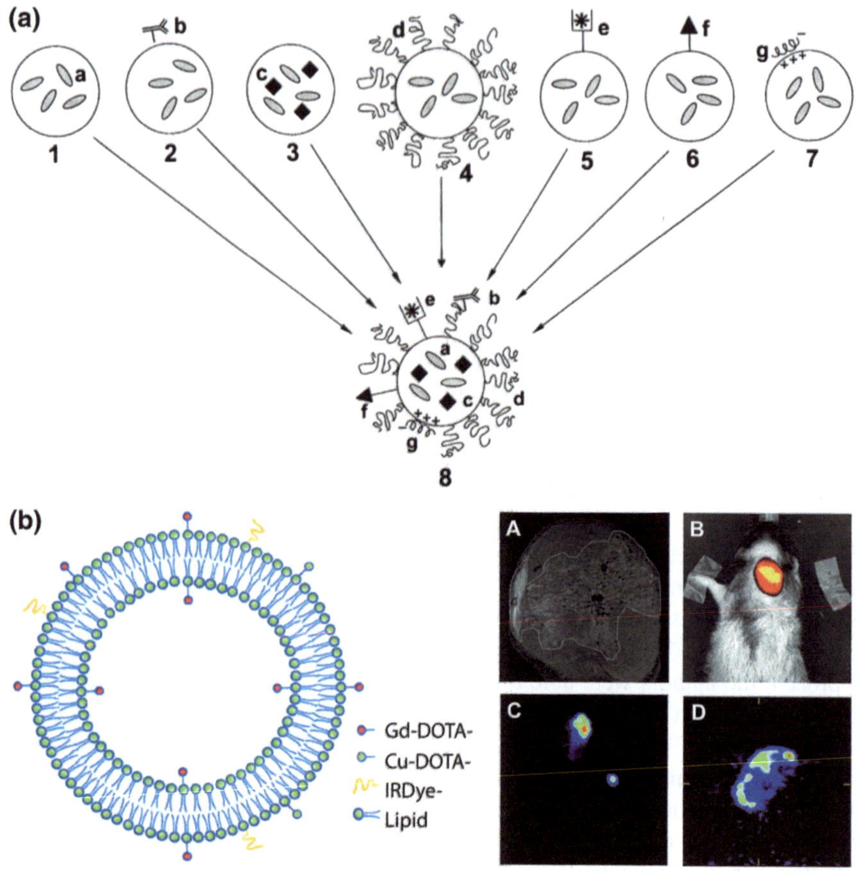

Fig. 3 Multimodal lipid nanotheranostic systems incorporating multiple single-modality agents. **a** Schematic representation of the "holy grail" for multifunctional liposome design, combining *a* small molecule drugs, *b* ligand targeting, *c* nanocrystals, *d* PEGylation, *e* chelated radionuclides, *f* cell-penetrating peptides, and *g* siRNA. Adapted with permission from Torchilin, copyright Elsevier (2006) [82]. **b** *Left* theranostic liposome with MRI, PET/SPECT, and fluorescence imaging modalities provided by inclusion of amphipathic chelators (DOTA-lipid) and fluorophores (IRdye-lipid) combined with core loading of hydrophilic agents. *Right* In vivo imaging of SCCHN tumour xenograft model by *A* MRI using Gd(III)-liposomes, *B* NIR fluorescence imaging using IRdye-Gd-liposomes, *C* SPECT using ^{99m}Tc-Gd-liposomes, and *D* PET using ^{64}Cu-Gd-liposomes. Adapted with permission from Li et al., copyright American Chemical Society (2012) [83]

agents can be achieved by core encapsulation, embedding in the lipid bilayer, or conjugation/adsorption to the liposome surface [84], thus enabling integration of multiple single-modality components for advanced nanotheranostics. Following this strategy, liposome theranostics has high potential for clinical translation as all of the assembled therapeutic and imaging agents, as well as the nanocarrier itself, are already approved by regulatory agencies and are well-established in the clinic. Below are highlighted the recent developments in liposome theranostics.

2.2.1 Building Imaging Functionality for Liposome Nanomedicine

Visualizing nanoparticle biodistribution and pharmacokinetics can help researchers better assess how effectively a nanocarrier accumulates in the tumour region and avoids RES clearance, and compare their design with other formulations. For example, Grange et al. recently reported liposomal dox labeled with an amphipathic form of gadolinium (Gd(III)-DOTAMA($C18)_2$) for MRI [85]. The authors prepared a targeted formulation using a peptide ligand that binds to the vascular factor neural cell adhesion molecule (NCAM), which is upregulated in Kaposi's sarcoma, a highly angiogenic disease. Using MRI, the authors showed that the NCAM-targeted liposome exhibits approximately 50 % overall enhanced tumour accumulation compared to the untargeted form, which is primarily retained in the blood. Ex vivo TEM and in vitro confocal imaging indicate that this result is due to enhanced cell uptake by NCAM targeting. The lipophilic Gd(III) remained sequestered in the tumour cell membrane, enabling monitoring of the tumour response longitudinally. Using MRI, the authors were able to conclude that despite faster clearance, NCAM ligand targeting enhances liposome accumulation in Kaposi's sarcoma and improves tumour response.

Following administration of a chemotherapeutic, patient response is assessed to optimize the subsequent dose or, in the case of a poor tumour response, to consider alternate options. For example, in breast cancer treatment, tumour response is primarily determined based on tumour staging and hormonal status. In nanomedicine, however, individual differences in tumour vasculature significantly impact delivery of the nanocarrier and must be considered. Poor vascular permeability limits tumour uptake and intratumoral distribution and is associated with faster tumour growth and poor therapeutic outcome. To provide clinicians with a tool to assess vascular permeability, Karathanasis et al. co-encapsulated dox and the CT contrast agent iodixanol in a liposome to be used for mammography [86]. In an animal study, the authors administered the liposome, once a week for 2 weeks, and found that the sum of the CT signal enhancement within the tumour region over the 2 weeks was correlated with overall survival. This relationship was present despite large heterogeneity in the two CT signals within the same animal. Using this nanotheranostic, the authors were able to predict the therapeutic outcome based on nanoparticle accumulation, and could in the future adjust the administered dose to accommodate for individual differences in vascular permeability.

MRI "smart delivery" systems are the main class of activatable theranostic liposomes reported for tracking intracellular uptake. For MRI labels such as Gd(III), the MR signal (T_1 relaxivity) changes based on its environment and interaction with water molecules. Liposome-encapsulated gadolinium cannot interact with water molecules as water cannot easily permeate the lipid bilayer. In this case the T_1 relaxation enhancement is minimal. Following release of the gadolinium label, interaction with water molecules increases T_1 relaxivity and appears as positive contrast, or a bright spot, on the MR image. MRI contrast enhancement has been quantitatively correlated with therapeutic payload release from activatable liposome systems. Viglianti et al. co-loaded the hydrophilic MRI label $MnSO_4$ and dox in the liposome core [87]. Using T_1-weighted images, the authors developed an algorithm for quantifying dox release and MRI contrast in a rat oral fibrosarcoma model. The model was confirmed both ex vivo using liquid chromatography and by histology. A follow-up study was conducted by Tagami et al., replacing toxic $MnSO_4$ with gadolinium using diethylene triamine pentaacetic acid (DTPA) as a chelating agent [88]. As in the first study, the release profile of dox (measured by fluorescence) and MRI contrast agent were identical, and MRI could be used to monitor drug release. Both of these studies envisaged activatable MRI as a tool for treatment planning during delivery of temperature-sensitive liposomes (TSL). TSLs such as ThermoDox® are under clinical study to selectively improve drug release at the tumour site [89–91]. The lipid content is chosen to provide a melting temperature of 40–43 °C. Co-delivery of heat (e.g., by focused ultrasound) destabilizes the lipid membrane leading to rapid release of the therapeutic agent. Due to variations in tissue structure and composition, thermal gradients can form leading to nonuniform drug distribution. Activatable MRI could be used to monitor the drug distribution in real time and adjust heat delivery to provide sufficient and uniform release.

Combining multiple imaging modalities significantly increases the power of nanotheranostics for drug development and treatment planning (i.e., multimodal imaging). As outlined in Table 1, each imaging modality suffers from technical disadvantages such as low resolution for PET/SPECT, sensitivity for MRI and CT, or tissue penetration depth for fluorescence imaging, and not all modalities are suitable for every desired application. Imaging both the systemic- and micro-distribution of nanocarriers requires the integration of noninterfering modalities to provide both high spatial resolution and sensitivity [92]. To achieve this goal, Li et al. reported a liposomal dox formulation with MRI, PET/SPECT, and fluorescence imaging functionality (Fig. 3b) [83]. During liposome synthesis, some of the lipid shell is replaced with an amphipathic chelator (DOTA-DSPE) and an MRI agent (Gd(III)-DOTA-DSPE). Postsynthesis the liposome is labeled with ^{64}Cu for PET though metal chelation with DOTA-DSPE. Fluorescence imaging functionality is provided by postinsertion of an amphipathic fluorophore (IRdye-DSPE) into the outer shell of the lipid bilayer. For SPECT, ^{99m}Tc is encapsulated in the aqueous core. Dox is also encapsulated postsynthesis, or alternatively a therapeutic radionuclide (^{186}Re/^{188}Re) can be encapsulated for radiotherapy. The authors found that this formulation and synthesis procedure was more robust than other multistep encapsulation methods (<20 nm size change with loading), and has a high loading capacity for both imaging and therapy (>90 % efficiency for all imaging components and 65 % for dox). With high-resolution MRI, the authors followed liposome biodistribution and intratumoral micro-distribution in a xenograft model of head-and-neck squamous cell carcinoma. Using sensitive PET/SPECT and fluorescence imaging, tumor distribution was monitored quantitatively and longitudinally, respectively.

3 Engineering Theranostic Lipid Nanoparticles Using Intrinsically Multifunctional Components—The "One-for-All" Approach

The "all-in-one" approach for combining multiple functionalities (Fig. 3a) has been described as the "holy grail" for lipid nanomedicine [82, 93, 94]. Despite the advantages of the many advanced nanotheranostics investigated in the lab, there is a disparity between these and the simplistic nanocarriers seen in the clinic [27, 28, 95]. While transferrin conjugated and thermally sensitive liposomes are undergoing clinical trials, the only theranostic nanoparticles currently under investigation are inorganic iron oxide nanoparticles for magnetic heating and MRI contrast [96, 97].

Transitioning theranostic nanoparticles from the lab to the clinic is hindered by the current paradigm of the "all-in-one" approach [98]. Inclusion of functional units increases the complexity and cost of synthesis and purification, and may be justified due to the benefits they provide. However, physical entrapment of therapeutic and imaging agents generally leads to an increase in particle polydispersity, and this random effect is amplified if more than one agent is encapsulated. Polydispersity raises regulatory concerns because nanoparticle size and composition is uncontrolled, leading to unpredictable in vivo behavior [99, 100]. Furthermore, adding imaging agents to lipid nanoparticle drug formulations does not generally provide imaging capabilities above the threshold required for high-quality images. It is particularly difficult to achieve a therapeutic payload in lipid-nanoparticle complexes. Radiolabeling has been recognized as a possible solution because PET is a highly sensitive technique [101]. With a careful design and controlled synthesis, most regulatory challenges can be overcome.

Complementary to the "all-in-one" approach is the emerging paradigm of the "one-for-all" approach where nanoparticle components are intrinsically multifunctional. We will discuss the "one-for-all" approach using porphysomes as a representative "one-for all" nanoparticle, and how this concept might overcome the regulatory hurdles facing theranostic lipid nanomedicine.

3.1 Porphysome Theranostics Using Porphyrin-Lipid Technology

Porphyrins are chromophores widely studied for their versatility and numerous applications in medicine and technology [102, 103]. Porphyrins have long been utilized clinically for their theranostic capabilities including fluorescence imaging and photosensitization [104, 105]. Porphyrins used in photodynamic therapy generate cytotoxic singlet oxygen as well as fluorescence upon excitation, which can be used to monitor tumor accumulation and intracellular uptake for treatment planning. In addition, photobleaching is a surrogate measurement for therapeutic activity and has been used to quantify the results of adjusting treatment parameters such as light dosimetry [106]. Encapsulated porphyrins form nonfluorescent aggregates which emit absorbed light as heat, enabling their use for hyperthermia and photothermal ablation as well as PAI [107]. Depending on their molecular structure, porphyrins can be stable metal chelators for MRI and PET/SPECT [108]. Metal chelation can also affect the photonic properties, including shifting the absorption spectrum or stabilizing thermal generation.

Porphysomes are a novel class of liposome-like nanocarriers, self-assembled from porphyrin-lipid building blocks (Fig. 4a). Porphyrin-lipid is formed by conjugation of a single-chained phospholipid with either pyropheophorbide (derived from *Spirulina pacifica*) or bacteriochlorophyll (derived from *Rhodobacter sphaeroides*) [109]. The hydrophobic porphyrin is attached to the lipid chain such that the structure remains amphipathic, allowing for self-assembly in aqueous buffer. Other types of lipids, such as PEG-lipid, which is used to enhance in vivo pharmacokinetics, and cholesterol, which is used to enhance loading and circulation half-life, are also involved in the formulation to improve in vivo function. Using TEM, the self-assembled nanoparticles were shown to be spherical vesicles with diameters of 100–150 nm. The wall of the vesicle was shown to be a bilayer of high-density material separated by a 2 nm gap, corresponding to two monolayers of porphyrin–lipid. A stable nanostructure is formed with an extremely high density of porphyrin molecules (up to 100 mol% porphyrin-lipid of the total lipid content), forming a liposome-like nanocarrier offering novel biophotonic functions beyond those of molecular porphyrins. For example, the high porphyrin density drives "super" absorption (extinction coefficient $\varepsilon_{680\ \mathrm{nm}} = 2.9 \times 10^9\ \mathrm{M}^{-1}\ \mathrm{cm}^{-1}$) and "super" photoactivity quenching, which in turn converts light energy to heat with extremely high efficiency, providing photothermal and photoacoustic properties unprecedented for organic nanoparticles. Due to its natural chlorophyll origin, porphyrin-lipid itself is biodegradable and has very low toxicity in vivo (1000 mg/kg i.v. in mice caused no detectable functional, hematological or histological effects). Meanwhile, the aqueous core of porphysome can be actively or passively loaded with chemotherapeutics, such as dox. In addition, porphyrin-lipid retains the natural ability to chelate metals for multimodal imaging and manipulation of its photonic properties. Therefore, the nanoparticle built from this intrinsically multifunctional porphyrin building block provides "one-for-all" theranostics (multimodality imaging, phototherapy, and drug transport). This simplifies the composition and synthetic complexity of the "all-in-one" multifunctional liposome while maintaining the desired properties [95, 111]. Further, each subunit acts as structural component, therapeutic agent, and imaging agent, eliminating the compromise between imaging sensitivity and drug payload.

3.1.1 Intrinsic Multimodal Imaging Capabilities

With "super" absorption and "super" photoactivity quenching, porphysomes efficiently convert absorbed light energy to heat instead of generating fluorescence and singlet oxygen. Upon laser irradiation, the thermal expansion of porphysomes in the tissue generates a strong PA signal. The first application of porphysomes for PAI was for sentinel lymph node mapping [109]. After 15 min intradermal injection of porphysomes in rats at a dose as low as 2.3 pmol, the local lymphatic network was clearly detectable, showing the first draining lymph node, the inflowing lymph vessels, and

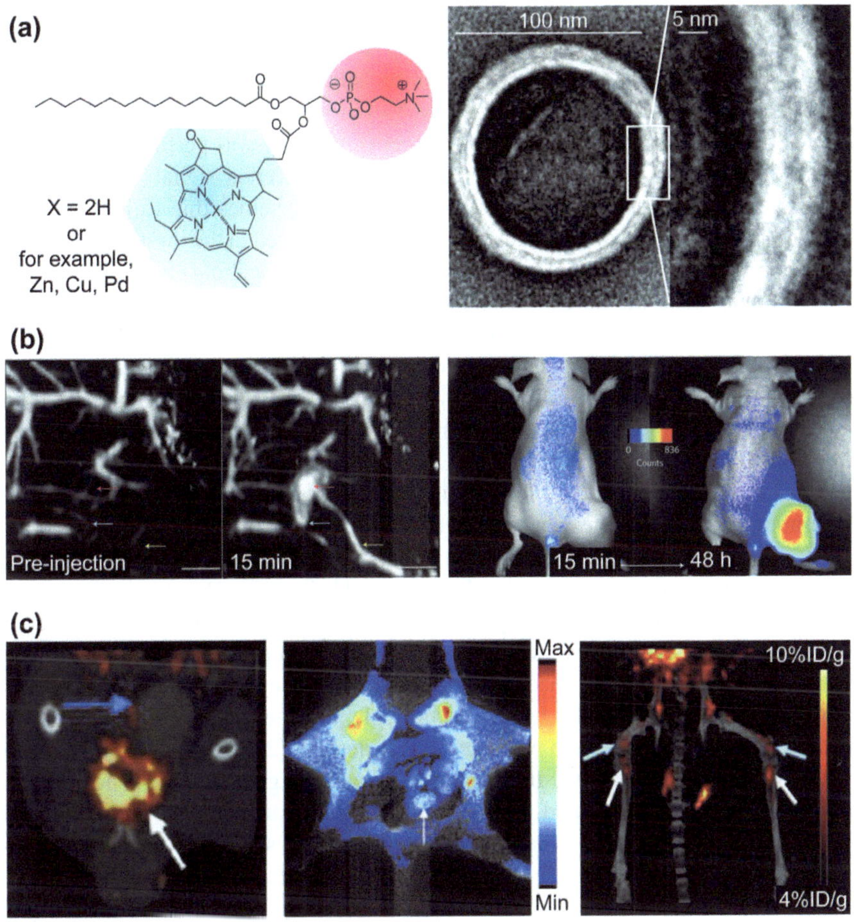

Fig. 4 Porphysomes are intrinsically multimodal, lipid nanoparticles formed from the self-assembly of porphyrin-lipid. **a** *Left* pyropheophorbide-lipid. *Right* electron micrographs of negatively stained porphysomes. **b** Porphysomes as dual imaging modality "smart delivery" systems. *Left* lymphatic mapping by PAI before and after intradermal injection of porphysomes (5 mm scale bar). *Right*: fluorescence activation after intravenous injection of porphysomes in a KB xenograft-bearing mouse. (**a**–**b**) Adapted with permission from Lovell et al., copyright Nature Publishing Group (2011) [109]. **c** In vivo multimodal imaging of ^{64}Cu-porphysomes in an orthotopic prostate cancer model. *Left* MicroPET/CT coronal image demonstrating tumour accumulation (*white arrow*). *Middle* Composite fluorescence and white-light image demonstrating porphysome disruption in the tumour. *Right* MicroPET/CT imaging of femur metastases. Adapted with permission from Liu et al., copyright American Chemical Society (2013) [110]

the surrounding lymph vessels (Fig. 4b). More recently, a temperature stimuli-responsive photoacoustic nanoswitch was developed by introducing 15 mol% of bacteriochlorophyll-lipid (Bchl-lipid) into a TSL [107]. At the melting temperature of the host lipid, membrane fluidity increases and the ordered packing of the Bchl-lipid is disrupted, resulting in a reversible loss of aggregate PA signal and an increase in monomer fluorescence. This parameter may be useful for precisely monitoring thermal therapies such as hyperthermia. Due to "super" quenching, porphysomes are a "smart delivery system," providing low background fluorescence imaging. The first application of porphysomes for fluorescence imaging was performed on a subcutaneous mouse model [109]. No significant fluorescence signal was observed immediately following intravenous injection of porphysomes due to its high self-quenching. After 48 h, enhanced fluorescence was observed in the tumour region as a result of porphysome accumulation and subsequent unquenching (Fig. 4b).

Additional imaging functionalities for whole-body imaging can be provided without adding any complexity to the system. Radioactivity is achieved by directly chelating a ^{64}Cu radioisotope in preformed porphysomes, while MRI is achieved by assembling particles with a paramagnetic ion (Mn(II))-labeled porphyrin-lipid building block. In both cases, stable chelation could be achieved using a simple, robust method: Preformed porphysomes or porphyrin molecules are mixed with the metal label and heated at 60 °C for 30 min. The long radioactive life-time (>48 h) of ^{64}Cu-porphysomes matches the nanoparticle circulation time (mouse half-time is 18 h). This is an ideal pairing for noninvasive, highly sensitive, and accurate real-time assessment of porphysome biodistribution [110, 112]. Due to the high sensitivity of PET, only 5 % of porphyrin-lipid must be labeled with ^{64}Cu, which does not impact the photophysical properties. Using ^{64}Cu-porphysomes, diseased tissue was delineated at both the macro (PET imaging) and microscopic level (fluorescence imaging) in a clinically relevant orthotopic prostate tumour model, and bony metastases as small as 1.5 mm were sensitively detected 24 h postinjection (Fig. 4c) [110]. With Mn(II) labeling, the fluorescence and singlet oxygen generation of porphyrin-lipid are quenched, improving the photostability and photothermal function of the Mn-porphysome. Moreover, Mn-porphysomes are capable of generating MR contrast at a level comparable to the clinically used agent Gd-DTPA [113]. ^{64}Cu/Mn-porphysomes comprised of a single functional unit that intrinsically integrates PET/MRI contrast generation, photothermal efficiency, and excellent photostability, have high potential for cancer theranostics and clinical translation.

3.1.2 Intrinsic Therapeutic Capabilities

Photothermal therapy (PTT) and PDT have emerged as viable clinical approaches, possessing advantages over conventional cancer therapies including improved selectivity achieved by local laser irradiation. PTT dissipates absorbed light energy as heat to destroy targeted tissues through a necrosis pathway, while PDT generates singlet oxygen as the predominant cytotoxic agent to damage the tissues in an area restricted to the region of photosensitizer accumulation. Porphysomes display structure-dependent photonic properties useful for both PTT and PDT (Fig. 5). Jin et al. reported the first in vivo comparison of porphysome-mediated PTT and PDT in hypoxic and hyperoxic conditions [114]. Porphysomes generated a rapid increase in temperature and induced obvious tissue damage under both tumour conditions, whereas PDT activity was eliminated. More recently, an activatable strategy enabled application of the PDT mechanism by introducing targeting ligands such as

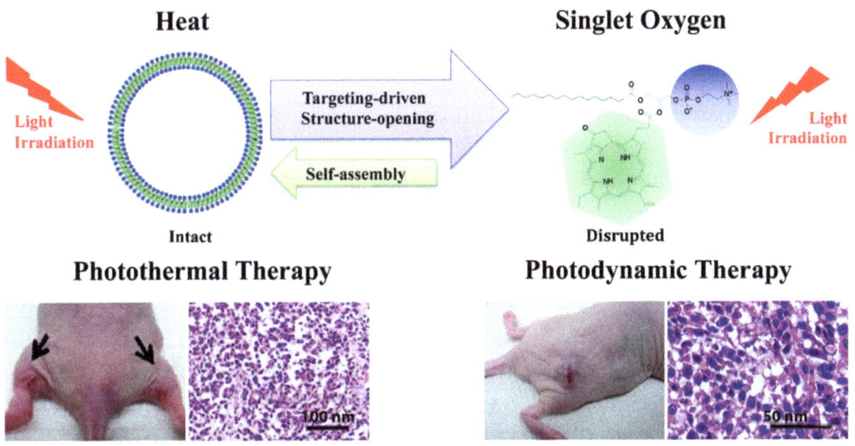

Fig. 5 Schematic illustration of structure-dependent porphysome PTT and PDT activity. Untargeted porphysomes remain intact and are effective PTT agents in vivo, while folate-porphysomes are internalized by folate-expressing KB tumour cells and are effective for PDT. Adapted with permission from Jin et al., copyright American Chemical Society (2013) [114] and Wiley-VCH (2014) [115]

folatc-lipid into the porphysome formulation [115]. The folate-porphysome exhibited enhanced intracellular uptake and efficient disruption of the nanostructure, releasing the photodynamic activity of the densely packed porphyrins for effective PDT, thus closing the loop for structure-dependent conversion between PTT and PDT. Replacing a fraction of the porphyrin-lipid with nonfluorescent lipids reduces the quenching efficiency, enabling fluorescence imaging and PDT in the intact state as well. These structure-dependent photonic properties are useful for monitoring porphysome stability and drug release, as well as for guiding administration of PTT and PDT.

Beyond these diverse intrinsic imaging modalities, the vesicular nature of porphysomes allows for loading of drugs into the core or bilayer of the nanoparticle. Lovell et al. demonstrated that dox could be actively loaded into porphysomes with 90 % loading efficiency [109]. To improve the delivery specificity, targeted porphysomes can be easily formed by inclusion of various targeting moieties such as folate-PEG-lipid (1 mol% lipid) for folate receptor-mediated targeting [109, 115]. These studies show great promise for porphysomes as a targeted nanocarrier for cancer theranostics.

3.2 Convergence of Porphysome and Lipoprotein Theranostics

Porphysomes in the 100–150 nm size range exhibit preferential accumulation in malignant tumors through the EPR effect, but may encounter significant impedance when diffusing through the tumour collagen network. Many studies have demonstrated that nanoparticles less than 40 nm exhibit more effective penetration in fibrous tumors compared to their larger counterparts [116–118]. Ng et al. developed a sub-40 nm activatable porphyrin nanodisc by using apolipoprotein to constrict the particle size [75]. The compact discoidal structure demonstrates similar structure-dependent photonic properties as porphysomes, including photoactivity quenching (99 % fluorescence quenching) and recovery of photodynamic activity upon cell internalization. In addition, the nanodiscs display a five fold increase in their collagen diffusion coefficient compared to porphysomes, indicating a greater potential for deep tissue penetration. The size-control strategy of using apolipoprotein could be extended to the synthesis of small spherical porphysomes by fusing lipoprotein and porphyrin-lipid technology. This strategy could offer the ability to incorporate large amounts of photosensitizers within a compact structure that allows for phototherapeutic action, fluorescence imaging, and the potential to effectively deliver therapeutics deep into poorly permeable tumors.

4 Conclusion

Lipid theranostics offer researchers and clinicians new tools to visualize nanoparticle fate, tumour accumulation, and intracellular uptake. The impact of these technological advancements on drug development and treatment planning represent significant achievements for personalized medicine. To date, experimental progress in lipid nanotheranostics has focused on methods and techniques to robustly provide imaging modalities and incorporate multiple functionalities in a single platform. However, few reports demonstrate use of the information provided by imaging to alter the treatment plan. Looking forward, nanotheranostics will benefit from practical studies demonstrating utility of imaging functionality, particularly to assess the advantages of multimodal theranostics. Through these studies the practical limitations of assembling multiple, single-modality theranostic agents may become more evident. The emerging paradigm of inherently multifunctional components is expected to play a large role in shaping the future of nanotheranostics. This concept can be extended beyond lipid nanomedicine to other theranostic platforms including porphyrin polymers [119] and cyanine micelles [120].

References

1. Baker M (2010) Whole-animal imaging: the whole picture. Nature 463(7283):977–980. doi:10.1038/463977a
2. Naumova AV, Modo M, Moore A, Murry CE, Frank JA (2014) Clinical imaging in regenerative medicine. Nat Biotechnol 32(8):804–818. doi:10.1038/nbt.2993
3. Sakamoto JH, van de Ven AL, Godin B, Blanco E, Serda RE, Grattoni A, Ziemys A, Bouamrani A, Hu T, Ranganathan SI, De Rosa E, Martinez JO, Smid CA, Buchanan RM, Lee SY, Srinivasan S, Landry M, Meyn A, Tasciotti E, Liu X, Decuzzi P, Ferrari M (2010) Enabling individualized therapy through nanotechnology. Pharmacol Res Official J Ital Pharmacol Soc 62(2):57–89. doi:10.1016/j.phrs.2009.12.011

4. Lammers T, Kiessling F, Hennink WE, Storm G (2010) Nanotheranostics and image-guided drug delivery: current concepts and future directions. Mol Pharm 7(6):1899–1912. doi:10.1021/mp100228v
5. Rizzo LY, Theek B, Storm G, Kiessling F, Lammers T (2013) Recent progress in nanomedicine: therapeutic, diagnostic and theranostic applications. Curr Opin Biotechnol 24 (6):1159–1166. doi:10.1016/j.copbio.2013.02.020
6. Phillips WT, Bao A, Brenner AJ, Goins BA (2014) Image-guided interventional therapy for cancer with radiotherapeutic nanoparticles. Adv Drug Deliv Rev 76:39–59
7. Lovell JF, Liu TW, Chen J, Zheng G (2010) Activatable photosensitizers for imaging and therapy. Chem Rev 110(5):2839–2857. doi:10.1021/cr900236h
8. Lin Q, Chen J, Ng KK, Cao W, Zhang Z, Zheng G (2014) Imaging the cytosolic drug delivery mechanism of HDL-like nanoparticles. Pharm Res 31(6):1438–1449. doi:10.1007/s11095-013-1046-z
9. Matsumura Y, Maeda H (1986) A new concept for macromolecular therapeutics in cancer chemotherapy: mechanism of tumoritropic accumulation of proteins and the antitumor agent smancs. Cancer Res 46(12 Pt 1):6387–6392
10. Maeda H (2001) The enhanced permeability and retention (EPR) effect in tumor vasculature: the key role of tumor-selective macromolecular drug targeting. Adv Enzyme Regul 41:189–207
11. Fang J, Nakamura H, Maeda H (2011) The EPR effect: unique features of tumor blood vessels for drug delivery, factors involved, and limitations and augmentation of the effect. Adv Drug Deliv Rev 63(3):136–151. doi:10.1016/j.addr.2010.04.009
12. Gabizon AA (1995) Liposome circulation time and tumor targeting: implications for cancer chemotherapy. Adv Drug Deliv Rev 16(2–3):285–294
13. Pegaz B, Debefve E, Ballini JP, Wagnieres G, Spaniol S, Albrecht V, Scheglmann DV, Nifantiev NE, van den Bergh H, Konan-Kouakou YN (2006) Photothrombic activity of m-THPC-loaded liposomal formulations: pre-clinical assessment on chick chorioallantoic membrane model. Eur J Pharm Sci Official J Eur Fed Pharm Sci 28(1–2):134–140. doi:10.1016/j.ejps.2006.01.008
14. Ernsting MJ, Murakami M, Roy A, Li SD (2013) Factors controlling the pharmacokinetics, biodistribution and intratumoral penetration of nanoparticles. J Controlled Release Official J Controlled Release Soc 172(3):782–794. doi:10.1016/j.jconrel.2013.09.013
15. Albanese A, Tang PS, Chan WC (2012) The effect of nanoparticle size, shape, and surface chemistry on biological systems. Annu Rev Biomed Eng 14:1–16. doi:10.1146/annurev-bioeng-071811-150124
16. Kaasgaard T, Andresen TL (2010) Liposomal cancer therapy: exploiting tumor characteristics. Expert Opin Drug Deliv 7(2):225–243. doi:10.1517/17425240903427940
17. Lammers T, Kiessling F, Hennink WE, Storm G (2012) Drug targeting to tumors: principles, pitfalls and (pre-) clinical progress. J Controlled Release Official J Controlled Release Soc 161(2):175–187. doi:10.1016/j.jconrel.2011.09.063
18. Ganta S, Devalapally H, Shahiwala A, Amiji M (2008) A review of stimuli-responsive nanocarriers for drug and gene delivery. J Controlled Release Official J Controlled Release Soc 126(3):187–204. doi:10.1016/j.jconrel.2007.12.017
19. Gao W, Chan JM, Farokhzad OC (2010) pH-Responsive nanoparticles for drug delivery. Mol Pharm 7(6):1913–1920. doi:10.1021/mp100253e
20. Rai P, Mallidi S, Zheng X, Rahmanzadeh R, Mir Y, Elrington S, Khurshid A, Hasan T (2010) Development and applications of photo-triggered theranostic agents. Adv Drug Deliv Rev 62(11):1094–1124. doi:10.1016/j.addr.2010.09.002
21. Huang Y, Jiang Y, Wang H, Wang J, Shin MC, Byun Y, He H, Liang Y, Yang VC (2013) Curb challenges of the "Trojan Horse" approach: smart strategies in achieving effective yet safe cell-penetrating peptide-based drug delivery. Adv Drug Deliv Rev 65(10):1299–1315. doi:10.1016/j.addr.2012.11.007

22. Mura S, Couvreur P (2012) Nanotheranostics for personalized medicine. Adv Drug Deliv Rev 64(13):1394–1416. doi:10.1016/j.addr.2012.06.006
23. Murakami M, Ernsting MJ, Li SD (2013) Theranostic nanoparticles for cancer imaging and therapy. In: Tiwari A, Tiwari A (eds) Nanomaterials in drug delivery, imaging, and tissue engineering. Wiley, New York, pp 363–383
24. Muthu MS, Leong DT, Mei L, Feng SS (2014) Nanotheranostics—application and further development of nanomedicine strategies for advanced theranostics. Theranostics 4(6): 660–677. doi:10.7150/thno.8698
25. Svenson S (2014) What nanomedicine in the clinic right now really forms nanoparticles? Wiley Interdisc Rev Nanomed Nanobiotechnol 6(2):125–135. doi:10.1002/wnan.1257
26. Kraft JC, Freeling JP, Wang Z, Ho RJ (2014) Emerging research and clinical development trends of liposome and lipid nanoparticle drug delivery systems. J Pharm Sci 103(1):29–52. doi:10.1002/jps.23773
27. Yang F, Jin C, Jiang Y, Li J, Di Y, Ni Q, Fu D (2011) Liposome based delivery systems in pancreatic cancer treatment: from bench to bedside. Cancer Treat Rev 37(8):633–642. doi:10.1016/j.ctrv.2011.01.006
28. Paliwal SR, Paliwal R, Agrawal GP, Vyas SP (2011) Liposomal nanomedicine for breast cancer therapy. Nanomedicine (London, England) 6(6):1085–1100. doi:10.2217/nnm.11.72
29. Zhao Y, Imura T, Leman LJ, Curtiss LK, Maryanoff BE, Ghadiri MR (2013) Mimicry of high-density lipoprotein: functional peptide-lipid nanoparticles based on multivalent peptide constructs. J Am Chem Soc 135(36):13414–13424. doi:10.1021/ja404714a
30. Li H, Marotta DE, Kim S, Busch TM, Wileyto EP, Zheng G (2005) High payload delivery of optical imaging and photodynamic therapy agents to tumors using phthalocyanine-reconstituted low-density lipoprotein nanoparticles. J Biomed Opt 10(4):41203. doi:10.1117/1.2011429
31. Zheng G, Chen J, Li H, Glickson JD (2005) Rerouting lipoprotein nanoparticles to selected alternate receptors for the targeted delivery of cancer diagnostic and therapeutic agents. Proc Natl Acad Sci USA 102(49):17757–17762. doi:10.1073/pnas.0508677102
32. McMahon KM, Mutharasan RK, Tripathy S, Veliceasa D, Bobeica M, Shumaker DK, Luthi AJ, Helfand BT, Ardehali H, Mirkin CA, Volpert O, Thaxton CS (2011) Biomimetic high density lipoprotein nanoparticles for nucleic acid delivery. Nano Lett 11(3):1208–1214. doi:10.1021/nl1041947
33. Damiano MG, Mutharasan RK, Tripathy S, McMahon KM, Thaxton CS (2013) Templated high density lipoprotein nanoparticles as potential therapies and for molecular delivery. Adv Drug Deliv Rev 65(5):649–662. doi:10.1016/j.addr.2012.07.013
34. Marotta DE, Cao W, Wileyto EP, Li H, Corbin I, Rickter E, Glickson JD, Chance B, Zheng G, Busch TM (2011) Evaluation of bacteriochlorophyll-reconstituted low-density lipoprotein nanoparticles for photodynamic therapy efficacy in vivo. Nanomedicine (London, England) 6 (3):475–487. doi:10.2217/nnm.11.8
35. Ng KK, Lovell JF, Zheng G (2011) Lipoprotein-inspired nanoparticles for cancer theranostics. Acc Chem Res 44(10):1105–1113. doi:10.1021/ar200017e
36. Jin H, Chen J, Lovell JF, Zhang Z, Zheng G (2012) Synthesis and development of lipoprotein-based nanocarriers for light-activated theranostics. Isr J Chem 52(8–9):715–727
37. Rensen PC, de Vrueh RL, Kuiper J, Bijsterbosch MK, Biessen EA, van Berkel TJ (2001) Recombinant lipoproteins: lipoprotein-like lipid particles for drug targeting. Adv Drug Deliv Rev 47(2–3):251–276
38. Lacko AG, Nair M, Prokai L, McConathy WJ (2007) Prospects and challenges of the development of lipoprotein-based formulations for anti-cancer drugs. Expert Opin Drug Deliv 4(6):665–675. doi:10.1517/17425247.4.6.665
39. Masquelier M, Vitols S, Peterson C (1986) Low-density lipoprotein as a carrier of antitumoral drugs: in vivo fate of drug-human low-density lipoprotein complexes in mice. Cancer Res 46(8):3842–3847

40. Lundberg B (1987) Preparation of drug-low density lipoprotein complexes for delivery of antitumoral drugs via the low density lipoprotein pathway. Cancer Res 47(15):4105–4108
41. McConathy WJ, Nair MP, Paranjape S, Mooberry L, Lacko AG (2008) Evaluation of synthetic/reconstituted high-density lipoproteins as delivery vehicles for paclitaxel. Anticancer Drugs 19(2):183–188. doi:10.1097/CAD.0b013e3282f1da86
42. Kader A, Pater A (2002) Loading anticancer drugs into HDL as well as LDL has little affect on properties of complexes and enhances cytotoxicity to human carcinoma cells. J Controlled Release Official J Controlled Release Soc 80(1–3):29–44
43. Yang M, Chen J, Cao W, Ding L, Ng KK, Jin H, Zhang Z, Zheng G (2011) Attenuation of nontargeted cell-kill using a high-density lipoprotein-mimicking peptide–phospholipid nanoscaffold. Nanomedicine (London, England) 6(4):631–641. doi:10.2217/nnm.11.10
44. Huntosova V, Buzova D, Petrovajova D, Kasak P, Nadova Z, Jancura D, Sureau F, Miskovsky P (2012) Development of a new LDL-based transport system for hydrophobic/amphiphilic drug delivery to cancer cells. Int J Pharm 436(1–2):463–471. doi:10.1016/j.ijpharm.2012.07.005
45. Zhang X, Chen B (2010) Recombinant high density lipoprotein reconstituted with apolipoprotein AI cysteine mutants as delivery vehicles for 10-hydroxycamptothecin. Cancer Lett 298(1):26–33. doi:10.1016/j.canlet.2010.05.023
46. Sabnis N, Nair M, Israel M, McConathy WJ, Lacko AG (2012) Enhanced solubility and functionality of valrubicin (AD-32) against cancer cells upon encapsulation into biocompatible nanoparticles. Int J Nanomed 7:975–983. doi:10.2147/ijn.s28029
47. Ghosh M, Ryan RO (2014) ApoE enhances nanodisk-mediated curcumin delivery to glioblastoma multiforme cells. Nanomedicine (London, England) 9(6):763–771. doi:10.2217/nnm.13.35
48. Duivenvoorden R, Tang J, Cormode DP, Mieszawska AJ, Izquierdo-Garcia D, Ozcan C, Otten MJ, Zaidi N, Lobatto ME, van Rijs SM, Priem B, Kuan EL, Martel C, Hewing B, Sager H, Nahrendorf M, Randolph GJ, Stroes ES, Fuster V, Fisher EA, Fayad ZA, Mulder WJ (2014) A statin-loaded reconstituted high-density lipoprotein nanoparticle inhibits atherosclerotic plaque inflammation. Nature Commun 5:3065. doi:10.1038/ncomms4065
49. Bricarello DA, Smilowitz JT, Zivkovic AM, German JB, Parikh AN (2011) Reconstituted lipoprotein: a versatile class of biologically-inspired nanostructures. ACS Nano 5(1):42–57. doi:10.1021/nn103098m
50. Sabnis N, Lacko AG (2012) Drug delivery via lipoprotein-based carriers: answering the challenges in systemic therapeutics. Ther Deliv 3(5):599–608
51. Lin Q, Chen J, Zhang Z, Zheng G (2014) Lipid-based nanoparticles in the systemic delivery of siRNA. Nanomedicine (London, England) 9(1):105–120. doi:10.2217/nnm.13.192
52. Lin Q, Jin CS, Huang H, Ding L, Zhang Z, Chen J, Zheng G (2014) Nanoparticle-enabled, image-guided treatment planning of target specific RNAi therapeutics in an orthotopic prostate cancer model. Small (Weinheim an der Bergstrasse, Germany) 10(15):3072–3082. doi:10.1002/smll.201303842
53. Krieger M, Smith LC, Anderson RG, Goldstein JL, Kao YJ, Pownall HJ, Gotto AM Jr, Brown MS (1979) Reconstituted low density lipoprotein: a vehicle for the delivery of hydrophobic fluorescent probes to cells. J Supramol Struct 10(4):467–478. doi:10.1002/jss.400100409
54. Song L, Li H, Sunar U, Chen J, Corbin I, Yodh AG, Zheng G (2007) Naphthalocyanine-reconstituted LDL nanoparticles for in vivo cancer imaging and treatment. Int J Nanomed 2(4):767–774
55. Chen J, Corbin IR, Li H, Cao W, Glickson JD, Zheng G (2007) Ligand conjugated low-density lipoprotein nanoparticles for enhanced optical cancer imaging in vivo. J Am Chem Soc 129(18):5798–5799. doi:10.1021/ja069336k
56. Corbin IR, Chen J, Cao W, Li H, Lund-Katz S, Zheng G (2007) Enhanced cancer-targeted delivery using engineered high-density lipoprotein-based nanocarriers. J Biomed Nanotechnol 3(4):367–376

57. Allijn IE, Leong W, Tang J, Gianella A, Mieszawska AJ, Fay F, Ma G, Russell S, Callo CB, Gordon RE, Korkmaz E, Post JA, Zhao Y, Gerritsen HC, Thran A, Proksa R, Daerr H, Storm G, Fuster V, Fisher EA, Fayad ZA, Mulder WJ, Cormode DP (2013) Gold nanocrystal labeling allows low-density lipoprotein imaging from the subcellular to macroscopic level. ACS Nano 7(11):9761–9770. doi:10.1021/nn403258w
58. Lee JY, Kim JH, Bae KH, Oh MH, Kim Y, Kim JS, Park TG, Park K, Lee JH, Nam YS (2014) Low-Density lipoprotein-mimicking nanoparticles for tumor-targeted theranostic applications. Small (Weinheim an der Bergstrasse, Germany). doi:10.1002/smll.201303277
59. Corbin IR, Li H, Chen J, Lund-Katz S, Zhou R, Glickson JD, Zheng G (2006) Low-density lipoprotein nanoparticles as magnetic resonance imaging contrast agents. Neoplasia (NY) 8 (6):488–498. doi:10.1593/neo.05835
60. Frias JC, Williams KJ, Fisher EA, Fayad ZA (2004) Recombinant HDL-like nanoparticles: a specific contrast agent for MRI of atherosclerotic plaques. J Am Chem Soc 126(50):16316–16317. doi:10.1021/ja044911a
61. Cormode DP, Skajaa T, van Schooneveld MM, Koole R, Jarzyna P, Lobatto ME, Calcagno C, Barazza A, Gordon RE, Zanzonico P, Fisher EA, Fayad ZA, Mulder WJ (2008) Nanocrystal core high-density lipoproteins: a multimodality contrast agent platform. Nano Lett 8(11):3715–3723. doi:10.1021/nl801958b
62. Bruns OT, Ittrich H, Peldschus K, Kaul MG, Tromsdorf UI, Lauterwasser J, Nikolic MS, Mollwitz B, Merkel M, Bigall NC, Sapra S, Reimer R, Hohenberg H, Weller H, Eychmuller A, Adam G, Beisiegel U, Heeren J (2009) Real-time magnetic resonance imaging and quantification of lipoprotein metabolism in vivo using nanocrystals. Nat Nanotechnol 4(3): 193–201. doi:10.1038/nnano.2008.405
63. Hill ML, Corbin IR, Levitin RB, Cao W, Mainprize JG, Yaffe MJ, Zheng G (2010) In vitro assessment of poly-iodinated triglyceride reconstituted low-density lipoprotein: initial steps toward CT molecular imaging. Acad Radiol 17(11):1359–1365. doi:10.1016/j.acra.2010.06.006
64. Skajaa T, Cormode DP, Falk E, Mulder WJ, Fisher EA, Fayad ZA (2010) High-density lipoprotein-based contrast agents for multimodal imaging of atherosclerosis. Arterioscler Thromb Vasc Biol 30(2):169–176. doi:10.1161/atvbaha.108.179275
65. Luthi AJ, Zhang H, Kim D, Giljohann DA, Mirkin CA, Thaxton CS (2012) Tailoring of biomimetic high-density lipoprotein nanostructures changes cholesterol binding and efflux. ACS Nano 6(1):276–285. doi:10.1021/nn2035457
66. Ponty E, Favre G, Benaniba R, Boneu A, Lucot H, Carton M, Soula G (1993) Biodistribution study of 99mTc-labeled LDL in B16-melanoma-bearing mice. Visualization of a preferential uptake by the tumor. Int J Cancer J Int du Cancer 54(3):411–417
67. Moerlein SM, Daugherty A, Sobel BE, Welch MJ (1991) Metabolic imaging with gallium-68- and indium-111-labeled low-density lipoprotein. J Nucl Med: Official Publ, Soc Nucl Med 32(2):300–307
68. Bünau VG (1970) J. B. Birks: photophysics of aromatic molecules. Wiley-Interscience, London 1970 (704 Seiten. Preis: 210 s. Berichte der Bunsengesellschaft für physikalische Chemie 74(12):1294–1295. doi:10.1002/bbpc.19700741223)
69. Jin H, Lovell JF, Chen J, Lin Q, Ding L, Ng KK, Pandey RK, Manoharan M, Zhang Z, Zheng G (2012) Mechanistic insights into LDL nanoparticle-mediated siRNA delivery. Bioconjug Chem 23(1):33–41. doi:10.1021/bc200233n
70. Zhang Z, Cao W, Jin H, Lovell JF, Yang M, Ding L, Chen J, Corbin I, Luo Q, Zheng G (2009) Biomimetic nanocarrier for direct cytosolic drug delivery. Angew Chem Int Ed Engl 48(48):9171–9175. doi:10.1002/anie.200903112
71. Zheng G, Li H, Zhang M, Lund-Katz S, Chance B, Glickson JD (2002) Low-density lipoprotein reconstituted by pyropheophorbide cholesteryl oleate as target-specific photosensitizer. Bioconjug Chem 13(3):392–396

72. Cao W, Ng KK, Corbin I, Zhang Z, Ding L, Chen J, Zheng G (2009) Synthesis and evaluation of a stable bacteriochlorophyll-analog and its incorporation into high-density lipoprotein nanoparticles for tumor imaging. Bioconjug Chem 20(11):2023–2031. doi:10.1021/bc900404y
73. Zhang Z, Chen J, Ding L, Jin H, Lovell JF, Corbin IR, Cao W, Lo PC, Yang M, Tsao MS, Luo Q, Zheng G (2010) HDL-mimicking peptide-lipid nanoparticles with improved tumor targeting. Small (Weinheim an der Bergstrasse, Germany) 6(3):430–437. doi:10.1002/smll.200901515
74. Epand RM, Gawish A, Iqbal M, Gupta KB, Chen CH, Segrest JP, Anantharamaiah GM (1987) Studies of synthetic peptide analogs of the amphipathic helix. Effect of charge distribution, hydrophobicity, and secondary structure on lipid association and lecithin: cholesterol acyltransferase activation. J Biol Chem 262(19):9389–9396
75. Ng KK, Lovell JF, Vedadi A, Hajian T, Zheng G (2013) Self-assembled porphyrin nanodiscs with structure-dependent activation for phototherapy and photodiagnostic applications. ACS Nano 7(4):3484–3490. doi:10.1021/nn400418y
76. Chen Y, Pullambhatla M, Banerjee SR, Byun Y, Stathis M, Rojas C, Slusher BS, Mease RC, Pomper MG (2012) Synthesis and biological evaluation of low molecular weight fluorescent imaging agents for the prostate-specific membrane antigen. Bioconjug Chem 23(12):2377–2385. doi:10.1021/bc3003919
77. Yang M, Jin H, Chen J, Ding L, Ng KK, Lin Q, Lovell JF, Zhang Z, Zheng G (2011) Efficient cytosolic delivery of siRNA using HDL-mimicking nanoparticles. Small (Weinheim an der Bergstrasse, Germany) 7(5):568–573. doi:10.1002/smll.201001589
78. Torchilin VP (2005) Recent advances with liposomes as pharmaceutical carriers. Nat Rev Drug Discov 4(2):145–160. doi:10.1038/nrd1632
79. Petros RA, DeSimone JM (2010) Strategies in the design of nanoparticles for therapeutic applications. Nat Rev Drug Discov 9(8):615–627. doi:10.1038/nrd2591
80. Wang AZ, Langer R, Farokhzad OC (2012) Nanoparticle delivery of cancer drugs. Annu Rev Med 63:185–198. doi:10.1146/annurev-med-040210-162544
81. Akbarzadeh A, Rezaei-Sadabady R, Davaran S, Joo SW, Zarghami N, Hanifehpour Y, Samiei M, Kouhi M, Nejati-Koshki K (2013) Liposome: classification, preparation, and applications. Nanoscale Res Lett 8(1):102. doi:10.1186/1556-276x-8-102
82. Torchilin VP (2006) Multifunctional nanocarriers. Adv Drug Deliv Rev 58(14):1532–1555. doi:10.1016/j.addr.2006.09.009
83. Li S, Goins B, Zhang L, Bao A (2012) Novel multifunctional theranostic liposome drug delivery system: construction, characterization, and multimodality MR, near-infrared fluorescent, and nuclear imaging. Bioconjug Chem 23(6):1322–1332. doi:10.1021/bc300175d
84. Al-Jamal WT, Kostarelos K (2011) Liposomes: from a clinically established drug delivery system to a nanoparticle platform for theranostic nanomedicine. Acc Chem Res 44(10):1094–1104. doi:10.1021/ar200105p
85. Grange C, Geninatti-Crich S, Esposito G, Alberti D, Tei L, Bussolati B, Aime S, Camussi G (2010) Combined delivery and magnetic resonance imaging of neural cell adhesion molecule-targeted doxorubicin-containing liposomes in experimentally induced Kaposi's sarcoma. Cancer Res 70(6):2180–2190. doi:10.1158/0008-5472.can-09-2821
86. Karathanasis E, Chan L, Balusu SR, D'Orsi CJ, Annapragada AV, Sechopoulos I, Bellamkonda RV (2008) Multifunctional nanocarriers for mammographic quantification of tumor dosing and prognosis of breast cancer therapy. Biomaterials 29(36):4815–4822. doi:10.1016/j.biomaterials.2008.08.036
87. Viglianti BL, Ponce AM, Michelich CR, Yu D, Abraham SA, Sanders L, Yarmolenko PS, Schroeder T, MacFall JR, Barboriak DP, Colvin OM, Bally MB, Dewhirst MW (2006) Chemodosimetry of in vivo tumor liposomal drug concentration using MRI. Magn Reson Med Official J Soc Magn Reson Med Soc Magn Reson Med 56(5):1011–1018. doi:10.1002/mrm.21032

88. Tagami T, Foltz WD, Ernsting MJ, Lee CM, Tannock IF, May JP, Li SD (2011) MRI monitoring of intratumoral drug delivery and prediction of the therapeutic effect with a multifunctional thermosensitive liposome. Biomaterials 32(27):6570–6578. doi:10.1016/j.biomaterials.2011.05.029
89. Li L, ten Hagen TL, Bolkestein M, Gasselhuber A, Yatvin J, van Rhoon GC, Eggermont AM, Haemmerich D, Koning GA (2013) Improved intratumoral nanoparticle extravasation and penetration by mild hyperthermia. J Controlled Release Official J Controlled Release Soc 167(2):130–137. doi:10.1016/j.jconrel.2013.01.026
90. Li L, ten Hagen TL, Haeri A, Soullie T, Scholten C, Seynhaeve AL, Eggermont AM, Koning GA (2014) A novel two-step mild hyperthermia for advanced liposomal chemotherapy. J Controlled Release: Official J Controlled Release Soc 174:202–208. doi:10.1016/j.jconrel.2013.11.012
91. Landon CD, Park JY, Needham D, Dewhirst MW (2011) Nanoscale drug delivery and hyperthermia: the materials design and preclinical and clinical testing of low temperature-sensitive liposomes used in combination with mild hyperthermia in the treatment of local cancer. Open Nanomed J 3:38–64. doi:10.2174/1875933501103010038
92. Lee DE, Koo H, Sun IC, Ryu JH, Kim K, Kwon IC (2012) Multifunctional nanoparticles for multimodal imaging and theragnosis. Chem Soc Rev 41(7):2656–2672. doi:10.1039/c2cs15261d
93. Puri A, Loomis K, Smith B, Lee JH, Yavlovich A, Heldman E, Blumenthal R (2009) Lipid-based nanoparticles as pharmaceutical drug carriers: from concepts to clinic. Crit Rev Ther Drug Carrier Syst 26(6):523–580
94. Elbayoumi TA, Torchilin VP (2010) Current trends in liposome research. Meth Mol Biol (Clifton, NJ) 605:1–27. doi:10.1007/978-1-60327-360-2_1
95. Cheng Z, Al Zaki A, Hui JZ, Muzykantov VR, Tsourkas A (2012) Multifunctional nanoparticles: cost versus benefit of adding targeting and imaging capabilities. Science (NY) 338(6109):903–910. doi:10.1126/science.1226338
96. Etheridge ML, Campbell SA, Erdman AG, Haynes CL, Wolf SM, McCullough J (2013) The big picture on nanomedicine: the state of investigational and approved nanomedicine products. Nanomed Nanotechnol Biol Med 9(1):1–14. doi:10.1016/j.nano.2012.05.013
97. Bealle G, Di Corato R, Kolosnjaj-Tabi J, Dupuis V, Clement O, Gazeau F, Wilhelm C, Menager C (2012) Ultra magnetic liposomes for MR imaging, targeting, and hyperthermia. Langmuir: ACS J Surf Colloids 28(32):11834–11842. doi:10.1021/la3024716
98. Huynh E, Zheng G (2013) Engineering multifunctional nanoparticles: all-in-one versus one-for-all. Wiley Interdisc Rev Nanomed Nanobiotechnol 5(3):250–265. doi:10.1002/wnan.1217
99. Wei A, Mehtala JG, Patri AK (2012) Challenges and opportunities in the advancement of nanomedicines. J Controlled Release Official J Controlled Release Soc 164(2):236–246. doi:10.1016/j.jconrel.2012.10.007
100. Svenson S (2013) Theranostics: are we there yet? Mol Pharm 10(3):848–856. doi:10.1021/mp300644n
101. Petersen AL, Hansen AE, Gabizon A, Andresen TL (2012) Liposome imaging agents in personalized medicine. Adv Drug Deliv Rev 64(13):1417–1435. doi:10.1016/j.addr.2012.09.003
102. Kaibara A, Matsumara G (2012) Handbook of porphyrins: Chemistry, properties, and applications. Handbook of Porphyrins: Chemistry, Properties and Applications
103. Carter KA, Shao S, Hoopes MI, Luo D, Ahsan B, Grigoryants VM, Song W, Huang H, Zhang G, Pandey RK, Geng J, Pfeifer BA, Scholes CP, Ortega J, Karttunen M, Lovell JF (2014) Porphyrin-phospholipid liposomes permeabilized by near-infrared light. Nat Commun 5:3546. doi:10.1038/ncomms4546
104. Zhang Y, Lovell JF (2012) Porphyrins as theranostic agents from prehistoric to modern times. Theranostics 2(9):905–915. doi:10.7150/thno.4908

105. Josefsen LB, Boyle RW (2012) Unique diagnostic and therapeutic roles of porphyrins and phthalocyanines in photodynamic therapy, imaging and theranostics. Theranostics 2(9): 916–966. doi:10.7150/thno.4571
106. Glidden MD, Celli JP, Massodi I, Rizvi I, Pogue BW, Hasan T (2012) Image-Based quantification of benzoporphyrin derivative uptake, localization, and photobleaching in 3D tumor models, for optimization of PDT parameters. Theranostics 2(9):827–839. doi:10.7150/thno.4334
107. Ng KK, Shakiba M, Huynh E, Weersink RA, Roxin A, Wilson BC, Zheng G (2014) Stimuli-responsive photoacoustic nanoswitch for in vivo sensing applications. ACS Nano. doi:10.1021/nn502858b
108. Ali H, van Lier JE (1999) Metal complexes as photo- and radiosensitizers. Chem Rev 99 (9):2379–2450
109. Lovell JF, Jin CS, Huynh E, Jin H, Kim C, Rubinstein JL, Chan WC, Cao W, Wang LV, Zheng G (2011) Porphysome nanovesicles generated by porphyrin bilayers for use as multimodal biophotonic contrast agents. Nat Mater 10(4):324–332. doi:10.1038/nmat2986
110. Liu TW, Macdonald TD, Jin CS, Gold JM, Bristow RG, Wilson BC, Zheng G (2013) Inherently multimodal nanoparticle-driven tracking and real-time delineation of orthotopic prostate tumors and micrometastases. ACS Nano 7(5):4221–4232. doi:10.1021/nn400669r
111. de Souza N (2011) One particle to rule them all? Nat Meth 8(5):370–371
112. Liu TW, MacDonald TD, Shi J, Wilson BC, Zheng G (2012) Intrinsically copper-64-labeled organic nanoparticles as radiotracers. Angew Chem Int Ed Engl 51(52):13128–13131. doi:10.1002/anie.201206939
113. MacDonald TD, Liu TW, Zheng G (2014) An MRI-sensitive, non-photobleachable porphysome photothermal agent. Angew Chem Int Ed Engl 53(27):6956–6959. doi:10.1002/anie.201400133
114. Jin CS, Lovell JF, Chen J, Zheng G (2013) Ablation of hypoxic tumors with dose-equivalent photothermal, but not photodynamic, therapy using a nanostructured porphyrin assembly. ACS Nano 7(3):2541–2550. doi:10.1021/nn3058642
115. Jin CS, Cui L, Wang F, Chen J, Zheng G (2014) Targeting-triggered porphysome nanostructure disruption for activatable photodynamic therapy. Adv Healthc Mater 3 (8):1240–1249. doi:10.1002/adhm.201300651
116. Pluen A, Boucher Y, Ramanujan S, McKee TD, Gohongi T, di Tomaso E, Brown EB, Izumi Y, Campbell RB, Berk DA, Jain RK (2001) Role of tumor-host interactions in interstitial diffusion of macromolecules: cranial versus subcutaneous tumors. Proc Natl Acad Sci USA 98(8):4628–4633. doi:10.1073/pnas.081626898
117. Cabral H, Matsumoto Y, Mizuno K, Chen Q, Murakami M, Kimura M, Terada Y, Kano MR, Miyazono K, Uesaka M, Nishiyama N, Kataoka K (2011) Accumulation of sub-100 nm polymeric micelles in poorly permeable tumours depends on size. Nat Nanotechnol 6 (12):815–823. doi:10.1038/nnano.2011.166
118. Sykes EA, Chen J, Zheng G, Chan WC (2014) Investigating the impact of nanoparticle size on active and passive tumor targeting efficiency. ACS Nano 8(6):5696–5706. doi:10.1021/nn500299p
119. Hsu CY, Nieh MP, Lai PS (2012) Facile self-assembly of porphyrin-embedded polymeric vesicles for theranostic applications. Chem Commun (Camb) 48(75):9343–9345. doi:10.1039/c2cc33851c
120. Zhang Y, Jeon M, Rich LJ, Hong H, Geng J, Zhang Y, Shi S, Barnhart TE, Alexandridis P, Huizinga JD, Seshadri M, Cai W, Kim C, Lovell JF (2014) Non-invasive multimodal functional imaging of the intestine with frozen micellar naphthalocyanines. Nat Nanotechnol 9(8):631–638. doi:10.1038/nnano.2014.130

Synthetic High-Density Lipoprotein-Like Nanoparticles as Cancer Therapy

Kaylin M. McMahon, Linda Foit, Nicholas L. Angeloni, Francis J. Giles, Leo I. Gordon and C. Shad Thaxton

Abstract

High-density lipoproteins (HDL) are diverse natural nanoparticles that carry cholesterol and are best known for the role that they play in cardiovascular disease. However, due to their unique targeting capabilities, diverse molecular cargo, and natural functions beyond cholesterol transport, it is becoming increasingly appreciated that HDLs are critical to cancer development and

Kaylin M. McMahon, Linda Foit and Nicholas L. Angeloni contributed equally to this work.

K.M. McMahon · L. Foit · N.L. Angeloni · C.S. Thaxton
Feinberg School of Medicine, Department of Urology, Northwestern University, Tarry 16-703, 303 E. Chicago Ave, Chicago, IL 60611, USA

K.M. McMahon · L. Foit · N.L. Angeloni · C.S. Thaxton
Simpson Querrey Institute (SQI), 303 E. Superior St, Chicago, IL 60611, USA

F.J. Giles
Northwestern Medicine Developmental Therapeutics Institute, Northwestern University, 645 N. Michigan Ave, Chicago, IL 60611, USA

L.I. Gordon
Department of Medicine, Division of Hematology/Oncology, Northwestern University Feinberg School of Medicine, Chicago, IL 60611, USA

L.I. Gordon · C.S. Thaxton
Robert H. Lurie Comprehensive Cancer Center, Northwestern University, Chicago, IL 60611, USA

C.S. Thaxton (✉)
International Institute for Nanotechnology (IIN), Northwestern University, 2145 Sheridan Rd, Evanston, IL 60208, USA
e-mail: cthaxton003@md.northwestern.edu

C.A. Mirkin et al. (eds.), *Nanotechnology-Based Precision Tools for the Detection and Treatment of Cancer*, Cancer Treatment and Research 166,
DOI 10.1007/978-3-319-16555-4_6

progression. Accordingly, this chapter highlights ongoing research focused on the connections between HDL and cancer in order to design new drugs and targeted drug delivery vehicles. Research is focused on synthesizing biomimetic HDL-like nanoparticles (NP) that can be loaded with diverse therapeutic cargo (e.g., chemotherapies, nucleic acids, proteins) and specifically targeted to cancer cells. Beyond drug delivery, new data is emerging that HDL-like NPs may be therapeutically active in certain tumor types, for example, B cell lymphoma. Overall, HDL-like NPs are becoming increasingly appreciated as targeted, biocompatible, and efficient therapies for cancer, and may soon become indispensable agents in the cancer therapeutic armamentarium.

Keywords
High-density lipoprotein · Cholesterol · Nanotechnology · siRNA · Delivery · Cancer therapy

Contents

1 Structure and Composition of Natural High-Density Lipoproteins (HDL)

Natural high-density lipoproteins (HDL) range in size from 7 to 13 nm in diameter. HDLs are dynamic nanostructures with regard to size, shape, and molecular composition. As HDLs biologically mature, they interact with cell receptors, enzymes, and other proteins. Apolipoprotein A-I (apoA-I) is the main protein associated with HDLs and represents approximately 70 % of the protein content associated with HDLs [1]. ApoA-I is an amphipathic scaffold protein, which binds lipids and defines the ultimate size and shape of HDL species [2]. ApoA-II, the second most common protein associated with HDL, makes up approximately 20 % of the total protein. Although less studied, multiple other apolipoproteins associate with HDLs, such as apoA-IV, apoC-I, apoC-II, apoC-III, apoD, apoE, apoJ, apoL, and apoM [3, 4]. Additionally, a number of other proteins, lipids (e.g., phosphatidylcholines), free cholesterol, and esterified cholesterol contribute to the heterogeneity of HDL species [5].

HDLs are the smallest and densest of the plasma lipoproteins. HDLs are constantly remodeled in the bloodstream through interaction with other lipoproteins, enzymes, and contact with target cells. These interactions result in significant particle heterogeneity. For example, HDLs can be classified into subpopulations by density, size, shape, composition, and surface charge. HDLs are classified into two main sub-fractions based on density: HDL_2 ($1.063 < d < 1.125$ g/mL), which are relatively large in size, lipid-rich, and more buoyant than HDL_3 species ($1.125 < d < 1.21$ g/mL), which are smaller in size [1, 6]. These sub-fractions can be further categorized into five distinct subpopulations by using methods such as electrophoresis and ultracentrifugation [1, 4]: HDL_{2b} (9.7–12.9 nm), HDL_{2a} (8.8–9.7 nm), HDL_{3a} (8.2–8.8 nm), HDL_{3b} (7.8–8.2 nm), and HDL_{3c} (7.2–7.8 nm) [6]. Further, two-dimensional electrophoresis can separate HDL subpopulations by charge and size, ultimately resulting in 5–10 distinguishable HDL species [1, 6]. Finally, a number of other techniques can be used to categorize HDL into subpopulations based on protein content. In short, HDLs are highly dynamic structures that have great variability with regard to size, shape, and surface chemistry. Each of these parameters is known to greatly modulate in vivo HDL function [7], and it is important to keep these parameters in mind when developing therapeutic agents based on HDLs.

HDL biosynthesis is initiated by the secretion of apoA-I from hepatocytes and enterocytes of the liver and small intestine, respectively [8]. Following synthesis, apoA-I is lipid-poor, but begins to sequester phospholipids and some free cholesterol. Acquisition of these molecular components results in nascent HDLs that have a discoidal shape, ≤8 nm in diameter. Nascent HDLs contain two anti-parallel molecules of apoA-I wrapped in a belt-like fashion around a solubilized central core of phospholipids oriented as a bilayer [9–11]. These nascent HDLs are the smallest of formed HDLs and only contain small amounts of cholesterol which is mainly interdigitated in the ~160 phospholipids present in the core of the disc [12]. Self-assembly of nascent HDL is moderated through the interaction of apoA-I with ATP binding cassette receptor A1 (ABCA1), a transmembrane protein that mediates the transfer of phospholipids and free cholesterol to apoA-I. Free cholesterol transferred to HDL becomes esterified by the serum enzyme lecithin cholesterol acyltransferase (LCAT). LCAT catalyzes the esterification of free cholesterol associated with HDL and increases its hydrophobicity. Cholesteryl esters (CE) are then driven into the lipid bilayer creating a core of hydrophobic CEs surrounded by a monolayer of phospholipids. The transfer of CE into the core results in an increase in HDL size and a change in morphology from the disc-like structure of nascent HDL into a maturing spherical nanostructure, while creating a gradient enabling more free cholesterol to move onto the particle surface. Further, as the particle matures it begins to interact with additional receptors known to participate in cellular cholesterol flux to HDLs, including ATP-binding cassette receptor G1 (ABCG1) and scavenger receptor type B-1 (SR-B1). Additionally, other lipoproteins aid in the maturation process of HDLs. For instance, low-density lipoproteins (LDLs) transfer triglycerides to HDLs in exchange for CE, a process catalyzed by the serum protein cholesteryl ester transfer protein (CETP). Together, these interactions increase the size of HDL and contribute to the heterogeneity of HDLs. However, HDLs do not consist solely of apolipoproteins, cholesterol, and phospholipids.

It is becoming increasingly appreciated that HDLs are highly diverse in their chemical composition with regard to phospholipids, small molecules, proteins [3, 13], small RNAs [14, 15], hormones, carotenoids, vitamins, and bioactive lipids [16].

Due to the complex synthesis and remodeling of HDL there are significant structural differences between HDL sub-species and individual particles. These differences result in a myriad of different HDL functions. Two key factors that determine the function of HDL in the human body are HDL size and composition. Below, we discuss the functions of natural HDLs to outline their natural mechanisms-of-action and also to highlight opportunities for targeting specific diseases, such as cardiovascular disease and cancer. Ultimately, designing synthetic HDL-like NPs with biologically functional modifications may provide targeting mechanisms to cell types critical to disease pathogenesis [17].

2 Cholesterol Transport-HDL and Coronary Heart Disease

Several epidemiological studies demonstrate that plasma concentrations of HDL-cholesterol (HDL-C) inversely correlate with the risk of coronary heart disease (CHD); however, emerging data show that measuring HDL-C as a biomarker for CHD may not be adequate [18]. Yet cholesterol associated with HDL has been termed "good" cholesterol and suggests that HDLs directly protect against CHD. The process of reverse cholesterol transport (RCT) is believed to be the primary explanation for this phenomenon. During RCT, cellular cholesterol is removed from lipid-laden macrophages (foam cells) in atherosclerotic lesions by HDLs via interactions with ABCA1, ABCG1, and SR-B1 [19]. Cholesterol-rich HDLs are then trafficked to the liver, delivering their cholesterol cargo to hepatocytes through SR-B1 in an apoA-I-dependent process. Cholesterol becomes incorporated into bile and excreted in the feces. In addition, HDL may transfer CE to LDLs in exchange for triglycerides. LDLs loaded with cholesterol are then taken up by hepatocytes through the LDL receptor further promoting RCT.

2.1 Non-cholesterol Transport-HDL and Coronary Heart Disease

While RCT is believed to be the primary means by which HDL reduces CHD risk, HDLs are implicated in multiple other cardio-protective mechanisms. For instance, HDLs reduce inflammation at the site of atherosclerotic lesions [4]. HDLs also have antioxidant and antithrombotic effects that maintain endothelial integrity and promote repair [1, 4, 20].

2.1.1 Anti-inflammatory Properties of HDLs

Chronic inflammation develops as atherosclerotic plaques mature. Pro-inflammatory stimuli contribute to the release of circulating cytokines, such as interleukin-1 and tumor necrosis factor alpha (TNF-α), which stimulate the expression of adhesion molecules such as intercellular adhesion molecule-1 (ICAM-1) and vascular cell adhesion molecule-1 (VCAM-1), P-selectin, and E-selectin. These molecules recruit and tether leucocytes and monocytes to endothelial cells overlying lipid-rich atheromas. Upon recruitment, the captured cells undergo molecular transformation and differentiate into macrophages that contribute to the growing atherosclerotic lesion. In vitro and in vivo studies demonstrate that HDLs reduce the expression of endothelial adhesion molecules and prevent the recruitment of monocytes to the arterial wall [21–23].

2.1.2 HDLs Combat Oxidative Damage

Oxidative stress is a risk factor associated with premature atherosclerosis and cardiovascular disease. Oxidized LDL (oxLDL) is the main mediator of oxidative arterial damage and promotes a pro-inflammatory and pro-atherogenic phenotype that contributes to endothelial cell dysfunction and apoptotic cell death. oxLDL contains a number of free radical-induced lipid hydroperoxides (LOOH), LOOH-derived short-chain oxidized phospholipids, and oxidized sterols [24, 25]. HDLs have antioxidant properties that can inhibit the accumulation of peroxidation products on the surface of LDLs and mitigate damage. This is accomplished by several mechanisms. First, methionine residues (112 and 148) of apoA-I reduce LOOH into redox-inactive lipids. In addition, apoA-I protein sequesters LOOHs from LDLs [26]. Third, HDLs, particularly the smaller HDL_3 subpopulations, have a unique proteome that plays a critical role in protection against LDL oxidative stress [26]. Also, the transfer of LOOH species from oxLDL to HDL_3 aids in the reduction of free radical species in atherosclerotic lesions [27]. Lastly, enzymes associated with HDLs, such as paraoxonase 1 (PON1), platelet activating factor-acetyl hydrolase (PAF-AH), and LCAT contribute to the antioxidant function of HDLs by hydrolyzing pro-inflammatory short-chain oxidized phospholipids [28].

2.1.3 Antithrombotic and Anticoagulant Activity of HDLs

Finally, HDLs demonstrate protective effects through antithrombotic and anticoagulant activity through direct and indirect interactions with endothelial cells. Endothelial-derived nitric oxide (NO) is critical for vasodilation and maintains the integrity of vascular endothelial and smooth muscle cells [29]. Reduced NO production is a hallmark of atherosclerosis and results in increased neutrophil adherence to the endothelium, smooth muscle cell proliferation, and enhanced platelet aggregation and adhesion. HDL is capable of activating NO synthesis through endothelial nitric oxide synthase (eNOS) upon binding to endothelial cell SR-B1 [30]. Endothelial cell apoptosis contributes to arterial atherothrombosis by the release of "microparticles" that carry pro-thrombotic factors, such as P-selectin [31]. Along with the release of microparticles, apoptotic endothelial cells enhance adhesion between unactivated platelets and leukocytes, promoting coagulation. HDLs directly contribute to endothelial protection by inhibiting oxLDL- and TNF-α-mediated apoptosis and by suppressing growth factor deprivation [32]. Furthermore, it has been demonstrated that HDLs enhance prostacyclin synthesis in endothelial cells. Prostacyclin acts synergistically with NO to induce vascular smooth muscle (VSM) relaxation [32], inhibits platelet activation, and represses the release of growth factors responsible for local proliferation of VSM cells [29]. In addition to HDLs acting on endothelial cells to reduce coagulation, an inverse correlation between platelet aggregation and HDL levels in humans has been reported [33]. This phenomenon is not limited to CHD as recombinant HDLs were shown to reduce platelet aggregation in individuals with type 2 diabetes mellitus [34].

Collectively, HDLs function in many ways to protect against cardiovascular disease. Developing a more complete understanding of HDL structure–function relationships will continue to provide new insights into the mechanisms of atheroprotection. However, the utility of HDLs extends beyond treating cardiovascular disease. It has recently been discovered that HDLs can bind, transport, and deliver a range of cargoes, including small molecules, photothermal reactive agents, and nucleic acids. Understanding the mechanisms by which HDLs perform this function will provide new opportunities to develop biomimetic, synthetic HDL-like NPs for targeted delivery of therapies for disease treatment. One such disease is cancer, as cancer cells increase cellular cholesterol pools not only by increasing cholesterol synthesis, but also by increasing HDL uptake. Thus HDL-based delivery methodologies have great potential for specific and effective delivery of cargo directly to cancer cells, with minimal off-target effects.

3 HDL and Cancer

A sustained and increased supply of cholesterol is essential for cancer cell proliferation and tumor progression [35–41]. Cancer-related anomalies of cholesterol metabolism have been implicated in angiogenesis, metastasis, and drug resistance [38]. On a cellular level, cholesterol is a crucial component of membranes and modulates their fluidity, stability, and overall architecture [42]. Additionally, cholesterol is known to accumulate in discrete regions of the membrane, termed lipid rafts, endowing these areas with very unique properties. Lipid rafts serve as assembly platforms for molecules involved in signaling cascades, including those associated with cancer development [35, 43–45]. On an organismal level, cholesterol serves as a precursor for steroid hormones, known regulators of cell proliferation and differentiation, and are critical in the progression of breast and prostate cancer [35].

Lipoproteins play a role in cancer progression through delivery of cholesterol to malignant cells. Consistently, many cancer patients exhibit reduced levels of cholesterol in the blood, which are restored to normal values upon successful cancer treatment [38, 46, 47]. Interestingly, it is the levels of HDL, not LDL, that are most affected in patients suffering from cancer [5, 38]. Exogenous addition of HDL has been shown to promote the growth of breast cancer cells in vitro and promote aggressiveness of tumors in vivo [48–50]. HDLs also exhibit anti-inflammatory, antioxidant, antimicrobial and pro-immunity properties [51, 52] and carry non-lipid cargo including microRNA, hormones, vitamins, and metabolites [14, 16]. All of these characteristics play an important functional role for HDL in cancer progression and tumor cell survival.

Many malignant cells have been shown to overexpress SR-B1 [5, 38, 42, 53–57]. For instance, among 50 human ovarian epithelial cancers, 96 % of the tumors highly expressed SR-B1 [54]. Further, prostate cancer cells have been shown to express SR-B1 for the uptake of cholesterol-rich HDLs to support endogenous androgen biosynthesis [53]. Also, HDL binding to SR-B1 can trigger downstream signaling cascades, such as through Akt, that promote breast cancer progression [57]. Thus, there is a great deal of evidence supporting increased HDL uptake via SR-B1 in cancer cells. This presents a unique opportunity to exploit this system for the targeted delivery of cancer therapeutics.

3.1 Cancer Therapies: Drug Hurdles and Emerging Drug Potential

Most anticancer drugs are inherently nonspecific in their biodistribution and diffuse into both healthy and cancer tissues. Further, the mechanism-of-action of most cancer drugs affects all rapidly diving cells rather than cancer cells specifically. Solubility is also a major limitation as ~40 % of new anticancer drugs are characterized by poor water solubility [58]. Targeted drug delivery vehicles soluble under physiologically relevant conditions have great potential to overcome these issues. By preferentially increasing accumulation of the therapeutic agent in malignant rather than healthy cells, the therapeutic index of a drug can be significantly improved and undesirable side effects reduced. HDLs are an attractive tool for the targeted delivery of antineoplastic substances due to their central role in promoting cancer proliferation by providing a constant supply of cholesterol to malignant cells. A number of research groups are therefore focused on the development of synthetic nanostructures that imitate natural HDLs in regard to function and structure. In the remainder of the chapter we will refer to these artificial structures as HDL-like nanoparticles (HDL-like NPs).

The preferential accumulation of such HDL-like NPs in cancer cells is achieved by both passive and active targeting. On the one hand, tumor tissues are characterized by a leaky vasculature and low lymphatic drainage, leading to differences in interstitial pressure between the center of a tumor and its periphery. This pressure difference allows for the preferential retention of particles between 10 and 100 nm in the tumor, a passive targeting phenomenon known as the enhanced permeability and retention (EPR) effect [59, 60]. On the other hand, synthetic HDL-like NPs are, like natural HDL, actively targeted to cancer cells by specific interaction with SR-B1 [42, 53–57]. Importantly, SR-B1 facilitates the uptake of cholesterol esters and anticancer drugs from spherical HDLs and HDL-like NP to the cytosol via a non-endocytic pathway [61, 62], avoiding lysosomal degradation of particle cargo.

3.2 Synthetic HDL-Like NP Composition

In contrast to natural HDL, synthetic HDL-like NPs are highly customizable, providing researchers with the unique opportunity to control many of the particles' structural and compositional features and to endow these particles with tailored and unique functions. Examples of the structural diversity of HDL-like NPs are provided in Table 1. Because HDL-like NPs are designed to mimic their natural

Table 1 Examples of the structural diversity of HDL-like NPs for drug delivery

Structural feature	Examples
Composition of particle core	Cholesteryl esters [54]
	Inorganic scaffolds (Au) [89]
Shape	Discoidal [80]
	Spherical [89]
Protein associated with particle	Full-length apoA-I [97]
	ApoA-I-mimetic peptide [99]
	ApoE [96]
Phospholipids	Phosphatidylcholine (PC) [69],
	1,2-dioleoyl-sn-glycero-3-phosphocholine (DOPC) [98]
	1,2-dimyris-toyl-sn-glycero-3-phosphocholine (DMPC) [98]
	1,2-dioleoyl-sn-gly- cero-3-((N(5-amino-1-carboxypentyl) iminodiacetic acid) succinyl)(- nickel salt) (NiLipid) [98]
	1,2-distearoyl-sn-glycero-3-phosphoethanolamine-N-[folate (polyethylene glycol)-2000] (DSPE-PEG2000-folate) (PF) [98]
	1,2-dioleoyl-*sn*-glycero-3-phosphoethanolamine-N-[3-(2-pyridyldithio)propionate] (PDP PE) [5, 91]
	Dipalmitoylphosphatidylcholine (DPPC) [5, 91]
	Pyropheo-phorbide conjugated to the glycerol backbone of lysophospholipids [80]
Lipid layer surrounding the particle	Monolayer [91]
	Bilayer [89]
Mechanism of drug loading	Covalent attachment [98]
	Encapsulation [102]
	Integration in lipid layer [73]
Additional targeting moieties	Folic acid, covalently linked to lysine residues of apoA-I [103]
	Lipid-conjugated folate (1,2-distearoyl-sn-glycero-3-phosphoethanolamine-N-[folate(polyethylene glycol)-2000] (DSPE-PEG2000-folate) (PF) [98]
Incorporated drug	Paclitaxel [83]
	Doxorubicin [70]
	Valrubicin [62]
	Fenretinide [74]
Incorporated fluorophore	Fluo-BOA [72]
	Lipid-conjugated Rhodamine B [104]

counterparts in composition, surface properties, and size, many of them were shown to be non-immunogenic, capable of avoiding clearance by the reticuloendothelial *system*, and to have relatively long circulation times [59, 63].

3.3 HDL-like Nanoparticles for Drug Delivery

3.3.1 Discoidal HDL NPs for Drug Delivery

Both discoidal and spherical HDL-like NPs have been shown to target SR-B1 [64] and have therefore both been employed for targeted drug delivery to cancer cells. While spherical HDL-like NPs are more similar in shape to the more mature versions of their natural counterparts, discoidal HDL-like NPs are more reminiscent of nascent HDL. Like natural, nascent HDL, discoidal HDL-like NPs are prone to undergo maturation in the bloodstream due to interaction with LCAT [65–68]. This biological maturation of discoidal HDLs can lead to unwanted leakage of drug cargo [69]. To combat this problem, researchers used monocholesterylsuccinate (CHS) instead of cholesterol to prevent particle interaction with LCAT, and anchored apoA-I on the particle by covalently linking it to CHS [70, 71]. The authors employed a CHS-modified discoidal HDL-like NPs, termed recombinant HDL (rHDL), and loaded it with paclitaxel, a mitotic inhibitor used to treat a variety of cancers. Studies in rats showed improved drug levels in the blood over extended periods of time both compared to drug-loaded, but unmodified particle (P-d-rHDL), as well as compared to free or liposome-loaded drug [71]. Further, paclitaxel-loaded CHS-modified particles (cP-d-rHDL) showed increased cytotoxicity and uptake in the human cancer cell line MCF-7 compared to P-d-rHDL. In breast tumor bearing mice, cP-d-rHDL were superior in regard to both tumor targeting as well as limiting tumor growth compared to P-d-rHDL, liposome-loaded drug or free drug [70].

3.3.2 Spherical HDL-like NPs for Drug Delivery

Delivery of hydrophobic antineoplastic agents by HDL-like NPs is often achieved by encapsulation of the drug in the hydrophobic core. Recently, Lacko and coworkers showed that inclusion of highly water-insoluble drugs, like valrubicin, into spherical HDL-like NPs composed of phosphatidyl choline, apoA-I, free cholesterol, and cholesteryl oleate, led to increased toxicity in SR-B1 expressing malignant prostate and ovarian cancer cell lines compared to the free drug [62, 72]. Further, off-target cytotoxic effects in non-malignant epithelial prostate and ovarian cell lines with low SR-B1 expression were decreased for the drug-containing particle compared to the free drug [62, 72]. Encapsulation of valrubicin into HDL-like NP may expand the therapeutic spectrum of the drug, which had previously been used exclusively for the treatment of bladder cancer. The authors also used a similar NP construct to target the therapeutic agent fenretinide to two different neuroblastoma cells lines in vitro [72]. Compared to the free drug, cytotoxicity in the malignant cell lines was significantly increased, while cytotoxicity in retinal pigment epithelial cells, a control cell line for off-target fenretinide toxicity, was reduced [62, 72]. Feng and co-workers used a similar strategy of drug encapsulation to target the chemotherapeutic doxorubicin to hepatocellular carcinoma and hepatoma cells lines [73, 74]. In this work, HDL-like NPs were composed of egg phospholipids, apoA-I, and doxorubicin and showed increased cytotoxicity, apoptosis induction, and conjugate accumulation in target cells compared with the free or liposome encapsulated drug [73]. The drug-loaded HDL-like NPs also decreased tumor growth in a metastatic model of human hepatocellular carcinoma (HCC) in nude mice and reduced hemolysis-related side effects [74].

To better understand the uptake of lipophilic cargo from spherical HDL-like NPs through SR-B1, Lin et al. developed multifluorophore-labeled HDL-mimicking peptide phospholipid scaffold (HPPS) nanoparticles [75]. Although apoA-I is critical for determining the shape of HDLs and allowing for specific interactions with cellular receptors, the protein can be replaced by short (<20 amino acids) peptides that show no sequence similarity to the full-length protein [76]. These peptides mimic the amphipathic helical structure of apoA-I, including its receptor and lipid binding abilities [76]. Lin et al. [75] generated a variety of different cholesteryl oleate/phosphocholine HDL-like NPs with either fluorescent compounds in the particle core, fluorescently labeled apoA-I peptides and phospholipids on the surface, or combinations thereof. Through sequential inhibition studies, they found that after initial interaction of the HDL-like NP with SR-B1, the particle bound to a specific sub-domain of the receptor, leading to particle dissociation and internalization of the hydrophobic cargo into the cytosol by a lipid-raft/caveolae-like mechanism. Phospholipids and apoA-I-mimetic peptides were mainly retained on the cell surface. It is important to note that for natural, mature HDL, selective influx of CE payload does not require particle catabolism [77, 78], suggesting that the cellular fate of specific HDL-like NPs might be dependent on their specific composition and determined by additional factors that warrant further investigation.

3.3.3 HDL-like NPs for the Delivery of Photothermal Therapeutics

HDL-like NPs have also been used for the delivery of photothermal agents, which facilitate infrared light-induced temperature changes in tumor tissues leading to tissue necrosis. Mathew et al. generated an HDL-like NP composed of phosphocholine, apoA-I fused to a trans-activating transcriptional activator peptide (TAT-peptide, for enhanced cell internalization), and a water-insoluble gadolinium bis (naphthalocyanine) sandwich complex (as a photothermal compound) [79]. Using this particle, the authors achieved photothermal killing of human lung cancer cells in a near infrared light-irradiation-dependent manner. In addition, delivery of the photothermal compound was dependent on the expression of SR-B1 by target cells. Also, by conjugating pyropheophorbide, a reduced porphyrin, to lysophospholipids, Ng et al. created a phototherapeutic fluorescent pyro-lipid, which self-assembled with apoA-I into nanodiscs [80]. The authors used these particles to treat Chinese hamster ovary (CHO) cells that were transfected to express SR-B1. The fluorescence of the photosensitizer was quenched in intact particles; however, upon cellular uptake fluorescence of the particle became un-quenched, likely due to disruption of particle structure. Importantly, transfected SR-B1-expressing cells exhibited a dose-dependent decrease in survival when treated with 660 nm light, whereas SR-B1 negative, untransfected CHO cells neither took up the particles nor were affected by light treatment [80]. To our knowledge, these and other constructs are awaiting efficacy testing in murine models.

3.3.4 Modification of HDL-like NP for Uptake via Alternate Receptors

HDL-like NPs can also serve as scaffolds for the attachment of alternative targeting ligands for binding receptors beyond SR-B1. Corbin et al. [81] exploited the fact that over 90 % of non-mucinous ovarian cancers overexpress folate receptor-α (FR-α), as previously reported [82]. The authors first synthesized HDL-like NPs composed of apoA-1, egg yolk phosphatidylcholine, and cholesteryl oleate and then covalently attached folate to the lysine residues of apoA-I. Folate modification abolished the ability of the particle to interact with SR-B1, therefore re-routing the HDL-like NP to FR-α. The researchers used this technique to specifically deliver a near-infrared fluorescent dye to ovarian tumors in mice, enabling in vivo tumor imaging [81, 83]. Current studies are underway to utilize these particles to deliver therapeutic antineoplastic agents as well.

3.3.5 Natural HDLs for Systemic Delivery of siRNA

The potential of HDLs and HDL-like NPs as delivery vehicles for nucleic acids emerged from a few key studies. The first data demonstrating the interaction between natural HDLs and siRNAs focused on siRNA sequences terminally modified with lipophilic moieties, like cholesterol. Addition of lipophilic siRNAs resulted in siRNA-specific gene silencing in cultured cells in vitro and in the liver following systemic administration [84–86]. These observations prompted investigations to better understand how lipophilic siRNAs were being productively delivered to target cells and the liver. Toward this end, Wolfrum et al. demonstrated that systemically injected lipid-modified siRNAs spontaneously bound lipoproteins in the serum (i.e., HDL and LDL). As such, lipoprotein-bound siRNAs were then delivered to tissues that express receptors for specific lipoproteins. In the case of HDL-bound siRNAs, delivery was mediated by SR-B1 [87].

In addition, upon extraction of natural HDLs from human plasma, Vickers et al. demonstrated that HDLs naturally bind and stabilize microRNAs [14]. Further, they found distinct microRNA profiles from HDL of healthy human samples compared to subjects with hypercholesterolemia. For example, the most abundant microRNA associated with HDLs in healthy subjects was hsa-miR-135a, whereas hsa-miR-223 was the most abundant microRNA found in hypercholesterolemic patients. To determine if HDL had the capacity to load microRNAs in vivo, rHDLs free of RNA were intravenously injected into wild-type or apoE null mice. After 6 h of incubation, rHDLs were isolated from mouse plasma. rHDLs isolated from healthy mice produced 110 unique miRNAs while rHDLs isolated from the apoE null mice were found to contain 162 microRNAs. These results suggest that HDL has the capacity to bind microRNAs in vivo. In addition, Vickers et al. determined that HDL transfers microRNA to recipient cells though ABCA1. By increasing the expression of ABCA1 in J774 murine macrophage cells using a liver-X-receptor-α agonist (LXRα), they were able to demonstrate an increase in abundance of miR-223 associated with rHDL in the presence of the LXRα. Finally, to determine the importance of SR-B1 mediated microRNA delivery, baby hamster kidney (BHK) cells were stably transfected with an inducible human SR-B1 vector to increase the expression of SR-B1. Upon treatment of induced BHK cells with native HDL and native HDL pre-complexed with miR-223, induced BHK cells resulted in a 69-fold increase in intracellular levels of hsa-miR-223 compared to induced BHK cells treated with native HDL. These data suggest that HDL delivered microRNA depends on SR-B1 expression. These data demonstrate that natural HDLs bind, stabilize, and productively deliver RNAs to cells that express SR-B1.

Collectively, it has been demonstrated that HDLs are natural nucleic acid delivery vehicles with high potential for overcoming hurdles associated with systemic siRNA delivery. As such, there has been a significant focus on the development of synthetic HDL-like vectors for the delivery of nucleic acids. Importantly, and described below, synthetic HDL-like platforms vary with regard to size, charge, and surface chemistry. The goal of these efforts is to understand the parameters of natural HDLs that lead to RNA binding, stabilization, and productive delivery to target cells, like cancer cells, that express receptors for HDL.

3.4 Synthetic HDL-like Nanoparticles for Nucleic Acid Delivery

3.4.1 Spherical HDL Gold Nanoparticles for Delivery of Nucleic Acids

Our research group pioneered the synthesis of HDL-like NPs using a gold nanoparticle (AuNP) core template. We demonstrated that these particles tightly bind cholesterol ($K_d = 3.8$ nM) [88] and function to modulate cholesterol metabolism in target cells through the same receptors as natural HDLs [89, 90]. Because our HDL-like AuNPs tightly bind cholesterol, and natural HDLs spontaneously associate with lipidated siRNAs after systemic administration, we hypothesized that our HDL-like AuNP platform could be utilized to adsorb cholesterol modified nucleic acids for delivery to target cells that express SR-B1. Initially, we tested this hypothesis using cholesterylated antisense DNAs (chol-DNA) for delivery to prostate cancer cells that express SR-B1 [91]. As in our previous work, we developed a synthetic HDL-like nanostructure that closely mimics the size, shape, and surface chemistry of natural, mature spherical HDLs [88]. A 5 nm AuNP was used as a scaffold to control the size, shape, and composition of synthetic, spherical HDL-like AuNPs. We termed these high-density lipoprotein-like gold nanoparticles (HDL-like AuNPs). They are similar in size to their natural mature spherical HDL counterparts and have a similar surface composition: approximately 3 copies of apoA-I and an outer phospholipid layer of zwitterionic 1-2-dipalmitoyl-sn-glycero-3-phosphocholine (DPPC) [88], [89]. Moreover, chol-DNAs bound to HDL-like AuNPs (chol-DNA-HDL-like AuNPs) can overcome many of the difficulties associated with in vitro nucleic acid delivery. For instance, HDL-like AuNPs stabilize chol-DNA against nuclease degradation. Further, chol-DNA-HDL-like AuNPs do not exhibit off-target cellular toxicity and are efficient conjugates for delivering nucleic acid to regulate target gene expression. Following cell treatment, transmission electron micrographs suggested that chol-DNA-HDL-like AuNPs bypass endolysosomal sequestration, a major biological hurdle in nucleic acid delivery. This manuscript was the first to demonstrate synthetic, biomimetic HDL-like AuNPs as a delivery vehicle for nucleic acids [91].

In addition, our group recently reported the ability of HDL-like AuNPs to deliver cholesteryl-modified single stranded RNA (RNAi-HDL-like AuNPs) for modulating gene expression in the context of neovascularization and angiogenesis, both in vitro and in vivo [92]. RNAi-HDL-like AuNPs were functionalized with antisense RNA targeting vascular endothelial growth factor receptor 2 (VEGFR2), the receptor for vascular endothelial growth factor (VEGF), a key regulator of neo-angiogenesis. The RNAi-HDL-like AuNPs were delivered to cultured vascular endothelial cells, which express SR-B1, and are responsible for the formation of new blood vessels. RNAi-HDL-like AuNPs effectively reduced VEGFR2 expression in target endothelial cells. In addition, treatment with RNAi-HDL-like AuNPs also reduced VEGF-induced endothelial cell survival and morphogenesis. Importantly, knockdown of SR-B1 expression by cultured endothelial cells reduced the uptake of RNAi-HDL-like AuNPs and drastically reduced VEGFR2 knockdown. Accordingly, SR-B1 expression, like for natural nucleic-acid-carrying HDLs, was shown to be required for the internalization and function of RNAi-HDL-like AuNPs. Most importantly, in vivo data showed that RNAi-HDL-like AuNPs reduce neovascularization induced by VEGF after local and systemic injection of the conjugates. Further, systemic administration of RNAi-HDL-like AuNPs targeting VEGFR2 significantly reduced target gene expression, tumor volume, and tumor weight in Lewis lung carcinoma tumor xenografts, a tumor model well known for its high degree of vascularity. Collectively, these data suggest that synthetic, spherical HDL-like AuNPs are an efficient delivery vehicle for systemic nucleic acid delivery to cells and tissues involved in carcinogenesis both in vitro and in vivo.

3.4.2 HDL-Mimicking Peptide-Phospholipid Scaffold (HPPS) Nanoparticles for siRNA Delivery

Other HDL-like NPs have been used to deliver nucleic acids. Yang et al. [93] designed an HDL-like NP using a peptide-phospholipid scaffold, termed HPPS particles. The particles consisted of phospholipids, cholesteryl oleate, and amphipathic α-helical peptides, which mimic apoA-I. The components self-assembled into structures similar to plasma-derived HDL. Direct incubation of HPPS particles with cholesteryl conjugated siRNAs (chol-siRNA) targeting the oncogene B-cell lymphoma-2 (chol-si-bcl-2) resulted in final constructs with a hydrodynamic diameter of 25.3 ± 1.2 nm and contained an average of 8 chol-siRNAs per particle. The surface charge of the particle shifted from −2.7 ± 1.9 mV to −15.2 ± 4.8 mV when chol-siRNA was added, which is consistent with RNA loading. Data showed that HPPS particles delivered siRNA cargo to the cytosol of SR-B1 expressing cells and regulated target gene expression. Treatment of KB cells expressing SR-B1 resulted in a 35 ± 9 % reduction of BCL-2 protein as compared to the control, and a 2.5-fold increase in apoptosis was measured when compared to chol-si-bcl-2 siRNA alone. Importantly, HPPS chol-si-bcl-2 particles were not effective in knocking down *Bcl-2* in HT1080 cells that express minimal SR-B1. Furthermore, and consistent with SR-B1 expression, HPPS particles efficiently deliver siRNA cargo to the cytoplasm of target cells, bypassing endolysosomal sequestration. These results suggest that HPPS chol-si-bcl-2 particles utilize SR-B1 for efficient siRNA delivery and target gene knockdown.

3.4.3 Reconstituted HDL for siRNA Delivery

As mentioned above, reconstituted HDLs (rHDLs, distinct from the HDL-like NPs described in Sect. 3.3.1), are synthetic forms of human HDL essentially composed of phospholipids, apoA-I, cholesterol, and esterified cholesterol [63]. For siRNA delivery, Shahzad et al. [94] incorporated siRNA into rHDL nanoparticles. The particles were approximately 10 nm in diameter, had a net neutral charge (−3.2 mV), and demonstrated efficient delivery of siRNA to target cells. Further, rHDL-siRNA conjugates were shown to regulate target gene expression after systemic administration in orthotopic ovarian and colorectal cancer models. This group found that rHDLs loaded with fluorescently labeled siRNA showed preferential uptake in mouse tissues that highly expressed SR-B1 (i.e., tumor and liver), with minimal fluorophore-labeled siRNA in other tissues such as the brain, heart, lung, kidney, and spleen. Importantly, this work provides insight into cell-specific targeting in vivo, especially in tumor bearing mice, suggesting efficient delivery to tumor cells that express SR-B1.

A similar approach to delivering siRNA using an rHDL nanoparticle has been demonstrated by Ding et al. [95]. Here, rHDL nanoparticles were synthesized using a mixture of phospholipids, apoA-I, cholesterol, and cholesteryl esters. In contrast to the previously cited work, Ding et al. incorporated cholesteryl modified siRNA sequences into the rHDL nanostructure. This approach yielded a rHDL/chol-siRNA complex that was ~90 nm in diameter with a near-neutral charge of −4.2 mV. Consistent with previous findings, the rHDL protected siRNA from nuclease degradation and rHDL/chol-siRNA was effective in silencing gene expression in vitro and in vivo. The siRNA sequence was targeted to Pokemon. Pokemon is a proto-oncogene overexpressed in a number of human cancers and induces carcinogenesis by repressing two key tumor suppressive pathways: one, the alterative reading frame (ARF)-multiple murine double minute gene 2 (HDM2)-p53 pathway, and two, the retinoblastoma (Rb)-early-region-2 transcription factor (E2F) pathway.

Systemic administration of rHDL/chol-siRNA particles complexed with siRNA targeting Pokemon demonstrated a reduction in tumor volume of HepG2 xenografts.

3.4.4 ApoE Lipopeptide Nanoparticles for siRNA Delivery

In addition to apoA-I, apoE has been used as a targeting moiety for HDL-like NPs. Dong et al. [96] generated a library of NPs composed of phospholipids, cholesterol, polyethylene glycol-lipid, siRNA, and different lipopeptides—lipids with constituent groups conjugated to different amino acids, peptides, and polypeptide head groups. The library was generated to evaluate combinations of lipoamino acid derivatives designed to mimic apolipoproteins associated with natural lipoproteins. They initially screened the lipopeptide nanoparticle's (LNP) capacity to silence Factor VII (a blood clotting factor) in mice following intravenous injection. Upon further screening, a lead compound, CKK-E12, was used to evaluate the silencing of phosphatase and tensin homolog (PTEN) in different tissues after intravenous administration in mice. Data showed a significant silencing of PTEN in the liver compared to the lung, spleen, kidney, heart, and brain. On further examination of the liver, CKK-E12 LNPs silenced over 80 % of PTEN in hepatocytes; however, it showed no significant silencing in endothelia cells and leukocytes in the liver with comparison made to a luciferase siRNA-formulated CKK-E12 LPN, as a control. To study the effects of cell uptake and gene silencing with regard to apolipoprotein species, CKK-E12 LNPs were mixed with 11 isoforms of apoA, apoB, apoC, apoE and apoH. Upon transfection of HeLa cells, results showed that apoA, apoC, and apoH did not show significant gene silencing; however, four apoE isoforms showed improved luciferase silencing and the apoE-3 isoform increased cellular uptake and endolysosomal escape when added to cultured cells on treatment with CKK-E12 LNPs. Collectively, these data demonstrate that particle optimization can significantly improve specificity and efficacy, and improvements such as these offer the potential to broaden the therapeutic application of siRNA therapeutics. Further, the incorporation of apoE, instead of apoA-I into HDL-like particles appears to be superior in silencing gene targets in the liver [97]. ApoE-conjugated HDL-like NPs more efficiently delivered chol-siRNA to hepatocytes compared to apoA-I-conjugated particles, presumably due to the employment of a different uptake mechanism (LDL receptor for apoE vs. SR-B1 for apoA-I) [97].

In addition, Fischer et al. [98] explored the biodistribution, toxicity, and potential of human cloned apoE4 containing NPs to deliver a variety of different compounds, including cholesterol modified single stranded DNA (ss-DNA) [98]. The authors used a series of modifications to conjugate additional cargo to their apoE-HDL-like NPs, namely nickel-chelating lipids for binding His-tagged proteins, highly hydrophobic molecules like cholesterol for covalently linkage to cargo (like chol-ssDNA), and covalent conjugation of proteins to lipophilic moieties for binding to the lipid bilayer. The resulting suite of HDL-like NPs was stable in complex biological matrices and were shown to be non-cytotoxic in vitro at concentrations up to 320 mg/ml. Administration of the particles in vivo did not cause weight loss, organ specific toxicity, or overt immunogenicity. Biodistribution of the particles varied by route of administration, with preferred accumulation of the particles in kidney and liver after intravenous, intraperitoneal, intramuscular, or subcutaneous administration.

3.5 HDL-like Nanoparticles: Intrinsically Therapeutic Agents

More recently, HDL-like AuNP constructs have shown intrinsic therapeutic activity without requiring additional molecular drug cargo [90]. The HDL-like AuNPs are similar to the nanoparticles our group employed for cholesterylated nucleic acid formulation, as previously discussed. The structural conformation and number of apoA-I molecules, as well as phospholipid number, particle size, and negative particle charge are consistent with natural HDLs [91, 92]. Our data demonstrate that the HDL-like AuNPs directly compete with natural HDLs to target SR-B1, a high-affinity HDL receptor expressed by lymphoma cells. Uniquely, the AuNP template at the core of the HDL-like AuNP enables differential modulation of cholesterol flux compared with natural HDL. Data show that binding of SR-B1 and the manipulation of cholesterol homeostasis in B lymphoma cells upon treatment with the HDL-like AuNP leads to a selective induction of apoptosis. Furthermore, upon HDL-like AuNP treatment to mice bearing B-cell lymphoma xenografts, data show inhibition of B cell growth compared to mice treated with human HDL or saline. Collectively, these data suggest that HDL-like AuNPs may provide opportunities as stand-alone therapies due to cooperative SR-B1 targeting and manipulation of cellular cholesterol homeostasis in B cell lymphoma [90].

In addition to our findings, Zheng et al. found that their SR-B1-targeted HPPS NP inhibited motility and colony formation of nasopharyngeal carcinoma (NPC) cells [71, 99, 100]. The NPs were active in nude mice bearing subcutaneous tumors. This effect involved neither tumor cell necrosis nor apoptosis, suggesting an alternative tumor-shrinking mechanism that is related to HPPS-induced inhibition of NPC cell motility and colony formation. ApoA-I-mimetic peptides have been shown to exhibit antitumor activities by reducing plasma levels of lysophosphatidic acid, a stimulator of cell migration, invasion, and colony formation [101]. While apoA-I-peptide may be involved in the antitumor mechanism of HPPS, the exact mechanism is still unknown.

In conclusion, a significant body of evidence demonstrates that biomimicry may play an important role for next generation cancer therapies. Synthetic HDL-like NP mimics are appealing as new treatments, due to their inherent active targeting to cancer cells and their ability to deliver diverse therapeutic cargo. Importantly, synthetic HDL-like NPs can be manipulated such that they are able to bind and deliver small molecules, photothermal reactive agents, nucleic acids, and may also be stand-alone entities that are intrinsically active. Furthermore, HDL-like NPs can also serve as delivery vehicles for imaging agents. Research directed at these intriguing nanostructure conjugates has only begun. Further, preclinical and clinical development of these new approaches based on HDL biology may offer tremendous new therapeutic opportunities for patients with cancer.

References

1. Camont L, Chapman MJ, Kontush A (2011) Biological activities of HDL subpopulations and their relevance to cardiovascular disease. Trends Mol Med 17(10):594–603. doi:10.1016/J. Molmed.05.013
2. Xu S, Laccotripe M, Huang X et al (1997) Apolipoproteins of HDL can directly mediate binding to the scavenger receptor SR-BI, an HDL receptor that mediates selective lipid uptake. J Lipid Res 38(7):1289–1298
3. Warnick GR, McNamara JR, Boggess CN et al (2006) Polyacrylamide gradient gel electrophoresis of lipoprotein subclasses. Clin Lab Med 26(4):803–846. doi:Doi 10.1016/J. Cll.2006.07.005
4. Tabet F, Rye KA (2009) High-density lipoproteins, inflammation and oxidative stress. Clin Sci 116(1–2):87–98. doi:10.1042/Cs20080106
5. Damiano MG, Mutharasan RK, Tripathy S et al (2013) Templated high density lipoprotein nanoparticles as potential therapies and for molecular delivery. Adv Drug Deliv Rev 65 (5):649–662. doi:10.1016/j.addr.2012.07.013
6. Rosenson RS, Brewer HB, Chapman MJ et al (2011) HDL measures, particle heterogeneity, proposed nomenclature, and relation to atherosclerotic cardiovascular events. Clin Chem 57 (3):392–410. doi:10.1373/Clinchem.2010.155333
7. Asztalos BF, Tani M, Schaefer EJ (2011) Metabolic and functional relevance of HDL subspecies. Curr Opin Lipidol 22(3):176–185. doi:10.1097/MOL.0b013e3283468061
8. Robert O, Bonow EA (eds) (2012) Braunwald's heart disease—a textbook of cardiovascular medicine, 9th edn. Elsevier Saunders, Philadelphia, PA
9. Ikonen E (2008) Cellular cholesterol trafficking and compartmentalization. Nat Rev Mol Cell Biol 9(2):125–138. doi:10.1038/nrm2336
10. Calabresi L, Gomaraschi M, Rossoni G et al (2006) Synthetic high density lipoproteins for the treatment of myocardial ischemia/reperfusion injury. Pharmacol Therapeut 111(3):836–854. doi:10.1016/J.Pharmthera.01.003
11. Silva RAGD, Huang R, Morris J et al (2008) Structure of apolipoprotein A-I in spherical high density lipoproteins of different sizes. Proc Nat Acad Sci USA 105(34):12176–12181. doi:10.1073/Pnas.0803626105
12. Segrest JP, Li L, Anantharamaiah GM et al (2000) Structure and function of apolipoprotein A-I and high-density lipoprotein. Curr Opin Lipidol 11(2):105–115. doi:10.1097/00041433-200004000-00002
13. Cheung MC, Albers JJ (1982) Distribution of high-density lipoprotein particles with different apoprotein composition—Particles with a-I and a-Ii and Particles with a-I but No a-Ii. J Lipid Res 23(5):747–753
14. Vickers KC, Palmisano BT, Shoucri BM et al (2011) MicroRNAs are transported in plasma and delivered to recipient cells by high-density lipoproteins. Nat Cell Biol 13(4):423–U182. doi:10.1038/Ncb2210
15. Liu X, Suo R, Xiong SL et al (2013) HDL drug carriers for targeted therapy. Clin Chim Acta; Int J Cin Chem 415:94–100. doi:10.1016/j.cca.2012.10.008
16. Vickers KC, Remaley AT (2013) Functional diversity of HDL cargo. J Lipid Res doi:10. 1194/jlr.R035964
17. Pirillo A, Norata GD, Catapano AL (2013) High-density lipoprotein subfractions—what the clinicians need to know. Cardiology 124(2):116–125. doi:10.1159/000346463

18. Feig JE, Hewing B, Smith JD et al (2014) High-density lipoprotein and atherosclerosis regression: evidence from preclinical and clinical studies. Circ Res 114(1):205–213. doi:10.1161/CIRCRESAHA.114.300760
19. Rader DJ, Alexander ET, Weibel GL et al (2009) The role of reverse cholesterol transport in animals and humans and relationship to atherosclerosis. J Lipid Res 50(Suppl):S189–S194. doi:10.1194/jlr.R800088-JLR200
20. Gomaraschi M, Ossoli A, Vitali C et al (2013) HDL and endothelial protection: examining evidence from HDL inherited disorders. Clin Lipidol 8(3):361–370. doi:10.2217/Clp.13.30
21. Nicholls SJ, Dusting GJ, Cutri B et al (2005) Reconstituted high-density lipoproteins inhibit the acute pro-oxidant and proinflammatory vascular changes induced by a periarterial collar in normocholesterolemic rabbits. Circulation 111(12):1543–1550. doi:10.1161/01.Cir.0000159351.95399.50
22. Cockerill GW, Rye KA, Gamble JR et al (1995) High-density-lipoproteins inhibit cytokine-induced expression of endothelial-cell adhesion molecules. Arterioscler Thromb Vasc 15 (11):1987–1994
23. Navab M, Imes SS, Hama SY et al (1991) Monocyte transmigration induced by modification of low-density-lipoprotein in cocultures of human aortic-wall cells is due to induction of monocyte chemotactic Protein-1 synthesis and is abolished by high-density-lipoprotein. J Clin Invest 88(6):2039–2046. doi:10.1172/Jci115532
24. Gordon LI, Bass J, Yachnin S (1980) Inhibition of human polymorphonuclear leukocyte chemotaxis by oxygenated sterol compounds. Proc Nat Acad Sci USA 77(7):4313–4316
25. Yachnin S, Streuli RA, Gordon LI et al (1979) Alteration of peripheral blood cell membrane function and morphology by oxygenated sterols; a membrane insertion hypothesis. Curr Top Hematol 2:245–271
26. Kontush A, Chapman MJ (2010) Antiatherogenic function of HDL particle subpopulations: focus on antioxidative activities. Curr Opin Lipidol 21(4):312–318. doi:10.1097/Mol.0b013e32833bcdc1
27. Davidson WS, Silva RAGD, Chantepie S et al (2009) Proteomic analysis of defined HDL subpopulations reveals particle-specific protein clusters relevance to antioxidative function. Arterioscler Thromb Vasc 29(6):870–U234. doi:10.1161/Atvbaha.109.186031
28. Subramanian VS, Goyal J, Miwa M et al (1999) Role of lecithin-cholesterol acyltransferase in the metabolism of oxidized phospholipids in plasma: studies with platelet-activating factor-acetyl hydrolase-deficient plasma. Biochim Biophys Acta 1439(1):95–109
29. Mineo C, Deguchi H, Griffin JH et al (2006) Endothelial and antithrombotic actions of HDL. Circ Res 98(11):1352–1364. doi:10.1161/01.RES.0000225982.01988.93
30. Yuhanna IS, Zhu Y, Cox BE et al (2001) High-density lipoprotein binding to scavenger receptor-BI activates endothelial nitric oxide synthase. Nat Med 7(7):853–857. doi:10.1038/89986
31. Furie B, Furie BC (2004) Role of platelet P-selectin and microparticle PSGL-1 in thrombus formation. Trends Mol Med 10(4):171–178. doi:10.1016/j.molmed.2004.02.008
32. Farmer JA, Liao J (2011) Evolving concepts of the role of high-density lipoprotein in protection from atherosclerosis. Curr atherosclerosis rep 13(2):107–114. doi:10.1007/s11883-011-0166-3
33. Naqvi TZ, Shah PK, Ivey PA et al (1999) Evidence that high-density lipoprotein cholesterol is an independent predictor of acute platelet-dependent thrombus formation. Am J Cardiol 84 (9):1011–1017
34. Calkin AC, Drew BG, Ono A et al (2009) Reconstituted high-density lipoprotein attenuates platelet function in individuals with type 2 diabetes mellitus by promoting cholesterol efflux. Circulation 120(21):2095–2104. doi:10.1161/CIRCULATIONAHA.109.870709
35. Simons K, Ikonen E (2000) How cells handle cholesterol. Science 290(5497):1721–1726. doi:10.1126/science.290.5497.1721

36. Clendening JW, Pandyra A, Boutros PC et al (2010) Dysregulation of the mevalonate pathway promotes transformation. Proc Nat Acad Sci USA 107(34):15051–15056. doi:10.1073/pnas.0910258107
37. Ginestier C, Monville F, Wicinski J et al (2012) Mevalonate metabolism regulates Basal breast cancer stem cells and is a potential therapeutic target. Stem Cells 30(7):1327–1337. doi:10.1002/stem.1122
38. Cruz PM, Mo H, McConathy WJ et al (2013) The role of cholesterol metabolism and cholesterol transport in carcinogenesis: a review of scientific findings, relevant to future cancer therapeutics. Front pharmacol 4:119. doi:10.3389/fphar.2013.00119
39. Brown MS, Goldstein JL (1986) A receptor-mediated pathway for cholesterol homeostasis. Science 232(4746):34–47. doi:10.1126/science.3513311
40. Olson RE (1998) Discovery of the lipoproteins, their role in fat transport and their significance as risk factors. J Nutr 128(2):439s–443s
41. Chapman MJ (1980) Animal lipoproteins: chemistry, structure, and comparative aspects. J Lipid Res 21(7):789–853
42. Gorin A, Gabitova L, Astsaturov I (2012) Regulation of cholesterol biosynthesis and cancer signaling. Curr Opin Pharmacol 12(6):710–716. doi:10.1016/j.coph.2012.06.011
43. Brown AJ (2007) Cholesterol, statins and cancer. Clin Exp Pharmacol Physiol 34(3):135–141. doi:10.1111/j.1440-1681.2007.04565.x
44. Simons K, Gerl MJ (2010) Revitalizing membrane rafts: new tools and insights. Nat Rev Mol Cell Biol 11(10):688–699. doi:10.1038/nrm2977
45. Simons K, Ikonen E (1997) Functional rafts in cell membranes. Nature 387(6633):569–572. doi:10.1038/42408
46. Niendorf A, Nagele H, Gerding D et al (1995) Increased LDL receptor mRNA expression in colon cancer is correlated with a rise in plasma cholesterol levels after curative surgery. Int J Cancer J Int Du Cancer 61(4):461–464
47. Solomon KR, Freeman MR (2011) The complex interplay between cholesterol and prostate malignancy. Urol Clin North Am 38(3):243–259. doi:10.1016/j.ucl.2011.04.001
48. Rotheneder M, Kostner GM (1989) Effects of low- and high-density lipoproteins on the proliferation of human breast cancer cells in vitro: differences between hormone-dependent and hormone-independent cell lines. Int J Cancer J Int Du Cancer 43(5):875–879
49. Uda S, Accossu S, Spolitu S et al (2012) A lipoprotein source of cholesteryl esters is essential for proliferation of CEM-CCRF lymphoblastic cell line. Tumour Biol: J Int Soc Oncodevelopmental Biol Med 33(2):443–453. doi:10.1007/s13277-011-0270-6
50. Danilo C, Frank PG (2012) Cholesterol and breast cancer development. Curr Opin Pharmacol 12(6):677–682. doi:10.1016/j.coph.2012.07.009
51. Navab M, Reddy ST, Van Lenten BJ et al (2009) The role of dysfunctional HDL in atherosclerosis. J Lipid Res 50(Suppl):S145–S149. doi:10.1194/jlr.R800036-JLR200
52. Eren E, Yilmaz N, Aydin O (2012) High density lipoprotein and it's dysfunction. Open Biochem J 6:78–93. doi:10.2174/1874091X01206010078
53. Leon CG, Locke JA, Adomat HH et al (2010) Alterations in cholesterol regulation contribute to the production of intratumoral androgens during progression to castration-resistant prostate cancer in a mouse xenograft model. Prostate 70(4):390–400. doi:10.1002/pros.21072
54. Shahzad MMK, Mangala LS, Han HD et al (2011) Targeted delivery of small interfering RNA using reconstituted high-density lipoprotein nanoparticles. Neoplasia 13(4):309–U142. doi:10.1593/Neo.101372
55. Llaverias G, Danilo C, Mercier I et al (2011) Role of cholesterol in the development and progression of breast cancer. Am J Pathol 178(1):402–412. doi:10.1016/j.ajpath.2010.11.005
56. Muntoni S, Atzori L, Mereu R et al (2009) Serum lipoproteins and cancer. Nutr Metab Cardiovasc Dis: NMCD 19(3):218–225. doi:10.1016/j.numecd.2008.06.002
57. Danilo C, Gutierrez-Pajares JL, Mainieri MA et al (2013) Scavenger receptor class B type I regulates cellular cholesterol metabolism and cell signaling associated with breast cancer development. Breast Cancer Res: BCR 15(5):R87. doi:10.1186/bcr3483

58. Farokhzad OC, Langer R (2009) Impact of nanotechnology on drug delivery. ACS Nano 3 (1):16–20. doi:10.1021/Nn900002m
59. Sabnis N, Lacko AG (2012) Drug delivery via lipoprotein-based carriers: answering the challenges in systemic therapeutics. Ther Deliv 3(5):599–608
60. Huynh E, Zheng G (2013) Engineering multifunctional nanoparticles: all-in-one versus one-for-all. Wiley Interdisc Rev Nanomed Nanobiotechnol 5(3):250–265. doi:10.1002/wnan.1217
61. Ng KK, Lovell JF, Zheng G (2011) Lipoprotein-inspired nanoparticles for cancer theranostics. Acc Chem Res 44(10):1105–1113. doi:10.1021/Ar200017e
62. Sabnis N, Nair M, Israel M et al (2012) Enhanced solubility and functionality of valrubicin (AD-32) against cancer cells upon encapsulation into biocompatible nanoparticles. Int J Nanomed 7:975–983. doi:10.2147/Ijn.S28029
63. Lacko AG, Nair M, Paranjape S et al (2002) High density lipoprotein complexes as delivery vehicles for anticancer drugs. Anticancer Res 22(4):2045–2049
64. Liadaki KN, Liu T, Xu S et al (2000) Binding of high density lipoprotein (HDL) and discoidal reconstituted HDL to the HDL receptor scavenger receptor class B type I. Effect of lipid association and APOA-I mutations on receptor binding. J Biol Chem 275(28):21262–21271. doi:10.1074/jbc.M002310200
65. Rader DJ (2006) Molecular regulation of HDL metabolism and function: implications for novel therapies. J Clin Invest 116(12):3090–3100. doi:10.1172/Jci30163
66. Ghosh M, Singh AT, Xu W et al (2011) Curcumin nanodisks: formulation and characterization. Nanomedicine 7(2):162–167. doi:10.1016/j.nano.2010.08.002
67. Singh AT, Evens AM, Anderson RJ et al (2010) All trans retinoic acid nanodisks enhance retinoic acid receptor mediated apoptosis and cell cycle arrest in mantle cell lymphoma. Br J Haematol 150(2):158–169. doi:10.1111/j.1365-2141.2010.08209.x
68. Singh AT, Ghosh M, Forte TM et al (2011) Curcumin nanodisk-induced apoptosis in mantle cell lymphoma. Leuk Lymphoma 52(8):1537–1543. doi:10.3109/10428194.2011.584253
69. Jia JT, Xiao Y, Liu JP et al (2012) Preparation, characterizations, and in vitro metabolic processes of paclitaxel-loaded discoidal recombinant high-density lipoproteins. J Pharm Sci-US 101(8):2900–2908. doi:10.1002/Jps.23210
70. Wang J, Jia JT, Liu JP et al (2013) Tumor targeting effects of a novel modified paclitaxel-loaded discoidal mimic high density lipoproteins. Drug Deliv 20(8):356–363. doi:10.3109/10717544.2013.834418
71. Zhang M, Jia J, Liu J et al (2013) A novel modified paclitaxel-loaded discoidal recombinant high-density lipoproteins: preparation, characterizations and in vivo evaluation. Asian J Pharm Sci 8 (1):11–18. doi:http://dx.doi.org/10.1016/j.ajps.2013.07.002
72. Sabnis N, Pratap S, Akopova I et al (2013) Pre-clinical evaluation of rHDL encapsulated retinoids for the treatment of neuroblastoma. Front Pediatr 1:6. doi:10.3389/fped.2013.00006
73. Wang BL, Yuan Y, Han L et al (2014) Recombinant lipoproteins reinforce cytotoxicity of doxorubicin to hepatocellular carcinoma. J Drug Target 22(1):76–85. doi:10.3109/1061186x.2013.839687
74. Yuan Y, Wang WN, Wang BL et al (2013) Delivery of hydrophilic drug doxorubicin hydrochloride-targeted liver using apoAI as carrier. J Drug Target 21(4):367–374. doi:10.3109/1061186x.2012.757769
75. Lin Q, Chen J, Ng KK et al (2013) Imaging the cytosolic drug delivery mechanism of HDL-like nanoparticles. Pharm Res. doi:10.1007/s11095-013-1046-z
76. Reddy ST, Navab M, Anantharamaiah GM et al (2014) Searching for a successful HDL-based treatment strategy. Biochim Biophys Acta 1841 1:162–167
77. Knecht TP, Pittman RC (1989) A plasma-membrane pool of cholesteryl esters that may mediate the selective uptake of cholesteryl esters from high-density lipoproteins. Biochim Biophys Acta 1002(3):365–375. doi:10.1016/0005-2760(89)90351-2

78. Pittman RC, Knecht TP, Rosenbaum MS et al (1987) A nonendocytotic mechanism for the selective uptake of high-density lipoprotein-associated cholesterol esters. J Biol Chem 262 (6):2443–2450
79. Mathew S, Murakami T, Nakatsuji H et al (2013) Exclusive photothermal heat generation by a gadolinium bis(naphthalocyanine) complex and inclusion into modified high-density lipoprotein nanocarriers for therapeutic applications. ACS Nano 7(10):8908–8916. doi:10.1021/Nn403384k
80. Ng KK, Lovell JF, Vedadi A et al (2013) Self-assembled porphyrin nanodiscs with structure-dependent activation for phototherapy and photodiagnostic applications. ACS Nano 7 (4):3484–3490. doi:10.1021/Nn400418y
81. Corbin IR (2013) Ligand-coupled lipoprotein for ovarian cancer-specific drug delivery. Methods Mol Biol 1049:467–480. doi:10.1007/978-1-62703-547-7_35
82. Kalli KR, Oberg AL, Keeney GL et al (2008) Folate receptor alpha as a tumor target in epithelial ovarian cancer. Gynecol Oncol 108(3):619–626. doi:10.1016/j.ygyno.2007.11.020
83. Corbin IR, Ng KK, Ding L et al (2013) Near-infrared fluorescent imaging of metastatic ovarian cancer using folate receptor-targeted high-density lipoprotein nanocarriers. Nanomedicine 8(6):875–890. doi:10.2217/nnm.12.137
84. Bijsterbosch MK, Rump ET, De Vrueh RLA et al (2000) Modulation of plasma protein binding and in vivo liver cell uptake of phosphorothioate oligodeoxynucleotides by cholesterol conjugation. Nucleic Acids Res 28(14):2717–2725. doi:10.1093/Nar/28.14.2717
85. Soutschek J, Akinc A, Bramlage B et al (2004) Therapeutic silencing of an endogenous gene by systemic administration of modified siRNAs. Nature 432(7014):173–178. doi:10.1038/Nature03121
86. Lorenz C, Hadwiger P, John M et al (2004) Steroid and lipid conjugates of siRNAs to enhance cellular uptake and gene silencing in liver cells. Bioorg Med Chem Lett 14 (19):4975–4977. doi:10.1016/J.Bmcl.07.018
87. Wolfrum C, Shi S, Jayaprakash KN et al (2007) Mechanisms and optimization of in vivo delivery of lipophilic siRNAs. Nat Biotechnol 25(10):1149–1157. doi:10.1038/nbt1339
88. Thaxton CS, Daniel WL, Giljohann DA et al (2009) Templated spherical high density lipoprotein nanoparticles. J Am Chem Soc 131(4):1384–1385. doi:10.1021/ja808856z
89. Luthi AJ, Zhang H, Kim D et al (2012) Tailoring of biomimetic high-density lipoprotein nanostructures changes cholesterol binding and efflux. ACS Nano 6(1):276–285. doi:10.1021/nn2035457
90. Yang S, Damiano MG, Zhang H et al (2013) Biomimetic, synthetic HDL nanostructures for lymphoma. Proc Nat Acad Sci USA 110(7):2511–2516. doi:10.1073/pnas.1213657110
91. McMahon KM, Mutharasan RK, Tripathy S et al (2011) Biomimetic high density lipoprotein nanoparticles for nucleic acid delivery. Nano Lett 11(3):1208–1214. doi:10.1021/nl1041947
92. Tripathy S, Vinokour E, McMahon KM et al (2014) High-density lipoprotein nanoparticles deliver RNAi to endothelial cells to inhibit angiogenesis. Part Part Syst Charact. doi:10.1002/ppsc.201400036
93. Yang M, Jin HL, Chen JA et al (2011) Efficient cytosolic delivery of siRNA using HDL-mimicking nanoparticles. Small 7(5):568–573. doi:10.1002/Smll.201001589
94. Shahzad MM, Mangala LS, Han HD et al (2011) Targeted delivery of small interfering RNA using reconstituted high-density lipoprotein nanoparticles. Neoplasia 13(4):309–319
95. Ding Y, Wang W, Feng MQ et al (2012) A biomimetic nanovector-mediated targeted cholesterol-conjugated siRNA delivery for tumor gene therapy. Biomaterials 33(34):8893–8905. doi:10.1016/J.Biomaterials.08.057
96. Dong Y, Love KT, Dorkin JR et al (2014) Lipopeptide nanoparticles for potent and selective siRNA delivery in rodents and nonhuman primates. Proc Nat Acad Sci USA 111(11):3955–3960. doi:10.1073/pnas.1322937111
97. Nakayama T, Butler JS, Sehgal A et al (2012) Harnessing a physiologic mechanism for siRNA delivery with mimetic lipoprotein particles. Mole Ther: J Am Soc Gene Ther 20 (8):1582–1589. doi:10.1038/mt.2012.33

98. Fischer NO, Weilhammer DR, Dunkle A et al (2014) Evaluation of nanolipoprotein particles (NLPs) as an in vivo delivery platform. Plos ONE 9(3). doi:ARTN e93342 DOI 10.1371/journal.pone.0093342
99. Zheng Y, Liu YY, Jin HL et al (2013) Scavenger receptor B1 is a potential biomarker of human nasopharyngeal carcinoma and its growth is inhibited by HDL-mimetic nanoparticles. Theranostics 3(7):477–486. doi:10.7150/Thno.6617
100. Zhang ZH, Chen J, Ding LL et al (2010) HDL-mimicking peptide-lipid nanoparticles with improved tumor targeting. Small 6(3):430–437. doi:10.1002/Smll.200901515
101. Su F, Grijalva V, Navab K et al (2012) HDL Mimetics inhibit tumor development in both induced and spontaneous mouse models of colon cancer. Mol Cancer Ther 11(6):1311–1319. doi:10.1158/1535-7163.Mct-11-0905
102. Rui MJ, Tang HL, Li Y et al (2013) Recombinant high density lipoprotein nanoparticles for target-specific delivery of siRNA. Pharm Res-Dordr 30(5):1203–1214. doi:10.1007/s11095-012-0957-4
103. Kim Y, Fay F, Cormode DP et al (2013) Single step reconstitution of multifunctional high-density lipoprotein-derived nanomaterials using microfluidics. ACS Nano 7(11):9975–9983. doi:10.1021/nn4039063
104. Lin QY, Chen J, Jin HL et al (2012) Efficient systemic delivery of siRNA by using high-density lipoprotein-mimicking peptide lipid nanoparticles. Nanomedicine 7(12):1813–1825. doi:10.2217/Nnm.12.73

Radiosensitization and Nanoparticles

Tatjana Paunesku, Stanley Gutiontov, Koshonna Brown and Gayle E. Woloschak

Abstract

Nanomaterials have been shown to have physical and chemical properties that have opened new avenues for cancer diagnosis and therapy. Nanoconstructs that enhance existing treatments for cancer, such as radiation therapy, are being explored in several different ways. Two general paths toward nanomaterial-enabled radiosensitization have been explored: (1) improving the effectiveness of ionizing radiation and (2) modulating cellular pathways leading to a disturbance of cellular homeostasis, thus rendering the cells more susceptible to radiation-induced damage. A variety of different agents that work via one of these two approaches have been explored, many of which modulate direct and indirect DNA damage (gold), radiosensitivity through hyperthermia (Fe), and different cellular pathways. There have been many in vitro successes with the use of nanomaterials for radiosensitization, but in vivo testing has been less efficacious, predominantly because of difficulty in targeting the nanoparticles. As improved methods for tumor targeting become available, it is anticipated that nanomaterials can become clinically useful radiosensitizers for radiation therapy.

Keywords

Ionizing radiation · Radiosensitization · Nanomaterials · Radiation oncology · Cancer therapy

T. Paunesku · S. Gutiontov · K. Brown · G.E. Woloschak (✉)
Department of Radiation Oncology, Tarry Building Room 4–713,
300 E Superior Street, Chicago, IL 60611, USA
e-mail: g-woloschak@northwestern.edu

C.A. Mirkin et al. (eds.), *Nanotechnology-Based Precision Tools for the Detection and Treatment of Cancer*, Cancer Treatment and Research 166,
DOI 10.1007/978-3-319-16555-4_7

Contents

1 Introduction

Physical and chemical peculiarities of nanomaterials and their distinct differences from biological molecules upon internalization by cells or tissues open new avenues for anticancer treatments in all fields of oncology including radiation oncology. Considering that radiation is used for treatment of over 50 % of all cancers, agents that could modulate the effects of radiotherapy generate much interest in the field. Probably the most drastic differences between biological molecules and nanomaterials can be observed with respect to their "response" to ionizing radiation. While chemical bonds of most organic polymers suffer radiation damage, most nanomaterials accept or release electrons and photons without significant structural changes following radiation exposure. Moreover, depending on their surface modifications, the same nanostructures sometimes have opposite effects on their immediate (bio) chemical surroundings. Thus, for example, fullerenes (C_{60}) increase reactive oxygen species (ROS) accumulation in cells in conjunction with ionizing radiation (e.g. [44]) leading to increased cell death, while hydroxyl group-modified fullerenes ($C_{60}OH_{24}$) have been found to be antioxidative and antiapoptotic (e.g. [55]).

In whole cells, ROS generated from the radiolysis of water is one of most pronounced, though not the only effect of ionizing radiation (e.g. [14]). ROS can cause damage to any one of the cellular components—nucleic acids, proteins, sugars and lipids, as well as direct energy deposition and ionization. Due to their short half-life in cells ROS traverse no more than 6 nm on average [53]; therefore ROS formation inside cell nuclei, close to the DNA is the most damaging for cells as it leads to the formation of DNA single-strand and double-strand breaks (DNA SSBs and DSBs). The extent of DNA damage in cells is most often measured by single cell gel electrophoresis (or "comet assay") which is extremely sensitive or, for "bulk" cells, with pulsed field electrophoresis, a technique much less sensitive to

DNA damage which requires high doses of radiation for detection [46]. Repair of DNA DSBs, completeness and accuracy of the repair make up the critical decision moment for irradiated cells. Misrepaired or unrepaired DNA damage can lead to cell death via apoptosis, autophagy, necrosis, or mitotic catastrophe. While cell death is often evaluated with clonogenic survival assays, the numbers and persistence of DNA DSB foci—complexes of proteins making up the DNA repair machinery, detected by immunocytochemistry—are used as a measure of the cellular ability to repair damage induced by irradiation [54]. The presence of nanomaterials in cells, in different subcellular compartments, can perturb either (or both) the physicochemical outcomes of radiation or the cellular capacity to repair radiation-induced damage. While some nanoconstructs may protect cells from the effects of radiation or have no "interactions" with radiation, nanoconstructs that increase cytotoxic effects of radiation —the radiosensitizing nanoconstructs, have opened a new area of study in the field of radiation oncology.

Two general paths for nanomaterial-enabled radiosensitization include: (a) the improvement of the effectiveness of ionizing radiation (e.g. by increasing energy deposition or by fortifying production of ROS) (Fig. 1) and (b) the modulation of cellular pathways and the disturbance of cellular homeostasis to render cells more susceptible to radiation (Fig. 2).

It should also be emphasized that for optimal nanoparticle-mediated radiosensitization, targeting of nanomaterials to specific cell types or subcellular organelles plays a significant role. While selection of best targeting moiety is in and of itself a complicated task, it should be remembered that surface modifications on nanomaterials are not static—nanomaterials accumulate and exchange components of tightly and loosely bound surface protein corona as they progress through cells and their environment (e.g. [40]). According to the "therapeutic ratio" concept, modulating radiation effects on all the cells in the organism equally would provide no advantage over radiation alone if radiation could not be targeted. Therefore, precise in vivo tissue-specific and subcellular delivery of nanomaterials plays a big role in the development of radiosensitizing nanoparticle constructs. However, while (some type) of targeting is absolutely essential for other anticancer nanoparticle formulations (for reviews see e.g. [47]), fortunately that is not so for radiosensitizing nanoconstructs. Ionizing radiation delivery has become so sophisticated that different treatment modalities enable dose delivery closely matching tumor topology. In such a scenario radiosensitizing nanoconstructs would further potentiate such dose differences. In any event, several other chapters of this book discuss nanoparticle targeting more directly, hence this topic will not be specifically addressed here. Rather, this chapter will focus on molecular and biochemical events that occur when nanoparticles are placed in the path of ionizing radiation quality photons. In the interest of brevity, we will not discuss either the basics of radiation physics and radiation biology or the molecular mechanisms relevant for cellular responses to radiation. Much more on these topics can be found in several radiation biology textbooks, e.g. [14, 25]. Finally, the use of nanoparticles for delivery of radionuclides (e.g. [45]) or the delivery of "small molecule" radiosensitizers (e.g. cisplatin [71], paclitaxel [61, 64], docetaxel [63], or curcumin [67]) will also not be

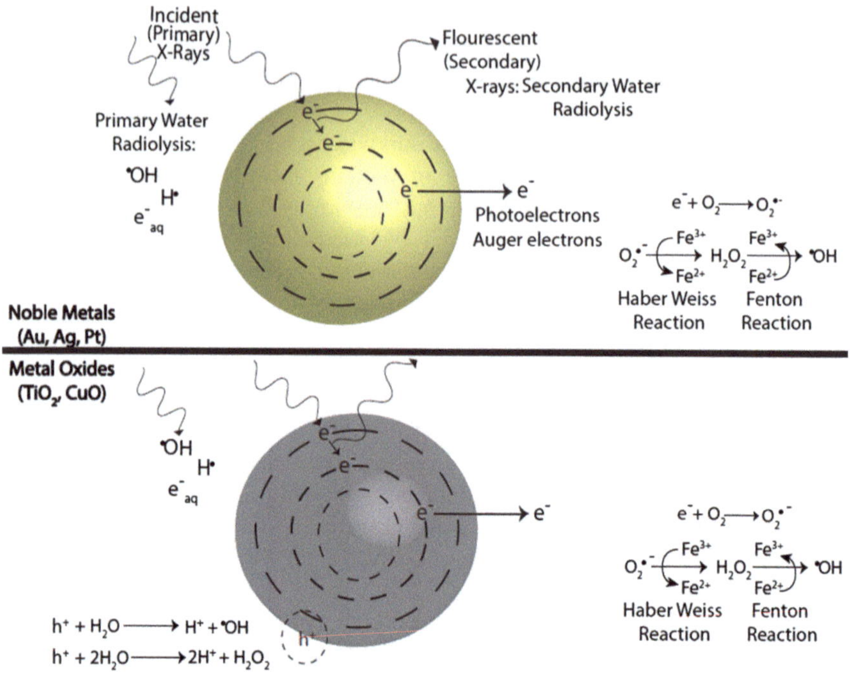

Fig. 1 Schematic drawing of radiocatalysis by nanoparticles. Exposure of nanoparticles to ionizing radiation (e.g. X-rays) causes Compton and Photoelectric effects—photoelectrons and Auger electrons (e^-) are ejected from the nanoparticles as well as photons of different energies, including secondary fluorescent X-rays. In addition, in semiconductor nanoparticles release of electrons leads to formation of reactive electropositive holes (h^+) on nanoparticle surface. Free electrons and electropositive holes on nanoparticle surface can damage cellular components directly or lead to production of reactive oxygen species (superoxide $O_2^{\cdot -}$, hydroxyl radical $^{\cdot}OH$, hydrogen peroxide H_2O_2) which are also formed by radiolysis of water (adapted from [2, 24, 39])

covered in this chapter, as this research belongs more appropriately into the topic of "cargo delivery" with nanoparticles, covered conceptually in other chapters of this book. We will touch only very briefly upon nanoconstruct-mediated nucleic acids delivery focusing on those cases where nucleic acids are especially relevant for regulating susceptibility toward ionizing radiation exposure.

Elemental and molecular compositions of nanoparticles discussed in this chapter will include gold, platinum, silver, cerium oxide nanoparticles, gadolinium-doped titanium dioxide and titanium dioxide, iron oxides, copper oxide, gadolinium-carrying nanoparticles, as well as a few polymeric nanoparticles made of elements with small *Z* values. In response to ionizing radiation elements with higher atomic numbers (greater *Z* value) produce energetic secondary X-rays, photoelectrons and Auger electrons, while lower *Z* elements release predominantly Auger electrons; the same is true for nanomaterials made of these materials [2]. Several high *Z* materials that have been tested as radiosensitizers, e.g. iodine [34] or germanium oxide [29],

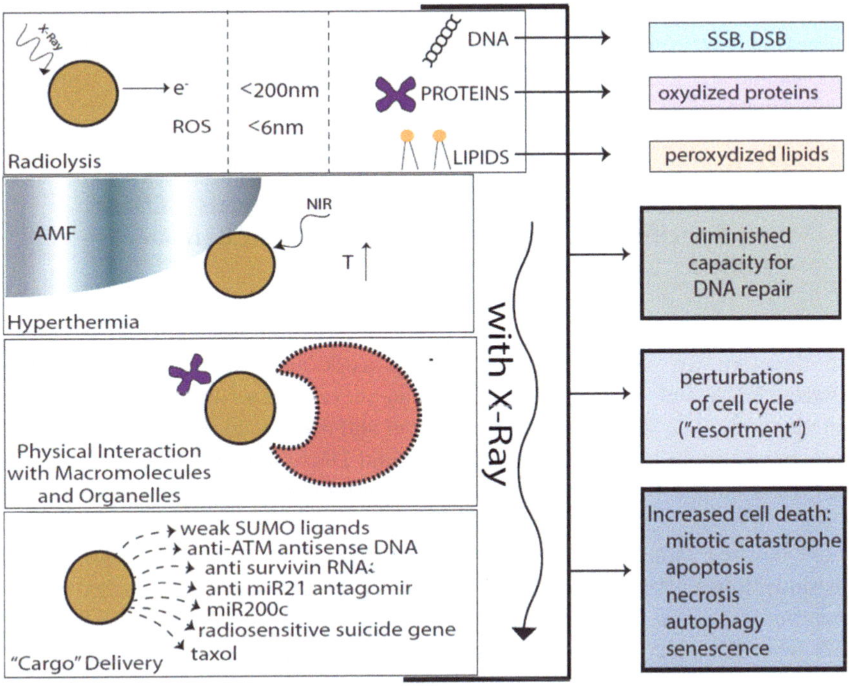

Fig. 2 Cellular effects of radiosensitizing nanomaterials. Several different "pathways" to radiosensitization by nanoparticles exist. In the first place, free electrons (released by nanoparticle ionization) are absorbed by surrounding molecules within 15–200 nm from particle surface (e.g. [2, 73]) and ROS (which on average traverse 6 nm distances in cytosol [53]) produced by nanoparticles damage cellular components in addition to incident ionizing radiation and products of water radiolysis. Free electrons and ROS oxidize proteins, peroxidize lipid membranes (cell and organelle membranes both), and produce SSBs and DSBs in DNA [14]. Secondly, nanomaterials may be used to develop hyperthermia under the influence of alternating magnetic field or near infrared light (e.g. [1, 33]). Thirdly, physical interactions between nanomaterials and cellular components, biomolecules, molecular assemblies, and organelles modulate behavior of these cellular elements in different ways leading to such outcomes as perturbations in cell cycle or DNA repair (e.g. [2, 69]). Fourthly, nanomaterials can be made of components (or carry a cargo) specifically designed to make the recipient cells more sensitive to radiation, such as, e.g. delivery of anisense oligonucleotides for DNA repair regulating genes, miRs, or delivery of weak ligands for UPS machinery (e.g. [3, 27, 74])

were found to work as radiosensitizers equally well as single molecules or as nanoparticle formulations.

Finally, we will also speculate about additional, potentially radiosensitizing (but not yet tested with radiation), nanoparticles. One can predict, for example, that nanoparticles that perturb the cell cycle may serve as radiosensitizers. Chemotherapeutic drugs with such properties have been in use for chemoradiotherapy for many years. For example, paclitaxel treatment favors accumulation of cancer cells in G2 and M phases, the most radiation-sensitive phases of the cell cycle, and for that reason it synergistically interacts with ionizing radiation [57]. Incidentally, poly (D,L-lactide-coglycolide)(PLGA) nanoparticles containing taxols are currently being developed as radiosensitizers [18].

1.1 Irradiation in the Presence of Nanomaterials—Effects on Specific (Sub)Cellular Structures and Chemical Milieu

Genomic DNA is the most important target of radiation-induced cellular killing, and therefore we will begin by reviewing work that investigated interactions between DNA and nanoparticles in vitro in the course of exposure to ionizing radiation [73]. The formation of single-strand and double-strand (SSBs and DSBs) DNA breaks in an in vitro setup, using plasmid DNA and gold nanoparticles was used as a proxy for the physicochemical processes governing gold-mediated increase in DNA damage. The basic mechanisms underlying this DNA sensitization were evaluated with electron bombardment with different electron energies [73]. In this study, low electron energies of 1, 10, and 100 eV as well as high energy 60 keV electrons were used. In addition to "standard" DNA lesions such as the SSB and DSB, loss of supercoiled DNA was also used as an index of DNA damage. In this work, considerations for the use of appropriate controls led the authors to compare pure plasmid DNA with "salted DNA" (three Na^+ per base) and 1:1 (one particle per plasmid) complexes of DNA with gold nanoparticles (solid gold particles 5 ± 2 nm in size). The yields of all three types of DNA damage (SSB, DSB, and supercoil loss) were enhanced by a factor of 2 or more when plasmid–nanoparticle complexes were compared to the salted DNA sample. This increase was the greatest when 10 eV electrons were used to irradiate DNA–nanoparticle complexes. This finding was explained by the fact that the damage to DNA with 10 eV electrons comes from transient anions forming in the DNA molecule (i.e., the bases, sugar, and phosphate groups) and the gold nanoparticles as well. These anions are resolved either by dissociative electron attachment (DEA), or by deposition of energy into local chemical bonds, resulting in bond breakage. Because DNA damage increased when electrons of 1 and 10 eV are used with nanoparticle–DNA complexes, the authors contended that the gold nanoparticles greatly increase the DEA cross section. Overall, the mechanisms responsible for plasmid DNA damage caused by gold nanoparticles and radiation included: [1] the production of short-range secondary electrons by the gold that has absorbed incoming electrons, with a high probability of interacting with the DNA if the gold particles are in complex with it, and [2] the efficient absorption of low energy electrons by the gold nanoparticles. Moreover, Sanche and collaborators consider these mechanisms critical for gold nanoparticle-dependent DNA damage caused by exposure to X-rays as well, especially for photon energies of 100–110 keV (produced at 100–250 kVp) [73]. It is worth noting, however, that any increase in the distance between the gold nanoparticle surface and plasmid DNA led to a decrease in DNA damage as well. For example, gold nanoparticles coated either with thiolated undecane, dithiolated diethylenetriaminepentaacetic or gadolinium chelating agents reduce (for the same nanoparticle size, particle:plasmid ratio and irradiation conditions) DNA damage enhancement from 2.3 to 1.6 [66].

Using a nucleic acid free system, i.e., gold and silver nanoparticles embedded in an alanine matrix, one recent study used electron spin resonance spectroscopy (ESR) to evaluate dose enhancement in the presence of gold and silver nanoparticles [11]. A conclusion from this work was that, irrespective of X-ray energy, larger and distantly spaced nanoparticles do not cause dose enhancement compared to smaller and closely clustered nanoparticles. Both silver and gold nanoparticles were used in numerous in vitro and in vivo assays (see Sect. 1.2).

In addition to low linear energy transfer (low LET) irradiation (X-rays and gamma rays and free electrons), heavy ion irradiation can be combined with the nanoparticles use as well. Two recent studies by Porcel and others [49, 50] with plasmid DNA and 3 nm platinum nanoparticles coated with polyacrylic acid compared effects of gamma rays with particulate, high linear energy transfer (high LET) irradiation. In this case SSBs and DSBs in plasmid DNA were caused by exposure to gamma rays (with LET = 0.2 keV/μm) and several heavy ions (He^{2+} LET = 2.3 keV/μm and C^{6+} LET = 13 keV/μm and LET = 110 keV/μm) alone or in combination with platinum nanoparticles. In all cases nanoparticle addition increased the effects of radiation damage. Use of high *Z* element (HZE) nanoparticles may be an interesting way to increase the efficiency of high LET radiation treatments.

A comprehensive cell-free study on gold nanoparticles in aqueous solution (without addition of nucleic acids) was also done recently by Misawa and Takahashi [39]. Diagnostic quality X-rays (with a dose rate of 1 Gy/min) in the presence of gold nanoparticles (5–250 nm in diameter) produced increased amounts of reactive oxygen species (ROS), compared to irradiation of aqueous media alone; this increase reached 1.46 for $^{\bullet}OH$ and 7.68 for O_2-. This finding was explained in part by electron emission of both photo- and Auger electrons and emission of photons of secondary fluorescent X-rays. The authors found that fluorescent X-ray production (which may lead to secondary excitations) and ionization along the X-ray path depended primarily on the volume of the gold particles. More dramatically, however, ROS generation in the presence of gold nanoparticles was inversely proportional to particle diameter indicating that the dominant parameter for ROS production is the surface area of nanoparticles, and not their weight percentage, possibly due to catalytic effects at the particle–water interface. The ROS species evaluated were $^{\bullet}OH$, O_2- and $^1O_2$, all of them biologically relevant, with diffusion lengths reaching no more than a few hundreds of nanometers in aqueous media. As mentioned earlier, in a more complex environment of cytosol or nucleosol most ROS progress no more than 6 nm on average [53]. In general, $^{\bullet}OH$ species interact with lipids, polysaccharides, proteins, and nucleic acids rapidly and locally. O_2- radicals react more slowly with lipids, however, in the presence of iron and hydrogen peroxide these radicals can lead to production of $^{\bullet}OH$ radicals (these chemical processes are described by Haber–Weiss and Fenton reactions). Finally, $^1O_2$ radicals react with unsaturated fatty acids, beta-carotene, amino acids, and methionine. Each one of these interactions damages biomolecules and causes stress than can lead to cell death or genomic instability.

1.2 Irradiation in the Presence of Nanomaterials—Effects on Cells in Vitro and/or Tumors in Vivo

While it is known that irradiation of gold nanoparticles in complexes with DNA leads to DNA damage, the effects of gold nanoparticles on whole cells are far more complex and have rarely been fully investigated in a single research paper. Most frequently, in cellular in vitro studies changes in cell viability are the primary endpoint of investigation, and the exact mode of action of nanoparticles has received relatively little attention. A few examples of such work (see Sect. 2) suggest that biochemical/biological effects of gold nanoparticles play perhaps an even greater role in radiosensitization than the production of secondary electrons.

In an effort to investigate the exact radiation parameters for increased radiation-dependent cell killing in vitro in the presence of gold nanoparticles, Rahman and others [51] used 1.9 nm gold nanoparticles (mixed at 0.197 mg/ml) with bovine aortic cells and exposed such cells to monochromatic, synchrotron-generated X-rays of 30, 40, 50, 60, 70, 81, and 100 keV. The greatest dose enhancement in cell killing was found for the combination of gold nanoparticles with 40 keV X-rays, followed by 50 and 60 keV dose enhancement values. Gold nanoparticles were also tested in combination with proton irradiation [48], and increased cell killing was found in samples of cells that had internalized the nanoparticles. (It should be noted that prolific ROS production outside of cells but close to cellular membrane can lead to cell death as well.)

Increased efficiency of radiation was also noted when radioactive seed therapy (I-125 brachytherapy seeds) was combined with gold nanoparticles [43]. Dose ranges varied between 2.1 and 4.5 cGy/h with an even dose distribution in adherent layers of HeLa cells. These cells were treated with 50 nm gold nanoparticles, and the number of gamma H2AX foci (signifying DNA DSBs) following irradiation in nanoparticle treated cells was 1.7–2.3 fold greater compared to controls.

Nanoconstructs prepared by simple one-pot synthesis as silver nanoparticles coated with egg white proteins were used to radiosenistize breast adenocarcinoma cells [30]; increasing concentration of nanoparticles with a fixed (4 Gy) irradiation led to increased cell killing even though no cytotoxicity was recorded in the absence of radiation, even at the maximal concentration of conjugates (12 mg/L).

Silver was also used to make the shell on the surface of carbon nanodots [24]. These nanoconstructs (5–100 nm in size including polyethylene glycol (PEG) coating) were tested in vitro both for photodynamic therapy and as radiosensitizers. Reduction in cell survival of Du145 prostate cancer cells with combination treatment, compared to radiation alone, was twofold with 15 Gy.

A growing body of literature shows promising data on nanoparticle-mediated radiosensitization in vivo. For example, syngeneic EMT-6 mammary tumors implanted subcutaneously into the hind limb of Balb/c mice and injected intravenously with 1.9 nm gold nanoparticles were irradiated 2 min later with 110 keV X-rays (produced at 250 KVp) by Hainfeld and others [12]. At gold concentration of 2.7 g/kg and 26 Gy dose complete tumor regression for 1 year was obtained for 86 % of mice compared to 20 % of mice given irradiation alone. In their more recent work, the same group of investigators used 11 nm gold nanoparticles administered intravenously at 4 g/kg to B6C3F1 mice bearing syngeneic orthotopic Tu-2449 brain tumors. Fifteen hours after nanoparticle administration, these mice were exposed to local 30 Gy (at 100 KVp) irradiation. A year after treatment 50 % of mice exposed to radiation in the presence of nanoparticles were still alive [13].

Among the nanomaterials that radiosensitize cells by potentiating the physicochemical effects of radiation is cerium oxide. Cerium oxide particles (5–8 nm) engage in different chemical reactions with different reactive oxygen species in a pH-dependent manner. At acidic pH, cerium oxide nanoparticles scavenge superoxide radical and produce hydrogen peroxide. At neutral pH values, cerium oxide nanoparticles scavenge H_2O_2 in a chemical reaction resembling the decomposition of hydrogen peroxide to water and oxygen mediated by catalase [62]. This pH-dependent behavior of cerium oxide leads to radioprotection of healthy tissues and cells living under "neutral pH" conditions, and confers radiosensitization of cancer cells that are characterized by an acidic intracellular milieu. This was shown not only with clonogenic assays in vitro, but also in vivo, using mice carrying orthotopic pancreatic tumor (human cell line L2.6pl) xenografts. While nanoparticles alone delivered intraperitonealy (IP) caused tumor volume reduction and apoptotic cell death as determined by TUNEL staining, administration of cerium oxide nanoparticles prior to ionizing radiation lead to increased tumor regression and enhanced apoptotic cell death [62]. As the production of ROS in acidic subcellular environment could lead to increased toxicity, one of the potentially interesting approaches to increase the usability of cerium oxide nanoparticles is to target them to acidic lysosomes by covering them with a negative surface charge [8].

Increased production of ROS and the subsequent increase of cell death monitored by in vitro assays were found both with gold nanoparticles and nanoparticles made of other materials. For example, lanthanide-doped titanium oxide nanoparticles increased radiation-induced cell death of several human cell lines [59]. Levels of radiosensitization differed from cell type to cell type (embryonal rhabdomyosarcoma cell lines RD, alveolar rhabdomyosarcoma RH30 and breast cancer cell line MCF7), suggesting that the cellular capacity for nanoparticle uptake as well the ability to cope with ROS burdens play very significant roles in any radiosensitization by means of nanomaterials. The same nanoparticle type (silica coated and rare earth doped TiO_2 crystals 3–10 nm in size) was injected into tumors in vivo. Tumor-bearing mice were irradiated with 10 fractions of 2.5 Gy followed by 3 fractions of 2 Gy over the same period [60]. Hind limb xenograft tumors grown in SCID mice were human lung adenocarcinoma cell line A459; in the presence of nanoparticles (three injections of 50 ml of 1 mg/ml nanoparticles each) an additional twofold reduction of tumor volume upon radiation was found.

An interesting research direction with regard to rare earth elements and nanoparticle-based radiosensitization is the use of gadolinium-based nanoparticles [5, 36, 37]. These 2.9 nm particles consist of a polysiloxane core and gadolinium–DTPA chelates (with cyanine 5.5 added to allow optical imaging) and have been tested first in orthotopically implanted 9L gliosarcoma (9LGS) brain tumors in Fisher 344 rats. Following intravenous delivery these nanoconstructs had good renal excretion and could be used both for magnetic resonance imaging and radiosensitization [5]. The importance of timing between nanoparticle delivery and irradiation (delivered with a microbeam) was very noticeable in this study; a 5 min delay increased radiation damage to normal tissue and decreased overall survival of rats compared to radiation alone, while a 20 min delay lead to improved survival of animals compared to radiation alone. A recent study using the same type of nanoparticles in xenograft head and neck cancer models (subcutaneous implantation of human head and neck cancer cell line SQ20B into flank of nude mice) showed that intratumor injection of gadolinium nanoparticles followed immediately with 10 Gy dose of ionizing radiation had a synergistic effect on tumor size reduction [37]. In vitro experiments with the same cell line as well as FaDu and Cal33 head and neck carcinoma cell lines, demonstrated that the use of gadolinium nanoparticles was associated with an increase in nonreparable DNA DSBs and the shortening of G2/M arrest, leading to increased genomic instability, appearance of polyploid and then hypoploid cell populations and increased apoptosis.

It should also be noted that radiation itself can alter delivery of nanomaterials. For example, Joh et al. [19] found that exposure to ionizing radiation increases the permeability of the neovasculature for nanoparticles. When nude mice orthotopically inocculated with the human glioma cell line U251 received 20 Gy (focused through a 1.5 cm collimator of the small animal irradiator) prior to injection of 22 nm PEG-coated gold nanoparticles, increased extravasation and in-tumor deposition of nanoparticles occurred, suggesting that radiation can improve the passive tumor tissue targeting of nanoparticles [19]. Radiation induced a three-fold increase of enhanced permeability retention in irradiated versus nonirradiated tumors.

2 Modulation of Radiosensitivity by Hyperthermia

Modulation of temperature conditions in the body (especially increased tumor temperature) at the time of radiation treatment has been noted to be a potential radiosensitizer, as reviewed by Wust and others [65]. Though hyperthermia was used in the treatment of various diseases including cancer since ancient times, one of the first carefully recorded local applications of hyperthermia alone for cancer treatment was performed in 1898 by Swedish gynecologist Westermark, who treated cervical cancer by running hot water through an intracavitary spiral tube and noted an excellent clinical response in seven patients [41].

On a molecular and cellular level, temperatures of 40–45 °C trigger various cellular responses, including protein denaturation [22], alterations in the cytoskeleton and membrane [15] and cell death and permanent arrest [22]. Temperatures equal to or higher than the transition temperature of 42.5 °C are considered to be optimal for generating protein damage to cells, but these temperatures are difficult to achieve in vivo [14]. In combination with ionizing radiation, hyperthermia is particularly potent predominantly because the two have different targets, with radiation damaging DNA and hyperthermia damaging proteins. For example, temperatures near the transition temperature (42 °C) increase tumor vascular perfusion [15] leading to an increase in tumor oxygenation, which increases the efficiency of radiation as formulated by oxygen enhancement ratio (OER). Protein denaturation and aggregation in the nucleus caused by hyperthermia perturb DNA synthesis and repair [23]. Regardless of the precise molecular mechanisms, chemo- and radiosensitizing effects of hyperthermia are significant, and hyperthermia treatment is considered to be among the most potent radiosensitizers discovered to date. Furthermore, thermal radiosensitization appears to be most pronounced in cells that are hypoxic and those that are in S phase, i.e., in those cells that tend to be the most resistant to radiotherapy alone [15].

Magnetic nanoparticles in combination with magnetic field exposure can be used to induce a site-specific hyperthermia [70]. After localizing in a tumor, magnetic nanoparticles can be activated by an alternating magnetic field (AMF), causing a local increase in temperature. Such nanoconstructs can heat their immediate submicron environment through eddy currents (circular electric currents induced within conductors by a changing magnetic field), magnetic hysteresis, and Neel or Brownian relaxation (complex physical phenomena that depend on the size of the particles). There is also evidence of additive magnetic effects via electromagnetic coupling through nanoconstruct aggregation either extracellularly or in endosomes [21, 56, 70]. This is most likely caused by the increase in "aggregated size" of magnetic material. Deliberately designed magnetic nanoparticle assemblies, such as, e.g. magnetic nanoparticle–adenovirus assemblies, have an increased capacity for thermal ablation [70].

Three recent studies illustrate the promise for magnetic nanoparticles-dependent hyperthermia-induced radiosensitization. The first study involved nanoparticles with a hematite core and hydrodynamic diameter of 117 nm injected into syngeneic mouse breast cancer tumors implanted into hind limb of C3H/He mice. Tumors were first subjected to AMF at 169 kHz or microwave exposure and then to irradiation—15 Gy of 6 MeV electrons within 30 min of hyperthermia. While tumor volume tripling time after the best single treatment was 18.7–25 days, combination therapy showed a tumor tripling time of 42.6 days [9]. A more recent study by Lin and others used $Mn_{0.5}Zn_{0.5}Fe_2O_4$ nanoparticles coated with poly(ethyleneimine) to deliver gene therapy [28]. The DNA construct was a combination of a promoter region of radiation responsive early growth response protein 1 (Egr-1) and the "suicide gene complex" HSV-TK/GCV. Expression of herpes simplex virus thymidine kinase (HSV-TK) converts the prodrug ganciclovir (GCV) into GCV triphosphate, a DNA synthesis and cell cycle inhibitor. In a hepatoma xenograft model (HepG2 tumors in BALB/c nude mice) use of these nanoconstructs with induction of hyperthermia and combined with irradiation lead to tumor volume decrease greater than 90 %, two times better than radiotherapy alone [28].

The final study to be mentioned here is a phase II clinical trial using magnetic nanoparticles in 66 human patients, 59 with recurrent glioblastoma multiforme. Nanoparticles consisted of 12 nm magnetite with aminosilane coating, at iron concentration of 112 mg/ml, injected directly into tumors. Six one-hour long hyperthermia sessions (induced by AMF) were done semiweekly, with median tumor temperature of 52 °C. These sessions were immediately preceded or followed with stereotactic beam radiotherapy (SBRT) with a median dose 30 Gy, delivered as five 2 Gy fractions each week. The primary study endpoint was overall survival (OS) after recurrence—it reached 13.4 months, which is a significant improvement compared to, e.g. 6.2 months recoded in a metastudy on temozolomide as a salvage treatment [33].

It should be noted that hyperthermia has also been combined with radiation treatments with the aid of nanoconstructs with optical properties [1, 4]. Gold nanoshells, for example, have tunable optical resonance which allows them to be thermally activated by near infrared (NIR) light and be appropriate for thermal ablative therapy [16]. Such nanoconstructs, prepared with a silica core coated with gold, at the size of 150 nm have plasmon resonance for NIR wavelengths of 808 nm. This leads to intense NIR absorption and conversion to thermal energy. In one of the first examples of in vivo study with these constructs, nanoconstruct delivery to tumors depended only on enhanced permeability retention (EPR) [4], and the tumor model was a xenograft (human colorectal cell line HCT 116 in nude mice). In this setup, local hyperthermia (lasting 20 min) was induced by NIR laser illumination 24 h after IV injection of nanoconstructs. Hyperthermia was followed 5 min later by irradiation with 10 Gy. In comparison with irradiation alone, combination treatment doubled the time of the tumor growth delay. Similarly, in in vivo orthotopic models of breast cancer (syngeneic mouse breast cancer cells p53/from GEM transgenic mice in immunocompetent mice and primary human breast tumor xenografts in nude mice) Krishnan and collaborators achieved reduction in tumor size [1]. In this case, irradiation with 6 Gy was immediately followed with local hyperthermia (20 min at 42 °C), using a local NIR illumination of the tumor after an intravenous administration of gold nanoshells. A twofold decrease in tumorigenicity (measured as the number of cells with purported stem-like properties and as a number of cells forming colonies in cell culture) of cells isolated from these tumors at 48 h after treatment was statistically significant in comparison with cells isolated from ionizing radiation only treated tumors.

3 Modulation of Cellular Biochemistry and Biochemical Homeostasis by Nanomaterials

3.1 DNA Repair Modulation

The effect of nanoparticles on cells is largely modulated by their acquisition of a protein corona and the subsequent intracellular modification of such corona through interactions with intracellular proteins. For example, an in vitro study found that several different sizes of polyethylene glycol and tetraethylene glycol-coated iron oxide and titanium dioxide nanoparticles interact with proteasome [6] and alter its activity. Similarly, corona of silica nanoparticles was found to contain proteasomal subunits as well [31]. Considering that ubiquitin proteasome system regulates DNA synthesis and repair as well as countless other processes relevant for radiation-induced stress [35], it is possible that the rate of DNA repair in irradiated cells may be altered in the presence of such nanoparticles.

DNA repair can also be deliberately modified by nanoparticles with surface modifications designed to do so. In their recent publication Li and others modified 2 nm gold nanoparticles by the addition of weak SUMO (Small Ubiquitin-like Modifier)-2/3 ligands, at a density of 100 per particle [27]. Through this approach, (because of high density of a weak ligand on the particle surface) nanoparticles became able to interact with multiple SUMO molecules in a poly-SUMO chain; inside cells these constructs could be found both in cytoplasm and nucleus. Because SUMOylation is a very important molecular modification pathway in the oxidative stress response, use of this type of nanoparticle, in comparison to same type of nanoparticles without SUMO ligand, led to an increase in radiation sensitivity in cancer cells in vitro as established by clonogenic assays. Comparisons between nanoparticles loaded with control scrambled molecules and nanoparticles loaded with SUMO interacting peptide demonstrated delayed DNA repair by comet assays in breast cancer MCF7 cells exposed to 4 Gy and incubated for 2 h.

Modulation of DNA repair with nanoparticles that release nucleic acid cargo is discussed in Sect. 4.

3.2 Nanoparticle Induced Cell Stress

Increased biocompatibility of PEG-coated gold nanoparticles compared to their "naked" counterparts (e.g. [42]) inspired the development of nanogel particles of about 100 nm in size, composed of a cross-linked poly(2-[*N,N*,-diethylamino]ethyl methacrylate) (PEAMA) core tethered with PEG and loaded with an average of 15 gold nanoparticles (8 nm in size) per each nanogel particle. The ability of these nanoconstructs to enhance radiation-induced tumor cell killing was investigated in several human and animal cell lines (human lung adenocarcinoma A549, Chinese hamster V79 and mouse squamous cell carcinoma SCCVII) [69]. Cells incubated for 14 h with nanogel particles (gold content in cell media varied between 20 and

50 μg/ml) were irradiated with doses between 2 and 15 Gy. Cell survival curves for each treatment were constructed; gold nanogel particles intensified cell killing by 1.2–1.5 fold across different irradiation doses. However, the initial DNA DSB damage in all cells depended only on the radiation dose, even though subsequently DNA DSB foci lasted longer in nanoparticle treated cells. Cellular uptake of nanogel particles as shown by TEM was associated with cytoplasmic vesicles. Starting with these findings, the authors investigated protein concentrations of critical DNA repair molecules (e.g. Rad51 and Ku70) and endoplasmic reticulum (ER) stress proteins (IRE1alpha, BiP, PERK, and calnexin) finding a reduction of the former and increase of the latter. In short, this study indicated that the 8 nm gold nanoparticles inside cells deregulate cellular homeostasis in ways that cause increased ER stress and decreased capacity for DNA damage repair [69]. Whether these changes are responsible for increased cell killing following ionizing radiation or not is not conclusive; nevertheless, the fact remains that the gold nanogels do act as radiosensitizers. It should be mentioned, however, that in another study (not including nanomaterials), Yamamori and others [68] found that ER stress is itself responsible for Rad51 degradation and general suppression of DNA DSB repair.

Cellular stress in many cases leads to cell death. Copper oxide, one of the most recent additions to the group of metal oxide nanoparticles, for example, was found to increase on its own both autophagy and apoptosis [26]. It is appealing, then, to speculate that CuO nanoparticles would have an additive or possibly even synergistic effect with ionizing radiation exposure. We can probably expect an increase in number of such studies in the future.

3.3 Cell Cycle

Using the prostate carcinoma cell line DU-145, Roa and others have documented changes in cell cycle distribution following treatment with ~10 nm gold nanoparticles coated with 6-deoxy-6-fluoro-1-thio-D-glucose [52]. As in most other studies, gold nanoparticles were found in cytoplasmic vesicles and it is difficult to decide which exact mechanism is responsible for cell cycle perturbation after nanoparticle treatment. Inhibition of glucose uptake by cytochalasin B and with it, inhibition of uptake of glucose-coated nanoparticles partially reversed the nanoparticle effects. Cell cycle perturbations were most noticeable in G2/M phase of cell cycle during which cells are the most radiation sensitive. It is, therefore, very possible that the increase in radiosensitivity noted in nanoparticle treated cells could be associated with cell cycle perturbations, especially when one considers that nanoparticles never came closer than 6 nm from the nuclear DNA.

Increased number of cells in G2/M phase of cell cycle was noted also in glioblastoma cells (SNB-19 and U87MG) in culture after treatment with titanium dioxide nanotubes [38]. Moreover, in this study it was noted that the DNA repair was reduced in glioblastoma cell lines after they engulfed TiO_2 nanorods (approximately 10 nm in diameter and up to 500 nm long). As nanorods remained in vesicles inside the cytoplasm in these cells, actual mechanism(s) responsible for these changes in cellular homeostasis are as of yet unclear. Increased numbers and persistence of DNA DSBs indicated by gamma H2AX foci (specifically in nanorods treated and irradiated cells) described in this study is similar to findings of others with other types of nanomaterials (e.g. [19, 20, 43, 69]). Many possible avenues for speculations about decrease of DNA repair in this work can be envisioned. Interaction of these nanoconstructs with the proteasome, for example, is quite possible considering that TiO_2 nanorods of 6 and 20 nm alter proteasomal activity [6], which, in turn, regulates DNA repair (e.g. [35]).

Perturbation of cell cycle by nanomaterials, however, does not always lead to increased numbers of cells in G2/M phase of the cell cycle. On the contrary, Mackey and El-Sayed used 30 and 15 nm gold nanoparticles (suitable for plasmon resonance and conjugated with nucleus targeting and cytoplasm-targeting peptides) in human oral squamous carcinoma cells (HSC-3) and found an accumulation of cells in the S phase with a depletion of cells in the G2/M phase [32]. These particles were not used for radiosenitization (where they could even have caused increased radioresistance); they have been tested in combination with 5-fluorouracil. Because this chemotherapeutic drug kills the cells in S phase most effectively (unlike ionizing radiation), drug and nanoparticle use had synergistic effects in this study.

4 Radiosensitization Through Modulation of Gene Expression by Nanomaterials

Polymeric nanoparticles were used as carriers for nucleic acids that can alter gene activity in cells; e.g. [74]. Ataxia-telangiectasia-mutated (ATM) protein is critical for DNA repair following ionizing radiation; inactivation of this gene increases radiation sensitivity, and a possible approach to achieve this goal is by delivering antisense oligonucleotides to cells using polymeric Poly(D,L-lactide-co-glycolide) (PLGA) nanoparticles. In mice, an initially spontaneously arising head and neck squamous cell carcinoma cell line SCCVII can be implanted into syngeneic C3H/HeJ mice and used as a model for human HNSCC. Such mice with tumors of about 200 mm^3 were treated with anti ATM oligonucleotides encapsulated in PLGA particles of approximately 100 nm in diameter. The nanoparticles (2.5 mg/kg) were administered 24 h before and immediately preceding irradiation; three pairs of intratumoral nanoparticle injections were followed with three 2 Gy fractions over a period of 15 days. Animals were sacrificed 15 days after the treatments; tumor volumes were significantly reduced in mice treated with PLGA particles with anti-ATM oligonucleotides compared to mice treated with "mutated" oligonucleotide nanoparticles.

Another protein playing a significant role in radiation resistance is survivin, and plasmids encoding an antisurvivin siRNA have been used for the development of radiosensitizing nanoconstructs. Monodisperse 180–220 nm nanoparticles composed of human serum albumin loaded with siRNA plasmids were decorated for targeting with an antibody against heat shock protein 70. In two glioblastoma cell lines U87MG and LN229 these nanoconstructs lead to radiosensitization of 1.64 and 1.25, respectively; accompanied by survivin suppression and induction of survivin regulated caspases 3 and 7 [7].

A more broad approach to gene expression modulation was employed by investigators who delivered microRNAs with pleiotropic effects into tumors using nanomaterials as a delivery system. For example, miR200c is considered to be a negative regulator of cancer stem cell behavior and epithelial—mesenchymal transition, and cancer stem cells are considered to be more radiation resistant than their "non-stem" counterparts. Nanoconstucts were prepared with miR200c loaded into PEG—peptide (gelatinase substrate)—poly (ε-caprolactone) copolymer nanoparticles and given to several gastric adenocarcinoma cell lines. Nanoparticle treatment combined with radiation decreased numbers of "stem-like" cells in the different cell populations. A decrease of CD44 and increase of E-cadherin were noted in cells treated with nanoconstructs in combination with irradiation, as well as changes in expression of many other apoptosis relevant proteins. In short, miR200c loaded copolymer nanoparticles radiosensitized human gastric adenocarcinoma cell lines BGC823, SGC7901, and MKN45 with 1.13–1.25 sensitization enhancement ratio [3].

A similar study with antagomirs against miR21 used locked nucleic acid—lipid nanocapsules to radiosensitize U87MG glioblastoma cells in culture [10]. In cells treated with these nanoconstructs viability was suppressed and reduced to 50 % after 4 Gy irradiation.

5 Conclusions

Despite impressive in vitro successes, nanomaterials are still showing only moderately interesting results as radiosensitizers in vivo. One reason for this lies in the fact that delivery of nanomaterials must be specific enough that the radiosensitizing nanoparticles do not accumulate in healthy tissues that need to be protected from (and not sensitized to) irradiation (e.g. see [5]). Nevertheless, the radiosensitizing nanomaterials are not as limited by this requirement as are chemotherapy and other anticancer therapies. Because radiation is targeted to the cancer with very precise delivery and dosimetry approaches, if the radiosensitizers are outside the radiation field, they will not be effective. Hence, targeting the cancer specifically is not a firm requirement of radiosensitizers. Second, to be good radiosensitizers nanoparticles must reach most of the cells that need to be radiosensitized. Considering that the most radiation resistant cells in solid tumors are the ones most distant from blood supply, it is very probable that the same cells will also be the most difficult to reach with nanomaterials. In addition to often sparse neovascularization, interstitial pressure in tumors and accumulation of collagen also present significant barriers to nanoparticle distribution in tumors in vivo. Collagen density was found to be a critical feature limiting delivery of 40 nm carboxylate modified FluoroSpheres to hind limb tumors in SCID mice [58]. Interestingly, radiation itself may be leveraged against nanoparticle delivery problems. For example, focused radiation delivery can permeabilize brain tumor blood barrier and increase tumor deposition of pegylated nanoparticles (e.g. [19]). Thirdly, nanoparticle uptake comparisons between cells grown under normoxia (21 % oxygen) versus hypoxia (0.1 % oxygen) show significant differences. Uptake of the same nanoparticles (1.9 nm "Auravist") by three different cell lines (DU145 prostate cancer, MDA-MB-231 breast cancer, and L132 lung epithelial cells) varied drastically under different oxygen conditions. These differences in uptake, combined with the importance of oxygen in production of reactive oxygen species, resulted in reduction of sensitizer enhancement ratio for Auravist from 1.41 for normoxia to 1.1 for hypoxia [17].

Finally, even if a nanoparticle formulation is found that is suitable for radiosensitization, general concerns regarding nanoparticle use may still prevent wide application of nanomaterials as radiosensitizers. Potential complications pertaining, e.g. to renal clearance of nanoparticles may obviate any potential benefits that would stem from radiosensitization. A recent review on renal clearance of nanomaterials [72], especially gold nanoparticles, suggests that both the size (currently "set" as less than 5.5 nm) and surface coating of nanoparticles dictates their elimination by kidneys. For example, while coating with PEG aids the nanoparticles to evade the cells of reticuloendothelial system (RES), it also prevents nanoparticle removal through kidneys regardless of their dimensions [72].

References

1. Atkinson RL, Zhang M, Diagaradjane P, Peddibhotla S, Contreras A, Hilsenbeck SG, Woodward WA, Krishnan S, Chang JC, Rosen JM (2010) Thermal enhancement with optically activated gold nanoshells sensitizes breast cancer stem cells to radiation therapy. Sci Trans Med 2:55ra79
2. Butterworth KT, McMahon SJ, Currell FJ, Prise KM (2012) Physical basis and biological mechanisms of gold nanoparticle radiosensitization. Nanoscale 4:4830–4838
3. Cui F, Liu Q, Li R, Shen J, Wu P, Yu L, Hu W, Wu F, Jiang C, Yue G et al (2014) Enhancement of radiotherapy efficacy by miR-200c-loaded gelatinase-stimuli PEG-Pep-PCL nanoparticles in gastric cancer cells. Int J Nanomed 13:2345–2358
4. Diagaradjane P, Shetty A, Wang JC, Elliott AM, Schwartz J, Shentu S, Park HC, Deorukhkar A, Stafford RJ, Cho SH et al (2008) Modulation of in vivo tumor radiation response via gold nanoshell-mediated vascular-focused hyperthermia: characterizing an integrated antihypoxic and localized vascular disrupting targeting strategy. Nano Lett 8:1492–1500
5. Le Duc G, Miladi I, Alric C, Mowat P, Bräuer-Krisch E, Bouchet A, Khalil E, Billotey C, Janier M, Lux F et al (2011) Toward an image-guided microbeam radiation therapy using gadolinium-based nanoparticles. ACS Nano 5:9566–9574
6. Falaschetti CA, Paunesku T, Kurepa J, Nanavati D, Chou SS, De M, Song M, Jang J-T, Wu A, Dravid VP et al (2013) Negatively charged metal oxide nanoparticles interact with the 20S proteasome and differentially modulate its biologic functional effects. ACS Nano 7:7759–7772
7. Gaca S, Reichert S, Multhoff G, Wacker M, Hehlgans S, Botzler C, Gehrmann M, Rödel C, Kreuter J, Rödel F (2013) Targeting by cmHsp70.1-antibody coated and survivin miRNA plasmid loaded nanoparticles to radiosensitize glioblastoma cells. J Control Release 172: 201–206
8. Gao Y, Chen K, Ma J-L, Gao F (2014) Cerium oxide nanoparticles in cancer. Onco Targets Ther 7:835–840

9. Giustini AJ, Petryk AA, Hoopes PJ (2011) Comparison of microwave and magnetic nanoparticle hyperthermia radiosensitization in murine breast tumors. Proc SPIE 7901:1–11
10. Griveau A, Bejaud J, Anthiya S, Avril S, Autret D, Garcion E (2013) Silencing of miR-21 by locked nucleic acid-lipid nanocapsule complexes sensitize human glioblastoma cells to radiation-induced cell death. Int J Pharm 454:765–774
11. Guidelli EJ, Baffa O (2014) Influence of photon beam energy on the dose enhancement factor caused by gold and silver nanoparticles: An experimental approach. Med Phys 41:032101
12. Hainfeld JF, Slatkin DN, Smilowitz HM (2004) The use of gold nanoparticles to enhance radiotherapy in mice. Phys Med Biol 49:N309–N315
13. Hainfeld JF, Smilowitz HM, O'Connor MJ, Dilmanian FA, Slatkin DN (2013) Gold nanoparticle imaging and radiotherapy of brain tumors in mice. Nanomed (Lond) 8: 1601–1609
14. Hall EJ, Giaccia AJ (2011) Radiobiology for the radiologist Lippincott Williams & Wilkins
15. Hildebrandt B, Wust P, Ahlers O, Dieing A, Sreenivasa G, Kerner T, Felix R, Riess H (2002) The cellular and molecular basis of hyperthermia. Crit Rev Oncol Hematol 43:33–56
16. Hirsch LR, Stafford RJ, Bankson JA, Sershen SR, Rivera B, Price RE, Hazle JD, Halas NJ, West JL (2003) Nanoshell-mediated near-infrared thermal therapy of tumors under magnetic resonance guidance. Proc Natl Acad Sci USA 100:13549–13554
17. Jain S, Coulter JA, Butterworth KT, Hounsell AR, McMahon SJ, Hyland WB, Muir MF, Dickson GR, Prise KM, Currell FJ et al (2014) Gold nanoparticle cellular uptake, toxicity and radiosensitisation in hypoxic conditions. Radiother Oncol 110:342–347
18. Jin C, Bai L, Wu H, Liu J, Guo G, Chen J (2008) Paclitaxel-loaded poly(D, L-lactide-co-glycolide) nanoparticles for radiotherapy in hypoxic human tumor cells in vitro. Cancer Biol Ther 7:911–916
19. Joh DY, Sun L, Stangl M, Al Zaki A, Murty S, Santoiemma PP, Davis JJ, Baumann BC, Alonso-Basanta M, Bhang D et al (2013) Selective targeting of brain tumors with gold nanoparticle-induced radiosensitization. PLoS ONE 8:e62425
20. Joh DY, Kao GD, Murty S, Stangl M, Sun L, Zaki AA, Xu X, Hahn SM, Tsourkas A, Dorsey JF (2013) Theranostic gold nanoparticles modified for durable systemic circulation effectively and safely enhance the radiation therapy of human sarcoma cells and tumors. Trans Oncol 6:722–IN32
21. Johannsen M, Thiesen B, Wust P, Jordan A (2010) Magnetic nanoparticle hyperthermia for prostate cancer. Int J Hyperth 26:790–795
22. Kampinga HH (2006) Cell biological effects of hyperthermia alone or combined with radiation or drugs: a short introduction to newcomers in the field. Int J Hyperth 22:191–196
23. Kampinga HH, Dikomey E (2001) Review: Hyperthermic radiosensitization: mode of action and clinical relevance. Int J Radiat Biol 77:399–408
24. Kleinauskas A, Rocha S, Sahu S, Sun Y-P, Juzenas P (2013) Carbon-core silver-shell nanodots as sensitizers for phototherapy and radiotherapy. Nanotechnology 24:325103
25. Van der Kogel A, Joiner M (2009) Basic clinical radiobiology. A Hodder Arnold Publication, London
26. Laha D, Pramanik A, Maity J, Mukherjee A, Pramanik P, Laskar A, Karmakar P (2014) Interplay between autophagy and apoptosis mediated by copper oxide nanoparticles in human breast cancer cells MCF7. Biochim Biophys Acta 1840:1–9
27. Li Y, Perkins A, Su Y, Ma Y, Colson L, Horne D, Chen Y (2012) Gold nanoparticles as a platform for creating a multivalent poly-SUMO chain inhibitor that also augments ionizing radiation. Proc Natl Acad Sci USA 109:4092–4097
28. Lin M, Huang J, Zhang J, Wang L, Xiao W, Yu H, Li Y, Li H, Yuan C, Hou X et al (2013) The therapeutic effect of PEI-Mn0.5Zn0.5Fe2O4 nanoparticles/pEgr1-HSV-TK/GCV associated with radiation and magnet-induced heating on hepatoma. Nanoscale 5:991–1000
29. Lin M-H, Hsu T-S, Yang P-M, Tsai M-Y, Perng T-P, Lin L-Y (2009) Comparison of organic and inorganic germanium compounds in cellular radiosensitivity and preparation of germanium nanoparticles as a radiosensitizer. Int J Radiat Biol 85:214–226

30. Lu R, Yang D, Cui D, Wang Z, Guo L (2012) Egg white-mediated green synthesis of silver nanoparticles with excellent biocompatibility and enhanced radiation effects on cancer cells. Int J Nanomed 7:2101–2107
31. Lundqvist M, Stigler J, Cedervall T, Berggård T, Flanagan MB, Lynch I, Elia G, Dawson K (2011) The evolution of the protein corona around nanoparticles: A test study. ACS Nano 5:7503–7509
32. Mackey MA, El-Sayed MA (2014) Chemosensitization of cancer cells via gold nanoparticle-induced cell cycle regulation. Photochem Photobiol 90:306–312
33. Maier-Hauff K, Ulrich F, Nestler D, Niehoff H, Wust P, Thiesen B, Orawa H, Budach V, Jordan A (2010) Efficacy and safety of intratumoral thermotherapy using magnetic iron-oxide nanoparticles combined with external beam radiotherapy on patients with recurrent glioblastoma multiforme. J Neurooncol 103:317–324
34. Matsudaira H, Ueno AM, Furuno I (1980) Iodine contrast medium sensitizes cultured mammalian cells to X-rays but not to gamma rays. Radiat Res 84:144–148
35. McBride WH, Iwamoto KS, Syljuasen R, Pervan M, Pajonk F (2003) The role of the ubiquitin/proteasome system in cellular responses to radiation. Oncogene 22:5755–5773
36. Miladi I, Duc G Le, Kryza D, Berniard A, Mowat P, Roux S, Taleb J, Bonazza P, Perriat P, Lux F et al (2013) Biodistribution of ultra small gadolinium-based nanoparticles as theranostic agent: application to brain tumors. J Biomater Appl 28:385–394
37. Miladi I, Aloy M-T, Armandy E, Mowat P, Kryza D, Magné N, Tillement O, Lux F, Billotey C, Janier M et al (2014) Combining ultrasmall gadolinium-based nanoparticles with photon irradiation overcomes radioresistance of head and neck squamous cell carcinoma. Nanomedicine 11:247–257
38. Mirjolet C, Papa AL, Créhange G, Raguin O, Seignez C, Paul C, Truc G, Maingon P, Millot N (2013) The radiosensitization effect of titanate nanotubes as a new tool in radiation therapy for glioblastoma: a proof-of-concept. Radiother Oncol 108:136–142
39. Misawa M, Takahashi J (2011) Generation of reactive oxygen species induced by gold nanoparticles under X-ray and UV Irradiations. Nanomedicine 7:604–614
40. Monopoli M, Åberg C, Salvati A, Dawson K (2012) Biomolecular coronas provide the biological identity of nanosized materials. Nat Nanotechnol 7:779–786
41. Moyer HR, Delman KA (2008) The role of hyperthermia in optimizing tumor response to regional therapy. Int J Hyperth 24:251–261
42. Naahidi S, Jafari M, Edalat F, Raymond K, Khademhosseini A, Chen P (2013) Biocompatibility of engineered nanoparticles for drug delivery. J Control Release 166: 182–194
43. Ngwa W, Korideck H, Kassis AI, Kumar R, Sridhar S, Makrigiorgos GM, Cormack RA (2013) *In vitro* radiosensitization by gold nanoparticles during continuous low-dose-rate gamma irradiation with I-125 brachytherapy seeds. Nanomedicine 9:25–27
44. Ni J, Wu Q, Li Y, Guo Z, Tang G, Sun D, Gao F, Cai J (2007) Cytotoxic and radiosensitizing effects of nano-C60 on tumor cells in vitro. J Nanopart Res 10:643–651
45. Nikolić N, Vranjes-Ethurić S, Janković D, Ethokić D, Mirković M, Bibić N, Trajković V (2009) Preparation and biodistribution of radiolabeled fullerene C60 nanocrystals. Nanotechnology 20:385102
46. Olive PL, Banáth JP (2006) The comet assay: a method to measure DNA damage in individual cells. Nat Protoc 1:23–29
47. Patra HK, Turner APF (2014) The potential legacy of cancer nanotechnology: cellular selection. Trends Biotechnol 32:21–31
48. Polf JC, Bronk LF, Driessen WHP, Arap W, Pasqualini R, Gillin M (2011) Enhanced relative biological effectiveness of proton radiotherapy in tumor cells with internalized gold nanoparticles. Appl Phys Lett 98:193702
49. Porcel E, Liehn S, Remita H, Usami N, Kobayashi K, Furusawa Y, Le Sech C, Lacombe S (2010) Platinum nanoparticles: a promising material for future cancer therapy? Nanotechnology 21:85103

50. Porcel E, Li S, Usami N, Remita H, Furusawa Y, Kobayashi K, Sech C Le, Lacombe S (2012) Nano-Sensitization under gamma rays and fast ion radiation. J Phys: Conf Ser 373:012006
51. Rahman WN, Corde S, Yagi N, Abdul Aziz SA, Annabell N, Geso M (2014) Optimal energy for cell radiosensitivity enhancement by gold nanoparticles using synchrotron-based monoenergetic photon beams. Int J Nanomed 9:2459–2467
52. Roa W, Zhang X, Guo L, Shaw A, Hu X, Xiong Y, Gulavita S, Patel S, Sun X, Chen J et al (2009) Gold nanoparticle sensitize radiotherapy of prostate cancer cells by regulation of the cell cycle. Nanotechnology 375101:9 pp
53. Roots R, Okada S (1975) Estimation of life times and diffusion distances of radicals involved in X-Ray-induced DNA strand breaks or killing of mammalian cells. Radiat Res 64:306–320
54. Rothkamm K, Löbrich M (2003) Evidence for a lack of DNA double-strand break repair in human cells exposed to very low X-ray doses. Proc Natl Acad Sci USA 100:5057–5062
55. Stankov K, Borisev I, Kojic V, Rutoljski L, Bogdanovic G, Djordjevic A (2013) Modification of antioxidative and antiapoptotic genes expression in irradiated K562 cells upon fullerenol C60 (OH) 24 nanoparticle treatment. J Nanosci Nanotechnol 13:105–113
56. Stigliano RV, Shubitidze F, Kekalo K, Baker I, Giustini AJ, Hoopes PJ (2013) Understanding mNP hyperthermia for cancer treatment at the cellular scale. Proc SPIE 8584:85840E
57. Tishler RB, Schiff PB, Geard CR, Hall EJ (1992) Taxol: a novel radiation sensitizer. Int J Radiat Oncol Biol Phys 22:613–617
58. Torosean S, Flynn B, Axelsson J, Gunn J, Samkoe KS, Hasan T, Doyley MM, Pogue BW (2013) Nanoparticle uptake in tumors is mediated by the interplay of vascular and collagen density with interstitial pressure. Nanomedicine 9:151–158
59. Townley HE, Rapa E, Wakefield G, Dobson PJ (2012) Nanoparticle augmented radiation treatment decreases cancer cell proliferation. Nanomedicine 8:526–536
60. Townley HE, Kim J, Dobson PJ (2012) In vivo demonstration of enhanced radiotherapy using rare earth doped titania nanoparticles. Nanoscale 4:5043–5050
61. Vinchon-Petit S, Jarnet D, Paillard A, Benoit J-P, Garcion E, Menei P (2010) In vivo evaluation of intracellular drug-nanocarriers infused into intracranial tumours by convection-enhanced delivery: distribution and radiosensitisation efficacy. J Neurooncol 97:195–205
62. Wason MS, Colon J, Das S, Seal S, Turkson J, Zhao J, Baker CH (2013) Sensitization of pancreatic cancer cells to radiation by cerium oxide nanoparticle-induced ROS production. Nanomedicine 9:558–569
63. Werner M, Copp J, Karve S (2011) Folate-targeted polymeric nanoparticle formulation of docetaxel is an effective molecularly targeted radiosensitizer with efficacy dependent on the Timing of Radiotherapy. ACS 5:8990–8998
64. Wiedenmann N, Valdecanas D, Hunter N, Hyde S, Buchholz TA, Milas L, Mason KA (2007) 130 Nm albumin-bound paclitaxel enhances Tumor radiocurability and therapeutic gain. Clin Cancer Res 13:1868–1874
65. Wust P, Hildebrandt B, Sreenivasa G, Rau B, Gellermann J, Riess H, Felix R, Schlag PM (2002) Hyperthermia in combined treatment of cancer. Lancet Oncol 3:487–497
66. Xiao F, Zheng Y, Cloutier P, He Y, Hunting D, Sanche L (2011) On the role of low-energy electrons in the radiosensitization of DNA by gold nanoparticles. Nanotechnology 22, 465101:10 pp
67. Yallapu MM, Maher DM, Sundram V, Bell MC, Jaggi M, Chauhan SC (2010) Curcumin induces chemo/radio-sensitization in ovarian cancer cells and curcumin nanoparticles inhibit ovarian cancer cell growth. J Ovarian Res 3:11
68. Yamamori T, Meike S, Nagane M, Yasui H, Inanami O (2013) ER stress suppresses DNA double-strand break repair and sensitizes tumor cells to ionizing radiation by stimulating proteasomal degradation of Rad51. FEBS Lett 587:3348–3353
69. Yasui H, Takeuchi R, Nagane M, Meike S, Nakamura Y, Yamamori T, Ikenaka Y, Kon Y, Murotani H, Oishi M et al (2014) Radiosensitization of tumor cells through endoplasmic reticulum stress induced by PEGylated nanogel containing gold nanoparticles. Cancer Lett 347:151–158

70. Yoo D, Lee J-H, Shin T-H, Cheon J (2011) Theranostic magnetic nanoparticles. Acc Chem Res 44:863–874
71. Zhang X, Yang H, Gu K, Chen J, Rui M, Jiang G-L (2011) In vitro and in vivo study of a nanoliposomal cisplatin as a radiosensitizer. Int J Nanomed 6:437–444
72. Zhang X-D, Yang J, Song S-S, Long W, Chen J, Shen X, Wang H, Sun Y-M, Liu P-X, Fan S (2014) Passing through the renal clearance barrier: toward ultrasmall sizes with stable ligands for potential clinical applications. Int J Nanomed 9:2069–2072
73. Zheng Y, Cloutier P, Hunting DJ, Sanche L (2008) Radiosensitization by gold nanoparticles: comparison of DNA damage induced by low and high-energy electrons. J Biomed Nanotechnol 4:469–473
74. Zou J, Qiao X, Ye H, Zhang Y, Xian J, Zhao H, Liu S (2009) Inhibition of ataxia-telangiectasia mutated by antisense oligonucleotide nanoparticles induces radiosensitization of head and neck squamous-cell carcinoma in mice. Cancer Biother Radiopharm 24:339–346

Hybrid Nanoparticles for Cancer Imaging and Therapy

Chunbai He and Wenbin Lin

Abstract

Hybrid nanoparticles, composed of both inorganic and organic components, have been exploited as promising platforms for cancer imaging and therapy. This class of nanoparticles can not only retain the beneficial features of both inorganic and organic materials, but also allow systematic fine-tuning of their properties through the judicious combination of functional components. This chapter summarizes recent advances in the design and synthesis of hybrid nanomaterials, with particular emphasis on two main categories of hybrid nanoparticles: Nanoscale metal-organic frameworks (also known as nanoscale coordination polymers) and polysilsesquioxane nanoparticles. Preliminary applications of these hybrid nanoparticles in cancer imaging and therapy are described.

Keywords

Hybrid Nanoparticles, Nanoscale Metal-Organic Frameworks, Nanoscale Coordination Polymers, Polysilsesquioxane Nanoparticles, Cancer Therapy, Cancer Imaging · Hybrid nanoparticles · Nanoscale metal-organic Frameworks · Nanoscale coordination polymers · Polysilsesquioxane nanoparticles · Cancer therapy · Cancer imaging

Contents

C. He · W. Lin (✉)
Department of Chemistry, University of Chicago, 929 E 57th St, Chicago, IL 60637, USA
e-mail: wenbinlin@uchicago.edu

C.A. Mirkin et al. (eds.), *Nanotechnology-Based Precision Tools for the Detection and Treatment of Cancer*, Cancer Treatment and Research 166,
DOI 10.1007/978-3-319-16555-4_8

1 Introduction

In spite of the remarkable progress in our understanding of fundamental cancer biology in the past few decades, we have yet to achieve comparable advances in the detection, diagnosis, and treatment of cancer [1, 2]. Current imaging and therapeutic agents suffer from nonspecific distribution throughout the body, rapid clearance, poor pharmacokinetics, and undesirable side effects. A variety of organic and inorganic nanoparticles have recently emerged as promising platforms for cancer diagnostic and therapeutic applications owing to their tunable sizes, high agent loadings, tailorable surface properties, controllable or stimuli-responsive drug release kinetics, improved biocompatibility, and passive tumor targeting via the enhanced permeability and retention (EPR) effect [3, 4]. Decorated with desired functional groups, nanoparticles allow the implementation of different molecular imaging techniques, such as computed tomography (CT), magnetic resonance imaging (MRI), single-photon emission tomography (SPECT), positron emission tomography (PET), ultrasound imaging, and optical imaging [5–10]. Their fascinating and unique properties have been exploited for a range of anticancer therapies, such as chemotherapy, photodynamic therapy, neutron capture therapy, thermal therapy, and magneto-therapy [11–13]. Further, nanoparticles may be engineered to combine two or more of these therapies, leading to synergistic anticancer efficacy [14–16].

Although numerous nanoparticles have been developed as carriers for imaging and therapeutic agents, the majority can be categorized into either pure inorganic [e.g., quantum dots (QDs) [17], metallic nanostructures (gold nanoparticles) [18], and metal oxides (particularly magnetic iron oxides [19], up-conversion nanophosphors [20], and zeolites [21])] or organic materials (e.g., liposomes [22], dendrimers [23], micelles [24], and polymeric hydrogel nanoparticles [25]). Each of these classes of nanoparticles has its own strengths and weaknesses. Hybrid nanoparticles, composed of both inorganic and organic components, can not only retain the beneficial features of both inorganic and organic nanomaterials, but also combine a multitude of organic and inorganic components in a modular fashion to allow for systematic tuning of the properties of the resultant nanoparticles [26]. This chapter will focus on the synthesis and applications of two major classes of hybrid nanoparticles that are closely investigated in our laboratory and in other research groups for cancer imaging and therapy, namely nanoscale metal-organic frameworks [NMOFs, also known as nanoscale coordination polymers (NCPs)] and polysilsesquioxane (PSQ) nanoparticles.

NMOFs, or NCPs, are a class of hybrid nanoparticles formed by the self-assembly of metal ions or clusters and polydentate bridging ligands [27–30]. We and others have demonstrated their potential applications in delivering imaging agents and drug molecules to cancer cells [26, 27, 31–33]. NMOFs possess several potential advantages over existing nanocarriers. NMOFs are compositionally and structurally diverse, allowing for the facile synthesis of NMOFs of different compositions, shapes, sizes, and chemical properties that are suited for different biomedical applications. NMOFs are intrinsically biodegradable as a result of relatively labile metal-ligand bonds.

Polysilsesquioxane (PSQ) nanoparticles are another class of hybrid nanomaterials, which are formed via condensation of silanol-based monomers and can be considered as a special class of NCPs with Si as the metal-connecting point. While PSQs offer similar biocompatibility to other silica-based materials, they allow much higher drug loadings than silica-based materials with grafted drugs on only their surfaces [34, 35].

In this chapter, we intend to summarize recent advances in the development of NMOF/NCP and PSQ nanoparticles as delivery vehicles for imaging agents and molecular therapeutics. Preliminary in vitro and in vivo imaging and therapy results will be described to highlight the potential of hybrid nanoparticles in oncology.

2 Nanoscale Metal-Organic Frameworks

Metal-organic frameworks (MOFs), also called coordination polymers or coordination networks, are a new class of highly tunable hybrid materials crafted from metal-connecting points and organic bridging ligands. They are typically synthesized under mild conditions via coordination-directed self-assembly processes. Given their periodic and porous structures as well as the ability to carry a wide variety of functionalities, MOFs have been investigated for many applications such as nonlinear optics, gas storage, catalysis, separation, light harvesting, and chemical sensing [36–47]. When MOFs are scaled to the nanoregime to form nanoscale metal-organic frameworks (NMOFs), they maintain the structural diversity and physicochemical properties of bulk MOFs. Further, the particle dimensions can range from 10 to 100 of nm, making them potential nanocarriers for imaging agents and drug molecules for cancer diagnosis and therapy.

2.1 Synthesis of Nanoscale Metal-Organic Frameworks

Four general methods have been utilized to synthesize NMOFs: nanoprecipitation (Fig. 1), solvothermal (Fig. 1), reverse microemulsion (Fig. 2), and surfactant-templated solvothermal reactions (Fig. 2). Among these four methods, nanoprecipitation

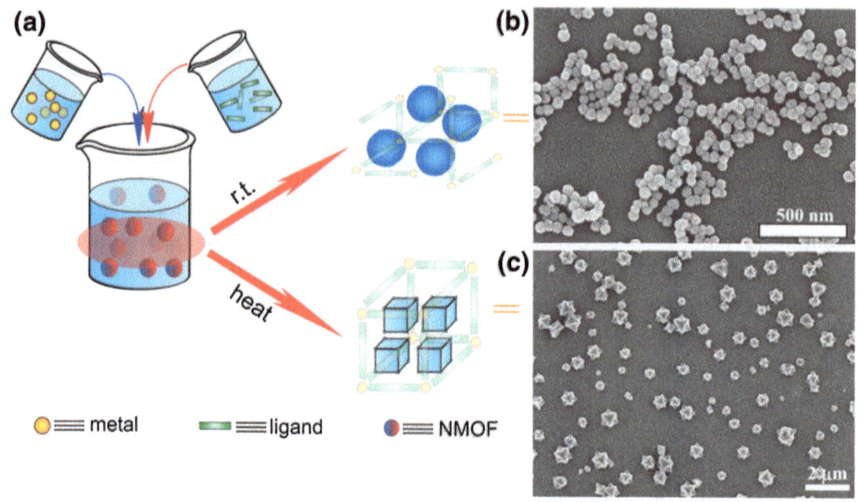

Fig. 1 **a** Surfactant-free synthesis of NMOFs [27]. **b**, **c** Representative SEM images of NMOFs synthesized by nanoprecipitation (**b**) [48] and solvothermal method (**c**) [31]. Reprinted with permission from [27]. Copyright (2014) American Chemical Society

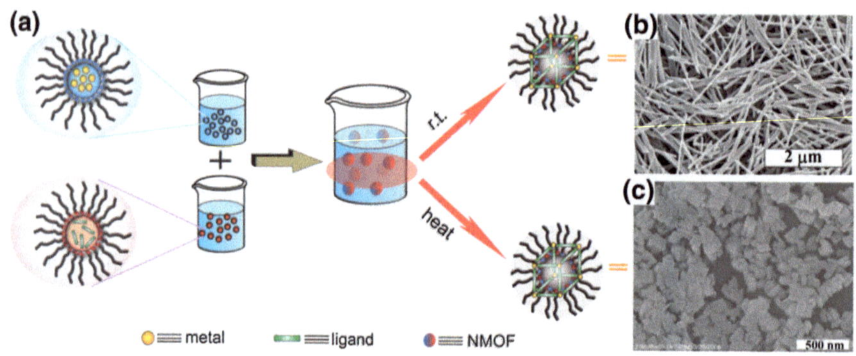

Fig. 2 **a** Surfactant-templated NMOF synthesis based on reverse microemulsions or surfactant-assisted solvothermal reactions [27]. **b**, **c** Representative SEM images showing NMOFs synthesized by surfactant-templated method [52, 53]. Reprinted with permission from [27]. Copyright (2014) American Chemical Society

typically yields amorphous materials and the other three methods can afford crystalline materials, owing to the ability to exert a better control on nanoparticle nucleation and growth kinetics. Surfactants are used in reverse microemulsions and surfactant-templated solvothermal reactions, which can better control the particle synthesis and stabilize the resultant particles.

For the nanoprecipitation method, precursor solutions are mixed to allow particle nucleation and growth. The resulting nanoparticles are insoluble in the solvent system whereas the individual precursors remain soluble; and therefore, the nanoparticles can be precipitated. This method was used to synthesize NMOFs composed of the anticancer prodrug *c,c,t*-$Pt(NH_3)_2Cl_2(succinate)_2$ (DSCP) and Tb^{3+} ions [48]. In this synthesis, the pH value of an aqueous solution of $TbCl_3$ and $[NMeH_3]_2DSCP$ was adjusted to 5.5, and methanol was added to the precursor solution to precipitate the nanoparticles. NMOFs formed by nanoprecipitation adopted spherical morphology with a diameter below 100 nm and carried an exceptionally high cisplatin loading compared to other nanocarriers. This method was used to synthesize a Zr-based NMOF containing DSCP by acetone-induced precipitation of a solution of $ZrCl_4$ and DSCP in *N,N*-dimethylforamide (DMF) [49].

Solvothermal synthesis of NMOFs can be carried out by either conventional heating or using a microwave. Temperatures and heating rates are important parameters that control the NMOF nucleation and growth. An Fe^{3+} NMOF with the formula of $Fe_3(\mu_3\text{-}O)Cl(H_2O)(BDC)_3$ was synthesized by heating a solution of $FeCl_3$ and terephthalic acid (BDC) with a microwave [31]. The resultant crystalline NMOF displayed octahedral morphology with a diameter of 200 nm and adopted the known MIL-101 structure. A solvothermal reaction between pamidronate (Pam) and $CaCl_2$ in diethylformamide (DEF)/H_2O afforded single crystals of Ca-Pam with rod-like morphology of $\sim 80 \times 80 \times 1000$ nm in dimensions [50]. Similarly, microwave heating of a solution of zoledronate (Zol) and $CaCl_2$ in DMF/H_2O led to crystalline particles of Ca-Zol that adopts a rod-like morphology of $\sim 70 \times 70 \times 1000$ nm in dimensions [50]. The solvothermal method was also used to synthesize a UiO NMOF with the formula of $Zr_6(\mu_3\text{-}O)_4(\mu_3\text{-}OH)_4$(amino-TPDC)$_6$. This NMOF with amino-triphenyldicarboxylate (amino-TPDC) bridging ligand was synthesized by heating a DMF solution of $ZrCl_4$ and amino-TPDC at 80 °C for 5 days, and exhibited a hexagonal-plate morphology with a diameter of ~ 100 nm and a thickness of ~ 30 nm [51].

Reverse microemulsions provide another method to control nucleation and growth kinetics of NMOFs because the building blocks for NMOFs are typically water-soluble. This surfactant-assisted synthesis can be carried out at room temperature or with heating. A crystalline $Gd(BDC)_{1.5}(H_2O)_2$ nanorod was synthesized with this method by mixing two separate microemulsions containing either $GdCl_3$ or $[NMeH_3]_2[BDC]$ [52]. The particle morphologies could be controlled by adjusting the water to surfactant molar ratio (*w* value) of the microemulsion. Nanorods of 100–1000 nm in length and 35–100 nm in diameter were synthesized in microemulsions with different *w* values.

Surfactant molecules can be used to template the NMOF synthesis under solvothermal conditions by coating the surfaces of growing NMOF particles. For example, a reverse microemulsion of $GdCl_3$ and $[NMeH_3]_6[BHC]$ (BHC = benzene hyxacarboxylic acid) was heated at 120 °C to afford Gd-BHC NMOFs, which exhibited a blocklike morphology with dimensions of $25 \times 50 \times 100$ nm and matched a previously reported lanthanide-BHC phase [40, 53]. This method was also used to synthesize NMOFs capped with lipid, such as 1,2-dioleoyl-*sn-glycero*-3-phosphate (DOPA). Several DOPA-coated NMOFs were synthesized by utilizing the surfactant-templated solvothermal reactions, including Zr-methotrexate (MTX) NMOFs, Zn-MTX NMOFs, La-DSCP NMOFs, and Zn bisphosphonate NCP containing platinum-based prodrugs [49, 54, 55].

The four general methods described above have been adopted to synthesize a large number of NMOFs. By independently adjusting NMOF precursors, reaction solvents, pH values, temperatures, surfactant, *w* values of microemulsions, a wide variety of NMOFs with well-defined morphologies and compositions can be readily synthesized.

2.2 Strategies to Incorporate Imaging or Therapeutic Agents Within Nanoscale Metal-Organic Frameworks

By taking advantage of the porous NMOF structure and flexibility of bridging ligands, several different methods have been developed to incorporate imaging or therapeutic agents within NMOFs with high loadings. These loading methods fall into two main categories: Direct incorporation during NMOF synthesis and post-synthetic loading [26, 27].

Imaging or therapeutic agents can be directly incorporated into the NMOF structure during synthesis as metal-connecting points/clusters or as the bridging ligand [33, 48, 52, 53, 56, 57]. Paramagnetic metal ions such as Gd^{3+}, Fe^{3+}, and Mn^{2+} serve as metal-connecting points in NMOF structures and can act as efficient magnetic resonance imaging (MRI) contrast enhancement agents. Platinum-based prodrugs (including DSCP, *cis,cis,trans*-$[Pt(NH_3)_2Cl_2(OCONHP(O)(OH)_2)_2]$, and $Pt(DACH)Cl_2[OCONHPO(OH)_2]_2)$, gemcitabine monophosphate, 2,3,5,6-tetraiodo-1,4-benzenedicarboxylic acid (I_4-BDC) were adopted as bridging ligands to form NMOFs to provide chemotherapeutics cisplatin, oxaliplatin, and gemcitabine for cancer therapy and high Z element iodine for CT imaging [48, 55, 56]. This strategy can lead to NMOFs possessing very high agent loadings with uniform distribution throughout the frameworks. For example, NCP-1 and NCP-2 containing bridging ligands of cisplatin or oxaliplatin prodrugs achieved as high as 48 wt% cisplatin loading and 45 wt% oxaliplatin loading, respectively (Fig. 3) [55].

The pores and channels in the NMOF structure allow postsynthetic loading of imaging or therapeutic agents. Biomedically relevant agents can be incorporated into the tunable pores of NMOFs by noncovalent or covalent interactions after the NMOF synthesis. A cisplatin prodrug, $Pt(NH_3)_2Cl_2(OH)(OEt)$, was loaded into UiO NMOFs via noncovalent encapsulation. Release studies showed sustained cisplatin release from the framework with negligible burst effects [51]. Férey and coworkers extended this strategy to NMOFs loaded with hydrophilic, amphiphilic, and hydrophobic agents [33]. Compared to noncovalent incorporation, postsynthesis covalent attachment of an agent offers a more robust approach in terms of preventing premature release of the agent. The agent covalently attached to the bridging ligand will only be released upon the decomposition of the NMOF structure. For example, Fe^{3+} NMOFs possessing 2-aminoterephallic acid (NH_2-BDC) as bridging ligands allowed for the covalent attachment of either an optical contrast agent or a chemotherapeutic [31, 58].

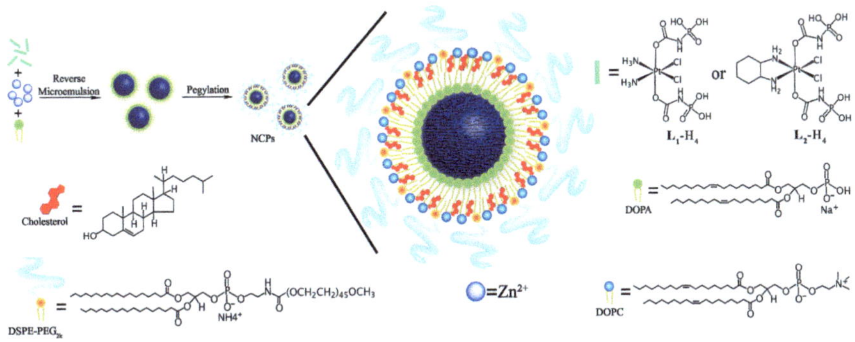

Fig. 3 General procedure of self-assembly of Zn bisphosphonate NCPs containing platinum-based prodrugs with lipid and PEG coatings [55]

2.3 NMOFs for Cancer Imaging

NMOFs have been evaluated for applications in magnetic resonance imaging (MRI), computed tomography (CT), and optical imaging [38, 59]. NMOFs show interesting and distinct advantages over other nanocarriers in delivering imaging contrast agents.

MRI is a noninvasive imaging technique based on the detection of nuclear spin reorientations in a magnetic field. MRI provides excellent spatial resolution, high soft tissue contrast, and unlimited penetration depth, but is intrinsically insensitive [32]. Therefore, large doses of contrast enhancement agents such as Gd(III) chelates are used in clinical scans to enhance the contrast between normal and diseased tissues [60]. The effectiveness of Gd(III)-containing NMOFs as T_1-weighted contrast agents was first demonstrated by Lin and coworkers [52, 53, 61]. For example, $Gd_2(BDC)_3(H_2O)_4$ nanorods gave an r_1 of 35.8 $mM^{-1}\ s^{-1}$ in an aqueous xanthan gum suspension, which is almost an order of magnitude higher than that obtained with the commercially available T_1-weighted contrast agent Omniscan [52]. Similar relaxivities were obtained for $Gd_2(BHC)(H_2O)_6$ NMOFs synthesized using a surfactant-mediated method. Many of the Gd(III)-based NMOFs can also act as efficient T_2-weighted contrast agents [53]. The relaxivity values of the NMOFs are dependent on nanoparticle size, with smaller NMOFs exhibiting larger r_1 relaxivities, which can be attributed to the fact that smaller particles possess higher surface to volume ratios to enhance the exchange between Gd(III) bound water and bulk water. Boyes and coworkers further tuned the relaxivities of Gd(III)-based NMOFs by surface modification with polymers, showing that hydrophilic polymers led to an increased r_1 relaxivity while hydrophobic polymers increased the r_2 relaxivity [62, 63]. Unfortunately, Gd-NMOFs leached significant amounts of Gd(III) ions in water, which limited their in vivo applications as MRI contrast agents, due to the toxicity of free Gd(III) ions [64]. To address the toxicity issue, the Lin group and others have synthesized Mn(II)-based NMOFs and evaluated their applications as MRI contrast agents [65, 66]. Mn(II) ions have been shown to be potent MRI contrast agents with lower toxicity compared to free Gd(III). Nanorods of Mn(BDC)(H_2O)$_2$ were found to have an r_1 and r_2 of 5.5 and 50.0 mM^{-1} s^{-1} on a per Mn(II) basis at 3 T, respectively, whereas nanorods of $Mn_2(BTC)_3(H_2O)_6$ exhibited an r_1 and r_2 of 7.8 and 70.8 mM^{-1} s^{-1} per Mn(II) basis, respectively. $Mn_2(BTC)_3(H_2O)_6$ NMOFs were further coated with a thin silica shell and functionalized with a cancer targeting peptide to afford NMOFs with active targeting capability. This NMOF was evaluated in vivo and demonstrated an enhancement in T_1-weighted signals in the liver, spleen, and aorta 1 h postinjection, which was attributed to the Mn(II) ions released from the NMOF. Horcajada and coworkers synthesized a series of iron-carboxylate NMOFs as T_2-weighted contrast agents with sizes ranging from 50 to 350 nm and crystalline structure matching the MIL series. Negative enhancement of the liver and spleen was observed after injection of these NMOFs to Wistar rats, which dissipated 3 months postinjection, suggesting the accumulation, degradation, and subsequent clearance of the NMOFs [33].

CT imaging is based on attenuation of X-rays by a specimen to provide 3D images with high spatial resolution [67]. High Z number elements such as iodine, barium, and bismuth are chosen as CT contrast agents; however, large doses (tens of grams) are typically required in the clinic to achieve adequate contrast [67]. Lin and coworkers developed NMOFs constructed from Cu(II) or Zn(II) and I_4-BDC as CT imaging contrast agents [56]. CT phantom studies showed that both NMOFs possessed slightly higher X-ray attenuation factors than commercially used contrast agent iodixanol. In another example, instead of incorporating the high Z element into the bridging ligand of the structure, Lin and coworkers incorporated high Z elements (Hf and Zr) into the $M_6(\mu_3\text{-}O_4)(\mu_3\text{-}OH)_4(RCO_2)_{12}$ (M = Zr or Hf) secondary building units of UiO-66 NMOFs with high Hf (57 wt%) and Zr (37 wt%) contents [68]. Hf-NMOFs of different sizes were coated with silica and poly(ethylene glycol) (PEG) to enhance biocompatibility, and were used for in vivo CT imaging of mice, showing increased attenuation in the liver and spleen (Fig. 4).

A number of luminescent NMOFs have been synthesized, but their unfavorable absorption properties and low quantum yields have precluded the applications in biomedical imaging [38]. Lin and coworkers synthesized NMOFs containing a phosphorescent $Ru(bpy)_3^{2+}$ derivative as the bridging ligand and Zn(II) or Zr(IV) as metal-connecting points [69]. Zr-based NMOFs with exceptionally high dye loadings (up to 57.4 wt%) were further coated with a layer of amorphous silica, functionalized with PEG, and an active targeting moiety for enhanced uptake in cancer cells. Confocal microscopy studies demonstrated an increased uptake of NMOFs in human lung cancer cells. Besides using optical imaging agents as bridging ligands for constructing NMOFs, optical dyes can also be incorporated within the frameworks post-synthetically. Kimizuka and coworkers created a series of NMOFs based on lanthanide ions and nucleotides [70–73]. Anionic dyes, such as perylene-3,4,9,10-tertacarboxylate, were incorporated within NMOFs of adenosine 5′-monophosphate and Gd^{3+}. Confocal microscopy showed the uptake of this NMOF into the lysosomes of HeLa cells and its biodistribution in a murine model was examined. Lanthanide-nucleotide NMOFs were also used to successfully encapsulate other anionic dyes and negatively charged quantum dots. Recently, Lin and coworkers

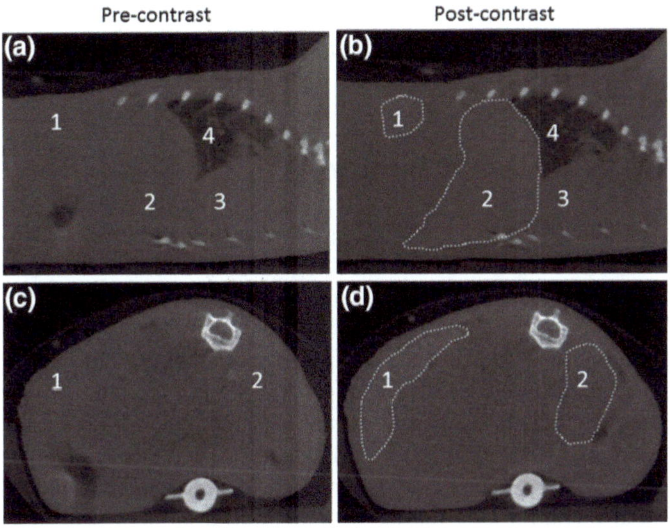

Fig. 4 Sagittal (**a** and **b**) and axial (**c** and **d**) CT slices of a mouse precontrast and 15 min after injection of Hf-NMOFs coated with silica and PEG. The areas of increased attenuation are outlined, and the labels are: 1-spleen (+131 HU), 2-liver (+86 HU), 3-heart, and 4-lungs [68]. Reproduced by permission of The Royal Society of Chemistry

incorporated fluorescein isothiocyanate (FITC) into UiO NMOFs by forming thiourea linkage between isothiocyanate group on FITC and amino group of bridging ligand to afford FITC conjugated UiO NMOFs (F-UiO) with exceptionally high FITC loadings, efficient fluorescence, and excellent ratiometric pH-sensing properties (Fig. 5) [74]. Real-time imaging studies in live cells revealed endocytosis and

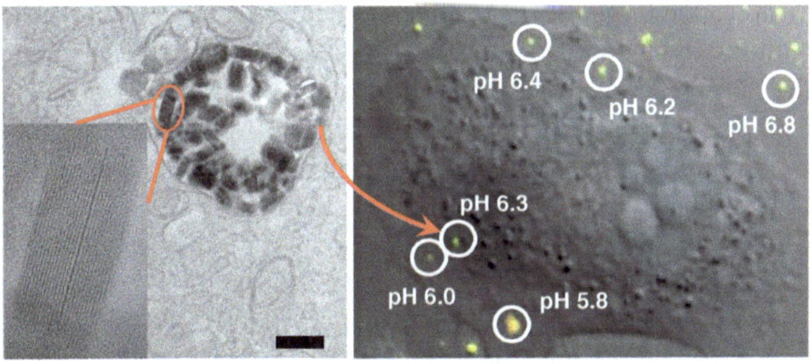

Fig. 5 High-resolution TEM image showing the distribution and structural integrity of UiO NMOFs in the endosomes of H460 cells. Inset is a zoomed-in view showing one intact UiO NMOF inside endosome. CLSM image was obtained from live cell imaging video, showing the overlay of *green* fluorescence (488 nm channel), *red* fluorescence (435 nm channel), and DIC. The pH of different endosomes or particles outside the cells was calculated based on the pH calibration curve obtained on live cells [74]. Reprinted with permission from [74]. Copyright (2014) American Chemical Society

exocytosis of F-UiO and endosome acidification in human lung cancer cells, which represents the first use of NMOFs for intracellular pH sensing and elucidating NMOF-cell interactions [74].

2.4 NMOFs for Cancer Therapy

Our understanding of cancer biology has progressed enormously in the past two decades, bringing a large number of anticancer therapeutics, including small molecule inhibitors, antibodies, chemotherapeutics, and nucleic acid drugs, to the clinic [3, 75–77]. However, current therapeutics are limited by their nonspecific distribution throughout the body, leading to high doses, rapid clearance, poor pharmacokinetics, and undesired side effects [3, 4, 78]. NMOFs are potential nanovectors for delivering therapeutic agents to targeted areas of the body to overcome these limitations. Nanoscale dimensions of NMOFs allow them to take advantage of the EPR effect to achieve specific and enhanced accumulation in the tumor site. NMOFs can also control drug release with their large surface areas, high porosity, and presence of functional groups to interact with loaded moieties.

Lin and coworkers have developed a series of NMOF platforms to deliver platinum-based chemotherapeutics. An amorphous NMOF was built from a cisplatin prodrug DSCP and Tb(III) ions by the nanoprecipitation method, and further coated with a thin layer of silica and a silyl-derived peptide to actively target cancer cells [48]. This NMOF decomposes in physiological media and thus releases the cisplatin prodrug by diffusing out of the silica shell in a controlled manner to induce cytotoxicity in human colon cancer cells. The Iron(III)-carboxylate NMOF was post-synthetically modified to carry a cisplatin prodrug, *c,c,t*-$Pt(NH_3)_2Cl_2$(succinate)(OEt), coated with silica, and targeted with RGD peptide targeting ligand [31]. Cytotoxicity was also observed for this NMOF. NMOFs constructed from a cisplatin prodrug DSCP and Zr(IV) or La(III) using surfactant-templated heating technique was capped with DOPA and modified with PEG [49]. These NMOFs showed cytotoxicity in human lung cancer cells. Recently, Lin and coworkers reported the self-assembly of zinc bisphosphonate NCPs carrying cisplatin and oxaliplatin prodrug with high drug loadings [55]. After PEGylation, these two NCPs showed minimal uptake by the mononuclear phagocyte system and excellent blood circulation half-lives of 16.4 and 12.0 h for NCPs carrying cisplatin and oxaliplatin, respectively. In multiple tumor xenograft murine models including colon, lung, and pancreatic cancer, NCPs exhibited superior potency and efficacy at very low drug dose compared with free drugs (Fig. 6).

NMOFs can be loaded with other chemotherapeutic agents including methotrexate (MTX), pamidronate (Pam), and zoledronate (Zol), and showed cytotoxicity in multiple human cancer cells in vitro [50, 54]. Horcajada and coworkers encapsulated busulfan within the iron-carboyxlate NMOFs and demonstrated their comparable cytotoxicity as the free drug against several cancer cell lines [33]. Maspoch and coworkers encapsulated several anticancer drugs as guest species within NMOFs synthesized from Zn^{2+} and 1,4-bis(imidazole-1-ylmethyl)benzene

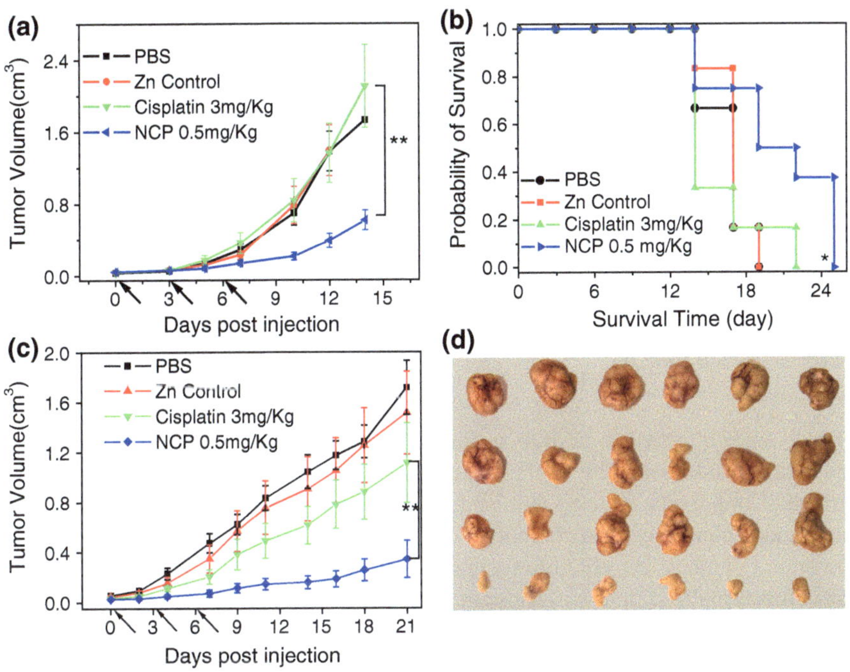

Fig. 6 In vivo antitumor activity of NCP carrying cisplatin. **a** In vivo tumor growth curves for NCP, free cisplatin, and other controls on the s.c. C26 tumor model. **b** Kaplan-Meier plots showing the percentage of animals remaining in the study with s.c. C26 tumor model. **c** In vivo tumor growth inhibition curves for NCP, free cisplatin, and other controls on the s.c. H460 tumor model. **d** Photos of resected H460 tumors from various groups. The last row is NCP group [55]

[79]. Huang and coworkers synthesized coordination polymer sphere by combining 1,1′-(1,4-butanediyl)bis(imidazole) (bbi) and ferrous ions and conjugated folic acid to the surface [80]. The authors demonstrated the cytotoxicity of this coordination polymer sphere loaded with doxorubicin in HeLa cells.

MOFs have also been studied as carriers for gaseous molecules for anticancer therapy such as nitric oxide (NO). NO has shown anticancer activity in high concentrations and other activities including antibacterial, antithrombotic, and wound-healing applications. Morris and coworkers synthesized two MOFs from either cobalt or nickel and 2,5-dihydroxyterephthalic acid, which can absorb seven times the amount of NO than any previously reported material on a per gram basis via ligation to coordinatively unsaturated metal centers, with little background release [81].

Most recently, NMOFs have been utilized for the delivery of nucleic acids. Lin and coworkers for the first time demonstrated the capability of a UiO NMOF to co-deliver cisplatin and small interfering RNA (siRNA) to drug-resistant ovarian cancer cells for overcoming drug resistance and enhanced anticancer efficacy [51]. A cisplatin prodrug was encapsulated into the channels of UiO NMOFs and

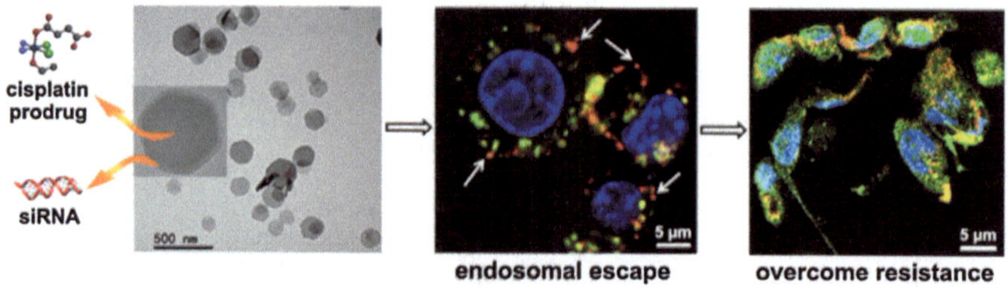

Fig. 7 UiO NMOFs carrying a cisplatin prodrug and siRNA exhibited capabilities to escape endosomal entrapment upon entering the cells and induce significant apoptosis in cisplatin-resistant ovarian cancer cells. [51] Reprinted with permission from [51]. Copyright (2014) American Chemical Society

siRNAs were attached to NMOF surface via multiple coordination bonds between phosphate residues on the siRNA backbone and vacant Zr sites on the NMOF surface. Co-delivery of cisplatin and siRNA with NMOFs led to an order of magnitude enhancement in chemotherapeutic efficacy in vitro (Fig. 7). Mirkin and coworkers synthesized UiO-66-N_3 ($Zr_6O_4OH_4(C_8H_3O_4\text{-}N_3)_6$) NMOFs and covalently functionalized their surface with oligonucleotides utilizing a strain promoted click reaction between DNA appended with dibenzylcyclooctyne and azide-functionalized UiO-66-N_3 [82]. When dispersed in aqueous solution, they exhibit increased stability and enhanced cellular uptake.

Several ongoing studies in the Lin group showed the potential of exploiting NMOFs as a versatile platform for combination therapies including chemotherapy, photodynamic therapy, and gene therapy. With more in-depth investigations, NMOFs are on the path through preclinical evaluations and hold great promise in anticancer therapy.

3 PSQ Nanoparticles

PSQ nanoparticles are formed via condensation of silanol-based monomers, yielding high drug loadings compared with other silica-based nanoparticles. PSQs have tunable structures and surface chemistry. PSQ particles can be considered as a special class of NMOFs/NCPs with Si as the metal-connecting point. Although not many examples are available for their synthesis and applications in the cancer imaging and therapy yet, the distinct properties of PSQ nanoparticles make them an interesting nanocarrier platform with significant potential in the clinic.

3.1 Synthesis of PSQ Nanoparticles

PSQs are hybrid materials composed of siloxane networks with organic or metal-organic bridging ligands. They are synthesized from bis(trialkoxysilanes) ((R'O)$_3$-

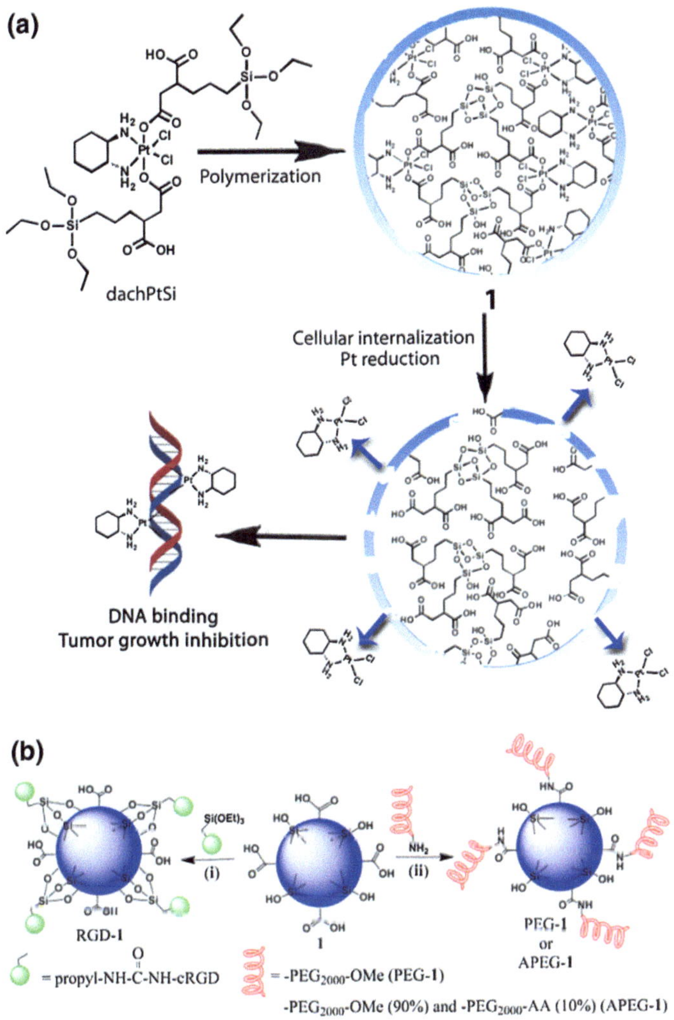

Fig. 8 Schematic representation of the synthesis of PSQ carrying a cisplatin prodrug [34]. Reproduced with permission from Ref. [34]

Si-R-Si-(OR')$_3$) via sol-gel reactions. As a homopolymer of (R'O)$_3$-Si-R-Si-(OR')$_3$, PSQs allow much higher drug loadings than other silica-based materials and their physicochemical properties can be easily tuned by changing the monomer properties than in a silica-based materials. Lin and coworkers constructed PSQs carrying a platinum complex, [Pt(dach)Cl$_2$(triethoxysilylpropyl succinate)$_2$], further surface-modified the PSQs with PEG and anisamide ligand for active targeting to cancer cells (Fig. 8) [34].

3.2 PSQ Nanoparticles for Cancer Imaging

The cancer imaging applications of PSQs have been demonstrated by the Lin group [35]. Gd(III) chelates were covalently linked to PSQ particles via a labile disulfide bond, and the Gd-PSQ nanoparticles were post-synthetically coated with PEG and anisamide ligand to enhance biocompatibility and cell uptake in cancer cells (Fig. 9). The effectiveness of this PSQ nanoparticle as efficient optical imaging and MRI contrast agents were demonstrated in human lung and pancreatic cancer cells (Fig. 9).

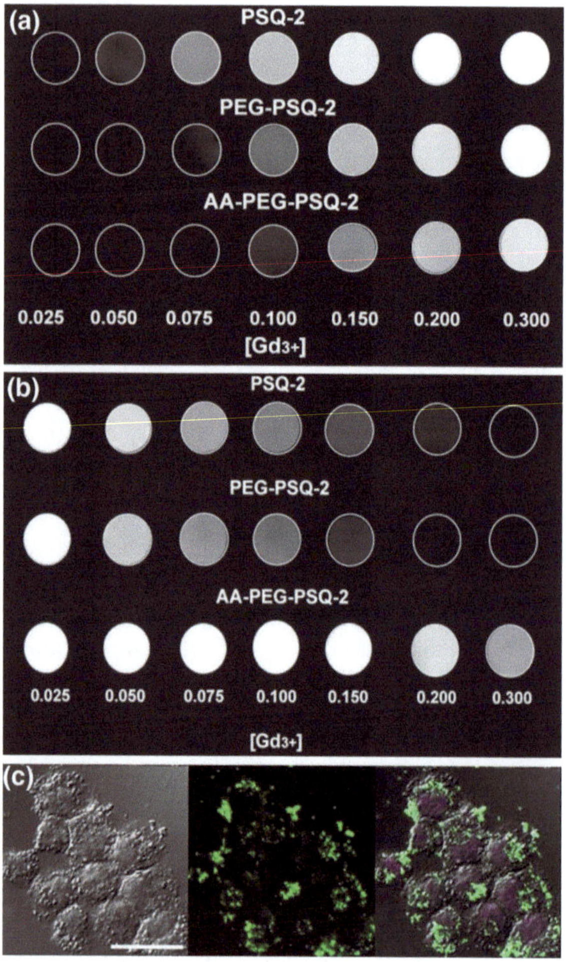

Fig. 9 Efficient MR and optical imaging of PSQ nanoparticles. T_1- (**a**) and T_2-weighted (**b**) MR phantom images of PSQ-**2**, PEG-PSQ-**2**, and AA-PEG-PSQ-**2** at various Gd(III) concentrations at 3 T. (**c**) Confocal microscope images of AsPC-1 cancer cells incubated with FITC-labeled anisamide-PEG-PSQ-2. The figures from *left* to *right* depict bright field images, *green* fluorescence of FITC-labeled PSQ nanoparticles, and merged images of bright field and *green* fluorescence with DRAQ5-stained nuclei, respectively. *Scale bar* is 20 μm. [35] Reproduced with permission from Ref. [35]

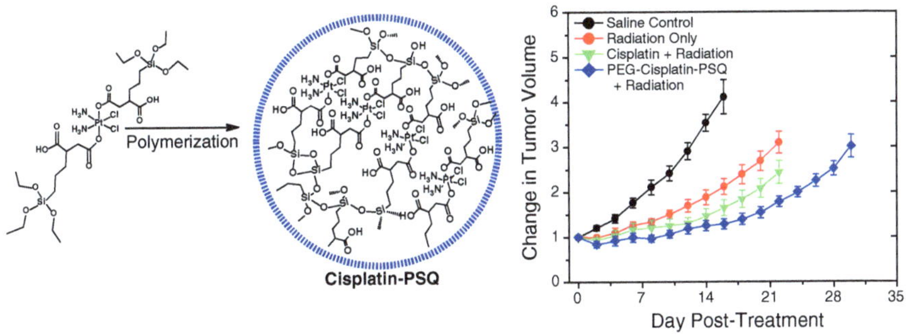

Fig. 10 Schematic showing the synthesis of PSQ nanoparticles carrying a cisplatin prodrug and their anticancer efficacy in human lung cancer xenograft murine models by chemoradiotherapy. [83] Reproduced with permission from Ref. [83]

3.3 PSQ Nanoparticles for Cancer Therapy

The Lin group constructed an oxaliplatin prodrug into PSQ nanoparticles and modified these particles with PEG and an active targeting ligand anisimide for enhanced accumulation in cancer cells [34]. These PSQ nanoparticles exhibited superior cytotoxicity to oxaliplatin free drug against four cancer cell lines in vitro and showed anticancer efficacy in human pancreatic xenograft murine models after intravenous injection. Most recently, Lin and coworkers reported a PSQ nanoparticle carrying a cisplatin prodrug and its utilization in chemoradiotherapy using human lung cancer as a disease model [83]. This PSQ nanoparticle has an exceptionally high loading of cisplatin and demonstrated significantly enhanced anticancer efficacy against free cisplatin both in vitro and in vivo with chemoradiotherapy (Fig. 10).

Cheng and coworkers developed PSQ nanoparticles with controllable sizes, drug loadings, and release profiles [84, 85]. The trialkoxysilane-containing drugs were synthesized by forming a degradable ester linker between drug and trialkoxysilane group and then condensed with tetraalkoxysilane to incorporate the drug into the resulting nanoparticles. Anticancer drugs including campothecin, paclitaxel, and docetaxel were loaded into the nanoparticles with precisely controlled sizes (ranging from 20 to 200 nm). The investigation on the correlation between the particle size and tumor accumulation/penetration revealed that particles with diameter of 20–50 nm exhibited enhanced anticancer efficacy, which was resulted from the faster cellular internalization and more efficient tumor accumulation/penetration [83].

4 Conclusions

Hybrid nanoparticles have emerged as powerful platforms for cancer imaging and therapy. Combining both organic and inorganic components within a single platform allows these materials to be functionalized for a multitude of applications. NMOFs possess nearly infinite tunability for specific applications and are intrinsically biodegradable. Although the development of NMOFs for cancer imaging and therapy applications is still at its early stage, preliminary results show that they hold great promise in clinical settings.

PSQ nanoparticles have been developed as imaging, therapeutic, and theranostic materials. These materials can be tailored to specific physical properties and applied to multiple biomedical applications. In vitro and in vivo evidences have demonstrated the advantages of PSQ nanoparticles over small molecule agents. Further improvements in their pharmacokinetics, biocompatibility, targeting efficiency, and biodegradability are required for their advancing into clinical use.

References

1. Vander Heiden MG (2011) Targeting cancer metabolism: a therapeutic window opens. Nat Rev Drug Discov 10(9):671–684. doi:10.1038/Nrd3504
2. Dobbelstein M, Moll U (2014) Targeting tumour-supportive cellular machineries in anticancer drug development. Nat Rev Drug Discov 13(3):179–196. doi:10.1038/Nrd4201
3. Davis ME, Chen Z, Shin DM (2008) Nanoparticle therapeutics: an emerging treatment modality for cancer. Nat Rev Drug Discov 7(9):771–782. doi:10.1038/Nrd2614
4. Kim CS, Duncan B, Creran B, Rotello VM (2013) Triggered nanoparticles as therapeutics. Nano Today 8(4):439–447. doi:10.1016/j.nantod.2013.07.004
5. Thorek DL, Ulmert D, Diop NF, Lupu ME, Doran MG, Huang R et al (2014) Non-invasive mapping of deep-tissue lymph nodes in live animals using a multimodal PET/MRI nanoparticle. Nat Commun 5:3097. doi:10.1038/ncomms4097
6. Lee N, Cho HR, Oh MH, Lee SH, Kim K, Kim BH et al (2012) Multifunctional Fe3O4/TaOx core/shell nanoparticles for simultaneous magnetic resonance imaging and X-ray computed tomography. J Am Chem Soc 134(25):10309–10312. doi:10.1021/Ja3016582
7. Rolfe BE, Blakey I, Squires O, Peng H, Boase NRB, Alexander C et al (2014) Multimodal polymer nanoparticles with combined F-19 magnetic resonance and optical detection for tunable, targeted, multimodal imaging in vivo. J Am Chem Soc 136(6):2413–2419. doi:10.1021/Ja410351h
8. Zhao YF, Sultan D, Detering L, Cho SH, Sun GR, Pierce R et al (2014) Copper-64-alloyed gold nanoparticles for cancer imaging: improved radiolabel stability and diagnostic accuracy. Angew Chem Int Edit 53(1):156–159. doi:10.1002/anie.201308494
9. Lee SB, Kim HL, Jeong HJ, Lim ST, Sohn MH, Kim DW (2013) Mesoporous silica nanoparticle pretargeting for PET imaging based on a rapid bioorthogonal reaction in a living body. Angew Chem Int Edit 52(40):10549–10552. doi:10.1002/anie.201304026
10. Nahrendorf M, Keliher E, Marinelli B, Waterman P, Feruglio PF, Fexon L et al (2010) Hybrid PET-optical imaging using targeted probes. Pro Nat Acad Sci USA 107(17):7910–7915. doi:10.1073/pnas.0915163107
11. Petros RA, DeSimone JM (2010) Strategies in the design of nanoparticles for therapeutic applications. Nat Rev Drug Discov 9(8):615–627. doi:10.1038/Nrd2591

12. Bardhan R, Lal S, Joshi A, Halas NJ (2011) Theranostic nanoshells: from probe design to imaging and treatment of cancer. Acc Chem Res 44(10):936–946. doi:10.1021/Ar200023x
13. Master A, Livingston M, Sen Gupta A (2013) Photodynamic nanomedicine in the treatment of solid tumors: perspectives and challenges. J Control Release 168(1):88–102. doi:10.1016/j.jconrel.2013.02.020
14. Szakacs G, Paterson JK, Ludwig JA, Booth-Genthe C, Gottesman MM (2006) Targeting multidrug resistance in cancer. Nat Rev Drug Discov 5(3):219–234. doi:10.1038/Nrd1984
15. Bock C, Lengauer T (2012) Managing drug resistance in cancer: lessons from HIV therapy. Nat Rev Cancer 12(7):494–501. doi:10.1038/Nrc3297
16. Pasparakis G, Manouras T, Vamvakaki M, Argitis P (2014) Harnessing photochemical internalization with dual degradable nanoparticles for combinatorial photo-chemotherapy. Nat Commun 5:3623. doi:10.1038/ncomms4623
17. Gao XH, Cui YY, Levenson RM, Chung LWK, Nie SM (2004) In vivo cancer targeting and imaging with semiconductor quantum dots. Nat Biotechnol 22(8):969–976. doi:10.1038/Nbt994
18. Huang XH, El-Sayed IH, Qian W, El-Sayed MA (2006) Cancer cell imaging and photothermal therapy in the near-infrared region by using gold nanorods. J Am Chem Soc 128(6):2115–2120. doi:10.1021/Ja057254a
19. Gupta AK, Gupta M (2005) Synthesis and surface engineering of iron oxide nanoparticles for biomedical applications. Biomaterials 26(18):3995–4021. doi:10.1016/j.biomaterials.2004.10.012
20. Shen J, Zhao L, Han G (2013) Lanthanide-doped upconverting luminescent nanoparticle platforms for optical imaging-guided drug delivery and therapy. Adv Drug Deliv Rev 65 (5):744–755. doi:10.1016/j.addr.2012.05.007
21. Li Z, Huve J, Krampe C, Luppi G, Tsotsalas M, Klingauf J et al (2013) Internalization pathways of anisotropic disc-shaped Zeolite L nanocrystals with different surface properties in HeLa cancer cells. Small 9(9–10):1809–1820. doi:10.1002/smll.201201702
22. Torchilin VP (2005) Recent advances with liposomes as pharmaceutical carriers. Nat Rev Drug Discov 4(2):145–160. doi:10.1038/Nrd1632
23. Lee CC, MacKay JA, Frechet JMJ, Szoka FC (2005) Designing dendrimers for biological applications. Nat Biotechnol 23(12):1517–1526. doi:10.1038/Nbt1171
24. Kataoka K, Harada A, Nagasaki Y (2012) Block copolymer micelles for drug delivery: design, characterization and biological significance. Adv Drug Deliv Rev 64:37–48. doi:10.1016/j.addr.2012.09.013
25. Hamidi M, Azadi A, Rafiei P (2008) Hydrogel nanoparticles in drug delivery. Adv Drug Deliv Rev 60(15):1638–1649. doi:10.1016/j.addr.2008.08.002
26. Taylor-Pashow KML, Della Rocca J, Huxford RC, Lin W (2010) Hybrid nanomaterials for biomedical applications. Chem Commun 46(32):5832–5849. doi:10.1039/C002073g
27. Della Rocca J, Liu D, Lin W (2011) Nanoscale metal-organic frameworks for biomedical imaging and drug delivery. Acc Chem Res 44(10):957–968. doi:10.1021/Ar200028a
28. Kitagawa S, Kitaura R, Noro S (2004) Functional porous coordination polymers. Angew Chem Int Edit 43(18):2334–2375. doi:10.1002/anie.200300610
29. Férey G (2008) Hybrid porous solids: past, present, future. Chem Soc Rev 37(1):191–214. doi:10.1039/B618320b
30. Yaghi OM, Li HL, Davis C, Richardson D, Groy TL (1998) Synthetic strategies, structure patterns, and emerging properties in the chemistry of modular porous solids. Acc Chem Res 31 (8):474–484. doi:10.1021/Ar970151f
31. Taylor-Pashow KML, Della Rocca J, Xie Z, Tran S, Lin W (2009) Postsynthetic Modifications of iron-carboxylate nanoscale metal-organic frameworks for imaging and drug delivery. J Am Chem Soc 131(40):14261–14263. doi:10.1021/Ja906198y
32. Della Rocca J, Lin W (2010) Nanoscale metal-organic frameworks: magnetic resonance imaging contrast agents and beyond. Eur J Inorg Chem (24):3725-3734. doi:10.1002/ejic.201000496

33. Horcajada P, Chalati T, Serre C, Gillet B, Sebrie C, Baati T et al (2010) Porous metal-organic-framework nanoscale carriers as a potential platform for drug delivery and imaging. Nat Mater 9(2):172–178. doi:10.1038/Nmat2608
34. Della Rocca J, Huxford RC, Comstock-Duggan E, Lin W (2011) Polysilsesquioxane nanoparticles for targeted platin-based cancer chemotherapy by triggered release. Angew Chem Int Edit 50(44):10330–10334. doi:10.1002/anie.201104510
35. Vivero-Escoto JL, Rieter WJ, Lau H, Huxford-Phillips RC, Lin W (2013) Biodegradable polysilsesquioxane nanoparticles as efficient contrast agents for magnetic resonance imaging. Small 9(20):3523–3531. doi:10.1002/smll.201300198
36. Zhang T, Lin W (2014) Metal-organic frameworks for artificial photosynthesis and photocatalysis. Chem Soc Rev 43(16):5982–5993. doi:10.1039/c4cs00103f
37. Jiang HL, Feng DW, Wang KC, Gu ZY, Wei ZW, Chen YP et al (2013) An exceptionally stable, porphyrinic Zr metal-organic framework exhibiting pH-dependent fluorescence. J Am Chem Soc 135(37):13934–13938. doi:10.1021/Ja406844r
38. Liu D, Lu K, Poon C, Lin W (2014) Metal-organic frameworks as sensory materials and imaging agents. Inorg Chem 53(4):1916–1924. doi:10.1021/Ic402194c
39. Rao XT, Song T, Gao JK, Cui YJ, Yang Y, Wu CD et al (2013) A highly sensitive mixed Lanthanide metal-organic framework self-calibrated luminescent thermometer. J Am Chem Soc 135(41):15559–15564. doi:10.1021/Ja407219k
40. Rieter WJ, Taylor KML, Lin W (2007) Surface modification and functionalization of nanoscale metal-organic frameworks for controlled release and luminescence sensing. J Am Chem Soc 129(32):9852–9853. doi:10.1021/Ja073506r
41. Xie Z, Ma L, deKrafft KE, Jin A, Lin W (2010) Porous phosphorescent coordination polymers for oxygen sensing. J Am Chem Soc 132(3):922–923. doi:10.1021/Ja909629f
42. Wanderley MM, Wang C, Wu C, Lin W (2012) A chiral porous metal-organic framework for highly sensitive and enantioselective fluorescence sensing of amino alcohols. J Am Chem Soc 134(22):9050–9053. doi:10.1021/Ja302110d
43. Harbuzaru BV, Corma A, Rey F, Jorda JL, Ananias D, Carlos LD et al (2009) A miniaturized linear pH sensor based on a highly photoluminescent self-assembled Europium(III) metal-organic framework. Angew Chem Int Edit 48(35):6476–6479. doi:10.1002/anie.200902045
44. Hirai K, Sumida K, Meilikhov M, Louvain N, Nakahama M, Uehara H et al (2014) Impact of crystal orientation on the adsorption kinetics of a porous coordination polymer-quartz crystal microbalance hybrid sensor. J Mater Chem C 2(17):3336–3344. doi:10.1039/C3tc32101k
45. Beauvais LG, Shores MP, Long JR (2000) Cyano-bridged Re(6)Q(8) (Q = S, Se) Cluster-Cobalt(II) framework materials: versatile solid chemical sensors. J Am Chem Soc 122 (12):2763–2772. doi:10.1021/Ja994186h
46. Lu G, Farha OK, Kreno LE, Schoenecker PM, Walton KS, Van Duyne RP et al (2011) Fabrication of metal-organic framework-containing silica-colloidal crystals for vapor sensing. Adv Mater 23(38):4449–4452. doi:10.1002/adma.201102116
47. Yoon M, Srirambalaji R, Kim K (2012) Homochiral metal-organic frameworks for asymmetric heterogeneous catalysis. Chem Rev 112(2):1196–1231. doi:10.1021/Cr2003147
48. Rieter WJ, Pott KM, Taylor KML, Lin W (2008) Nanoscale coordination polymers for platinum-based anticancer drug delivery. J Am Chem Soc 130(35):11584–11585. doi:10.1021/Ja803383k
49. Huxford-Phillips RC, Russell SR, Liu D, Lin W (2013) Lipid-coated nanoscale coordination polymers for targeted cisplatin delivery. Rsc Adv 3(34):14438–14443. doi:10.1039/C3ra42033g
50. Liu D, Kramer SA, Huxford-Phillips RC, Wang S, Della Rocca J, Lin W (2012) Coercing bisphosphonates to kill cancer cells with nanoscale coordination polymers. Chem Commun. 48(21):2668–2670. doi:10.1039/C2cc17635a
51. He C, Lu K, Liu D, Lin W (2014) Nanoscale metal-organic frameworks for the co-delivery of cisplatin and pooled siRNAs to enhance therapeutic efficacy in drug-resistant ovarian cancer cells. J Am Chem Soc 136(14):5181–5184. doi:10.1021/Ja4098862

52. Rieter WJ, Taylor KML, An HY, Lin W, Lin W (2006) Nanoscale metal-organic frameworks as potential multimodal contrast enhancing agents. J Am Chem Soc 128(28):9024–9025. doi:10.1021/Ja0627444
53. Taylor KML, Jin A, Lin W (2008) Surfactant-assisted synthesis of nanoscale gadolinium metal-organic frameworks for potential multimodal imaging. Angew Chem Int Edit 47 (40):7722–7725. doi:10.1002/anie.200802911
54. Huxford RC, deKrafft KE, Boyle WS, Liu D, Lin W (2012) Lipid-coated nanoscale coordination polymers for targeted delivery of antifolates to cancer cells. Chem Sci 3(1):198–204. doi:10.1039/C1sc00499a
55. Liu D, Poon C, Lu K, He C, Lin W (2014) Self-assembled nanoscale coordination polymers with trigger release properties for effective anticancer therapy. Nat Commun 5:4182. doi:10.1038/ncomms5182
56. Dekrafft KE, Xie ZG, Cao GH, Tran S, Ma L, Zhou OZ et al (2009) Iodinated nanoscale coordination polymers as potential contrast agents for computed tomography. Angew Chem Int Edit 48(52):9901 9904. doi:10.1002/anie.200904958
57. Miller SR, Heurtaux D, Baati T, Horcajada P, Greneche JM, Serre C (2010) Biodegradable therapeutic MOFs for the delivery of bioactive molecules. Chem Commun 46(25):4526–4528. doi:10.1039/C001181a
58. Nguyen JG, Tanabe KK, Cohen SM (2010) Postsynthetic diazeniumdiolate formation and NO release from MOFs. CrystEngComm 12(8):2335–2338. doi:10.1039/C000154f
59. Vivero-Escoto JL, Huxford-Phillips RC, Lin W (2012) Silica-based nanoprobes for biomedical imaging and theranostic applications. Chem Soc Rev 41(7):2673–2685. doi:10.1039/C2cs15229k
60. Lee GH, Chang Y, Kim TJ (2012) Blood-pool and targeting MRI contrast agents: from Gd-chelates to Gd-nanoparticles. Eur J Inorg Chem 12:1924–1933. doi:10.1002/ejic.201101137
61. Perrier M, Kenouche S, Long J, Thangavel K, Larionova J, Goze-Bac C et al (2013) Investigation on NMR relaxivity of nano-sized cyano-bridged coordination polymers. Inorg Chem 52(23):13402–13414. doi:10.1021/Ic401710j
62. Rowe MD, Chang CC, Thamm DH, Kraft SL, Harmon JF, Vogt AP et al (2009) Tuning the magnetic resonance imaging properties of positive contrast agent nanoparticles by surface modification with RAFT polymers. Langmuir 25(16):9487–9499. doi:10.1021/La900730b
63. Rowe MD, Thamm DH, Kraft SL, Boyes SG (2009) Polymer-modified gadolinium metal-organic framework nanoparticles used as multifunctional nanomedicines for the targeted imaging and treatment of cancer. Biomacromolecules 10(4):983–993. doi:10.1021/Bm900043e
64. Caravan P, Ellison JJ, McMurry TJ, Lauffer RB (1999) Gadolinium(III) chelates as MRI contrast agents: structure, dynamics, and applications. Chem Rev 99(9):2293–2352. doi:10.1021/Cr980440x
65. Taylor KML, Rieter WJ, Lin W (2008) Manganese-based nanoscale metal-organic frameworks for magnetic resonance imaging. J Am Chem Soc 130(44):14358–14359. doi:10.1021/Ja803777x
66. Paul G, Prado Y, Dia N, Riviere E, Laurent S, Roch M et al (2014) Mn-II-containing coordination nanoparticles as highly efficient T-1 contrast agents for magnetic resonance imaging. Chem Commun 50(51):6740–6743. doi:10.1039/C4cc01251h
67. Brenner DJ, Hall EJ (2007) Current concepts—computed tomography—an increasing source of radiation exposure. New Engl J Med 357(22):2277–2284. doi:10.1056/Nejmra072149
68. deKrafft KE, Boyle WS, Burk LM, Zhou OZ, Lin W (2012) Zr- and Hf-based nanoscale metal-organic frameworks as contrast agents for computed tomography. J Mater Chem 22 (35):18139–18144. doi:10.1039/C2jm32299d
69. Liu D, Huxford RC, Lin W (2011) Phosphorescent nanoscale coordination polymers as contrast agents for optical imaging. Angew Chem Int Edit 50(16):3696–3700. doi:10.1002/anie.201008277

70. Nishiyabu R, Aime C, Gondo R, Kaneko K, Kimizuka N (2010) Selective inclusion of anionic quantum dots in coordination network shells of nucleotides and lanthanide ions. Chem Commun 46(24):4333–4335. doi:10.1039/C001012j
71. Aime C, Nishiyahu R, Gondo R, Kimizuka N (2010) Switching on luminescence in nucleotide/lanthanide coordination nanoparticles via synergistic interactions with a cofactor ligand. Chem-Eur J 16(12):3604–3607. doi:10.1002/chem.201090007
72. Nishiyabu R, Hashimoto N, Cho T, Watanabe K, Yasunaga T, Endo A et al (2009) Nanoparticles of adaptive supramolecular networks self-assembled from nucleotides and lanthanide ions. J Am Chem Soc 131(6):2151–2158. doi:10.1021/Ja8058843
73. Aime C, Nishiyabu R, Gondo R, Kaneko K, Kimizuka N (2008) Controlled self-assembly of nucleotide-lanthanide complexes: specific formation of nanofibers from dimeric guanine nucleotides. Chem Commun 48:6534–6536. doi:10.1039/B815779k
74. He C, Lu K, Lin W (2014) Nanoscale metal-organic frameworks for real-time intracellular pH sensing in live cells. J Am Chem Soc 136(35):12253–12256. doi:10.1021/ja507333c
75. Duncan R (2006) Polymer conjugates as anticancer nanomedicines. Nat Rev Cancer 6(9):688–701. doi:10.1038/Nrc1958
76. Allen TM (2002) Ligand-targeted therapeutics in anticancer therapy. Nat Rev Cancer 2 (10):750–763. doi:10.1038/Nrc903
77. Harley CB (2008) Telomerase and cancer therapeutics. Nat Rev Cancer 8(3):167–179. doi:10.1038/Nrc2275
78. Li SD, Huang L (2008) Pharmacokinetics and biodistribution of nanoparticles. Mol Pharm 5 (4):496–504. doi:10.1021/mp800049w
79. Imaz I, Rubio-Martinez M, Garcia-Fernandez L, Garcia F, Ruiz-Molina D, Hernando J et al (2010) Coordination polymer particles as potential drug delivery systems. Chem Commun 46 (26):4737–4739. doi:10.1039/C003084h
80. Gao PF, Zheng LL, Liang LJ, Yang XX, Li YF, Huang CZ (2013) A new type of pH-responsive coordination polymer sphere as a vehicle for targeted anticancer drug delivery and sustained release. J Mater Chem B 1(25):3202–3208. doi:10.1039/C3tb00026e
81. McKinlay AC, Xiao B, Wragg DS, Wheatley PS, Megson IL, Morris RE (2008) Exceptional behavior over the whole adsorption-storage-delivery cycle for NO in porous metal organic frameworks. J Am Chem Soc 130(31):10440–10444. doi:10.1021/Ja801997r
82. Morris W, Briley WE, Auyeung E, Cabezas MD, Mirkin CA (2014) Nucleic acid-metal organic framework (MOF) nanoparticle conjugates. J Am Chem Soc 136(20):7261–7264. doi:10.1021/Ja503215w
83. Della Rocca J, Werner ME, Kramer SA, Huxford-Phillips RC, Sukumar R, Cummings ND, Vivero-Escoto JL, Wang AZ, Lin W (2014) Polysilsesquioxane nanoparticles for triggered release of cisplatin and effective cancer chemoradiotherapy. Nanomed Nanotechnol Biol Med 11(1):31–38. doi:10.1016/j.nano.2014.07.004
84. Tang L, Fan TM, Borst LB, Cheng JJ (2012) Synthesis and biological response of size-specific, monodisperse drug-silica nanoconjugates. Acs Nano 6(5):3954–3966. doi:10.1021/Nn300149c
85. Tang L, Gabrielson NP, Uckun FM, Fan TM, Cheng JJ (2013) Size-dependent tumor penetration and in vivo efficacy of monodisperse drug-silica nanoconjugates. Mol Pharmaceut 10(3):883–892. doi:10.1021/Mp300684a

Exploring the Tumor Microenvironment with Nanoparticles

Lei Miao and Leaf Huang

Abstract

Recent developments in nanotechnology have brought new approaches to cancer diagnosis and therapy. While enhanced permeability and retention effect (EPR) promotes nanoparticle (NP) extravasation, the abnormal tumor vasculature, high interstitial pressure and dense stroma structure limit homogeneous intratumoral distribution of NP and compromise their imaging and therapeutic effect. Moreover, heterogeneous distribution of NP in nontumor-stroma cells damages the nontumor cells, and interferes with tumor-stroma crosstalk. This can lead to inhibition of tumor progression, but can also paradoxically induce acquired resistance and facilitate tumor cell proliferation and metastasis. Overall, the tumor microenvironment plays a crucial, yet controversial role in regulating NP distribution and their biological effects. In this review, we summarize recent studies on the stroma barriers for NP extravasation, and discuss the consequential effects of NP distribution in stroma cells. We also highlight design considerations to improve NP delivery and propose potential combinatory strategies to overcome acquired resistance induced by damaged stroma cells.

Keywords

Nanoparticle · Tumor microenvironment · Extracellular matrix · Pericytes tumor-associated fibroblast

L. Miao · L. Huang (✉)
Division of Molecular Pharmaceutics and Center of Nanotechnology in Drug Delivery, Eshelman School of Pharmacy, University of North Carolina at Chapel Hill, Chapel Hill, NC 27599, USA
e-mail: leafh@unc.edu

C.A. Mirkin et al. (eds.), *Nanotechnology-Based Precision Tools for the Detection and Treatment of Cancer*, Cancer Treatment and Research 166,
DOI 10.1007/978-3-319-16555-4_9

Acronyms

EPR Enhanced Permeability and Retention Effect
ECM Extracellular Matrix
TME Tumor Microenvironment
BM Basement Membrane
IFP Interstitial Fluidic Pressure
TAF Tumor Associated Fibroblast
TAM Tumor Associated Macrophage
MMP Matrix Metalloproteinases
NP Nanoparticles

Contents

1 Introduction

Rapid development in nanotechnology allows the incorporation of multiple diagnostic and therapeutic agents (e.g., liposomes and quantum dots) into nanoparticles (NP) with a size range from 1 to 1000 nm [1–4]. These nanocarrier systems provide new approaches to diagnose, prevent, and treat aggressive malignancy. Nano-based delivery systems hold an advantage over traditional small molecule chemotherapy in that they can deliver drugs preferentially to tumors due to the enhanced permeability and retention (EPR) effect, sparing healthy tissues from dose-limiting side effects [5–7]. Since their size, shape, and surface properties can be tailored as needed, nanomedicine can enhance bioavailability, control drug release kinetics, improve pharmacokinetics, and provide a superior dosing schedule for better patient compliance [8–14]. Other advantages of nano-based delivery include the ability to simultaneously incorporate multiple therapeutic agents [15–20], or co-delivery a therapeutic agent along with an imaging agent for tumor visualization [21–23]. Nano-based delivery can also be designed to target tumors and other tumor-stroma components by surface modification with specific targeting ligands [24, 25]. Although many types of tumors have developed an innate resistance to chemotherapy, nanomedicine also has the potential to overcome resistance through an alternative path of cellular internalization [26, 27]. For example, members of the ATP binding cassette (ABC) family of influx transporters are defective in cisplatin-resistant cells [28]. NPs can bring cisplatin into cells via caveolin- or clathrin-mediated pathway, bypassing the defective transporter [29].

In the past two decades, over 20 nanotechnology-based therapeutic products have been approved for clinical use. Among these products, liposomal NP and polymer-drug conjugates are two major groups, accounting for more than 80 % of conjugates that have entered the clinic [30–32]. Although these nano-formulations are very efficient in decreasing adverse effects and inducing tumor regression, their actual clinical application, however, was limited to just a few types of tumors. A better understanding of the pathology of different tumor types and the common barriers that prevent intratumoral transport of NP is required to develop more broadly applicable strategies for the improvement of therapeutic outcomes across multiple cancer types.

Neoplastic epithelial cells co-exist in carcinomas with several different types of stromal cells and the extracellular matrix (ECM) that together create the tumor microenvironment (TME) [33]. To achieve an efficient tumor cell internalization, systemically delivered NP must first accumulate in tumors via blood flow, then extravasate from the microvessels, and finally pass through the ECM to reach the target cells [6, 31, 34, 35]. The first barrier is the extravasation of NP from tumor vasculature (Fig. 1). Although leaky and tortuous tumor vessels allow the accumulation of NP, mural wall cells, especially pericytes, and basement membrane (BM), paradoxically limit the penetration of NP through pole-opening on capillary walls [5, 34, 36, 37]. The uniformly elevated interstitial fluid pressure (IFP) resulting from vascular hyperpermeability and lymphatic malfunction further reduces convective transport [5, 14]. The second barrier leading to limited NP diffusion is the dense stroma (Fig. 1). Stroma consists of nontumor cells (e.g., fibroblast, tumor-associated fibroblast (TAF), epithelium, endothelium, muscle cells, and immune cells) and a highly cross-linked fibril-like ECM structure with collagen and glycosaminoglycan as major constituents [38–40]. The ECM not only supports tumor cell growth and metastasis, but also functions as a sieve with high osmotic pressure to inhibit passive diffusion of NP [14, 41–43].

Most nanotechnology research focuses on improving NP penetration and diffusion to tumor cells, and not on distribution of NP to other cellular components in stroma. Endothelial cells, TAFs, or tumor, associated macrophages (TAM), have been recognized as major components that promote cancer progression, therapy

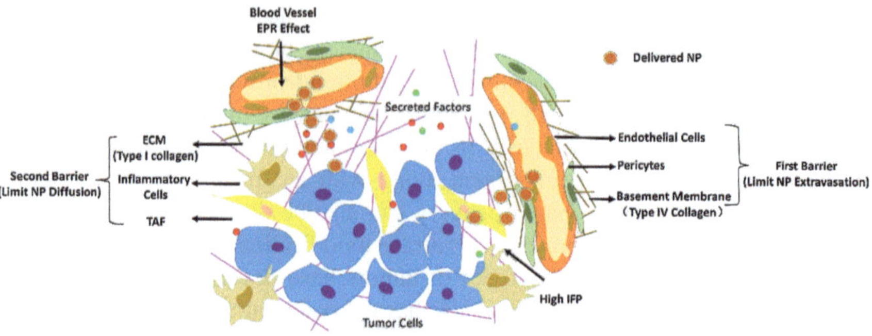

Fig. 1 Schematic illustration of the major cellular and noncellular components of the tumor microenvironment (TME). The EPR effect facilitates nanoparticle (NP) accumulation. However, high interstitial fluidic pressure (IFP), pericyte coverage, and the basement membrane limit NP extravasation from the blood vessels toward the interstitial space. When NP extravasate from the blood vessels, stroma cells and the extracellular matrix act as another barrier to further inhibit NP diffusion

resistance, and metastasis formation [44–47]. The modulation of these stromal cellular components by either small molecules or nanodrugs can facilitate the remodeling of tumor blood vessels or ECM. In addition, imaging of the TME by NP that target stroma cells increasingly contribute to accurate diagnosis, early response evaluation and treatment guidance. In this review, we discuss the barriers to NP delivery, and provide strategies to overcome these limitations. We also summarize NP that are designed to target stroma cells in vivo, discuss their diagnostic application in tumor imaging and the paradoxical therapeutic effects induced by stroma cell damage and depletion [48–50].

Finally, we propose possible strategies to overcome the heterogeneity of solid tumor patterns, the complexity of stroma cells in NP delivery, and also mention the NP-based diagnostic and therapeutic applications in metastatic tumors.

2 Paradoxical Features of the Tumor Microenvironment Impacting Nanoparticle Accumulation and Penetration

2.1 Abundant Neovasculature and EPR Effect Promote Nanoparticle Accumulation in Tumor

Nano-based formulations have shown promising antitumor effects compared to free drugs due to improved pharmacokinetic properties and preferential accumulation in the tumor with abnormal vasculature [51–53]. Intravenously injected nanodrugs are delivered into the pathological lesions through arterioles and released from capillaries. Therefore, the key mediators of nanomedicine intratumoral delivery are small vessels, especially capillaries [7]. Normal capillaries are lined by a tightly sealed endothelium, firmly attached and supported on the abluminal side by stellate-shaped pericytes, which are further enveloped in a thin layer of basement membrane (BM) [54]. In normal tissues, pericyte coverage of the endothelial abluminal surface varies among different organs and blood vasculatures, with a general range between 10 and 70 % [54, 55]. The vasculature BM, with major components of type IV collagen, laminin, entactin (nidogen), fibronectin, usually envelops blood vessels with a thickness ranging from 100–150 nm [56, 57]. Unlike normal blood vessels, tumor vasculatures usually have large pore openings (0.1–3 μm in diameter), leading to significantly higher vascular permeability and hydraulic conductivity [58, 59]. In addition, the extent of pericyte coverage on tumor vessels is typically diminished compared to normal tissues [54]. Both pericytes and BM are loosely associated with the endothelial cells and partially penetrate deep in the tumor parenchyma [6, 36, 54, 60]. This inherent leaky and loosely compacted vasculature tends to be abnormally permeable to macromolecules and NP (10–100 nm, in diameter). When coupled with impaired lymphatic drainage, the EPR effect brings several advantages in theranostic NP-based drug delivery (Fig. 1).

2.2 Functional Nanomaterials for Therapeutic and Diagnostic Applications in Cancer

Nanomedicine-based therapies refer to active pharmaceutic ingredients encapsulated into or conjugated with nano-based delivery vehicles, including liposomes, polymer micelles, polymer-drug conjugates, dendrimers, and macromolecule (Fig. 2) [6]. The particle size of these nanovectors ranges from 10 to 100 nm, which prevents first-pass elimination in kidneys, in turn allowing accumulation in tumors via the leaky vasculature. Biocompatible polymers such as poly-ethyleneglycol (PEG) and targeting ligands such as antibodies, peptides, and small molecules have been attached onto the surface of liposomes to achieve increased circulation time and cell internalization [61, 62]. Examples include the doxorubicin-liposome Doxil and the vincristine-liposome Onco TCSs [63, 64]. Doxil increases the half-life in the blood due to the chemical coating of PEG. It is effective in the treatment of hypervasculatured tumors, including Kaposi Sarcoma and Ovarian Cancers [63]. PEGylation has also been used to prepare polymer-drug conjugates. Another attractive polymer employed to formulate drug conjugates is N-(2-hydroxyproyl) methacrylamide (HPMA). A number of HPMA products are currently in clinical trials due to their desirable attributes, such as hydrophilicity, functionalizable side chain and biodegradability. Examples include a HPMA copolymer paclitaxel formulation in phase I trial for treating solid tumors [63]. Abraxane ($\sim$130 nm albumin-bound paclitaxel NP) is another paclitaxel formulation, which is one of the only two FDA approved nano-formulations in clinical trials besides Doxil. Though various NP formulations have been developed to target tumor cells inducing cell apoptosis, only modest survival benefits have been achieved. One possible reason is

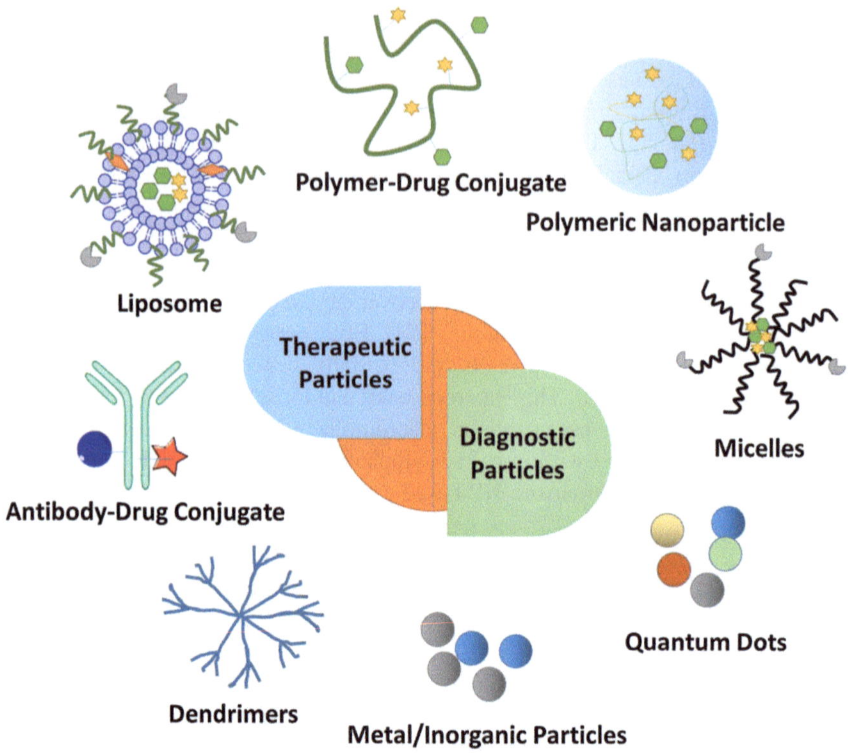

Fig. 2 Multi-functional nanoparticles (NP) for diagnostic and therapeutic effects in the treatment of tumors. Examples include: liposomes, polymer-drug conjugates, polymeric nanoparticles, micelles, antibody-drug conjugates, dendrimers, metal NP and quantum dots

that the abnormal tumor vasculature and the dense interstitial matrix hinder delivery of the drug throughout the entire tumor in sufficient concentration.

Another clinical application of NP is cancer imaging. Precise imaging of tumor cells and their microenvironment provides accurate diagnosis and treatment guidance [65]. In the past decades, several theranostic NPs have been designed to target tumor. These include magnetic and iron oxide NP for magnetic resonance imaging (MRI), surface modified iodine, gold, and bismuth NP for X-ray computed tomography (CT) and fluorescent labeled dextran NP, silica NP, and surfaces stabilized quantum dots for fluorescence imaging [65]. Preferential internalization of certain imaging NP by stroma cells provides the possibility of imaging the TME [65]. Visualization of the TME not only contributes to disease diagnosis, but also underlines the integral distribution pattern of NP, which can subsequently guide therapeutic NP treatment.

Although, more and more imaging agents and nanodrugs are emerging based on the EPR effect, there are still many obstacles to overcome for an effective tumor diagnosis and therapy. Limited intratumoral penetration, disparate stromal cell distribution and response are two major barriers for imaging and therapy.

2.3 Barriers for Extravasation of Nanoparticle from Blood Vessel into Extracellular Matrix

2.3.1 High Interstitial Fluid Pressure (IFP) Limits NP Convection

On one hand, high permeability of the tumor vessels and a lack of functional lymphatic vessels results in the EPR effect, driving NP extravasation; on the other hand, these phenomena lead to high IFP, limiting NP extravasation. In normal tissues, IFP is around 0 mm Hg; whereas tumors, exhibiting interstitial hypertension, have an IFP almost identical to the microvascular pressure (with a range of 10–40 mm Hg) [66–68]. High IFP limits convection of NP, paradoxically promoting passive diffusion [69]. Diffusion is a much slower transvascular process than convection, especially for the transport of large NP [14]. Moreover, stroma cells compress intratumoral blood and lymphatic vessels, which consequently impairs blood flow, leading to blood stasis, loss of function, and further inhibition of NP penetration [70]. The vascular abnormalities can also cause hypoxia and acidosis. Hypoxia renders tumor cells resistant to both cytotoxic drugs and radiation, while also inducing genetic instability and selecting for more malignant tumor cells with potentially metastatic properties [59, 71]. Finally, because of the steep drop in IFP on the edge of tumors, intratumoral fluid can escape from the tumor periphery into the surrounding tissue, expulsing therapeutic NP, and also excreting growth factors (e.g., VEGF-A, PDGF-C) to facilitate tumor progression [14]. Altogether, the high IFP and abnormal vasculature pose a formidable barrier to both the delivery and efficacy of nanodrugs.

2.3.2 Pericytes Coverage as One Factor to Explain Limited Nanoparticle Extravasation

Pericytes are a ubiquitous part of the TME [54, 72]. They were first identified in 1923 and named based on their function as a major constituent of mural cells lining against microvessels [73]. Although no specific molecular marker has been identified specifically to pericytes, alpha smooth muscle actin (α-SMA), PDGFR-β, NG2 proteoglycan, RGS5, and XlacZ4 are commonly used [54, 73, 74]. Signaling pathways implicated in the development of pericytes and their interactions with endothelial cells have recently been reviewed in detail [54]. Briefly, pericytes recruitment involves multiple pathways in a tumor type-specific manner. They are recruited mainly by endothelial cells through PDGF-BB/PDGFR-β signaling. Alternative recruitment signaling includes HB-EGF, pericyte-expressed EGFR and SDF-1α/CXCR4 [75–77]. VEGF-A, a potent mediator of endothelial sprouting and neovascularization, acts as a negative regulator of pericyte function and vessel maturation [78]. Therefore, VEGF-A and PDGF-BB coordinately regulate pericyte coverage. Ablation of pericytes by anti-PDGF antibody or VEGF has been reported to increase vascular tortuosity and tumor growth in low PDGF-BB tumor models [79, 80]. This is paradoxical, since one would expect that increased leakiness of blood vessel with low-pericyte coverage would severely facilitate NP extravasation and inhibition tumor cell proliferation.

The relationship between pericyte coverage and NP extravasation has been investigated in detail by various groups [1, 81–83]. Different from the initial assumption that all tumor microvessels have low and loose pericyte coverage [54], emerging evidence has demonstrated heterogeneous pericyte coverage within in one single tumor or with regards to different tumor types. By defining pericytes as α-SMA positive cells attached to endothelial cells, Kano et al. has classified malignant tumors into high pericyte coverage and low-pericyte coverage subtypes [81, 82]. They further established that more coverage relates to a worse prognosis and more fibrotic interstitium for pancreatic, diffuse-type gastric cancer, clear cell renal cell carcinomas, and glioblastoma (60–70 % coverage), when compared to low-coverage cancers with a better prognosis such as colon cancer and ovarian cancer (10–20 % coverage) [73]. Disparate intratumoral NP transport in response to cytokine-mediated modification of pericyte coverage has been observed in these two types of tumors in a series of work by Kano et al. [73, 81–83]. Murine colon cancer CT26 is an example of low-pericyte coverage tumor, while the BxPC3 pancreatic model has been characterized by hypovasculature with more than 70 %

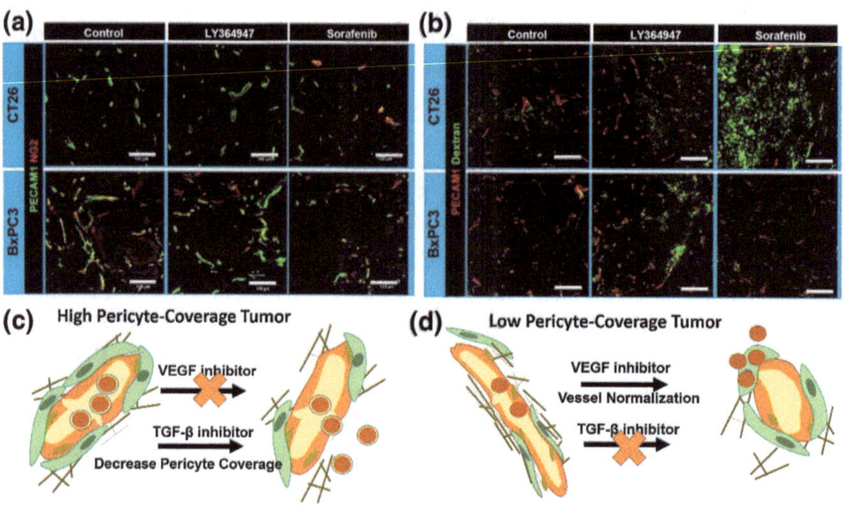

Fig. 3 Effects of VEGF inhibitor (Sorafenib) and TGF-β inhibitor (LY364947) in the CT26 or BxPC3 model. **a** Vascular phenotypes revealed by immunofluorescence staining. *Green*, CD31 or platelet endothelial cell adhesion molecule (PECAM)-1; *red*, NG2. **b** Extravasation of 2 MDa dextran from vasculature. Dextran is shown in *green* and CD31/PECAM-1 in *red*. Scale bars = 100 μm. **c** and **d** are schemes that explain the different effects of VEGF inhibitor and TGF-β inhibitor on high or low-pericyte coverage tumors (Reproduced from Kano 2009, copyright of Elsevier)

pericyte (Fig. 3). They compared the effects of three types of kinase inhibitors, including TGF-β inhibitor (LY364947), PDGF-B signaling inhibitor (imatinib), and VEGF inhibitor (Sorafenib) on extravasation of a modeled NP, 2 MDa dextran, and a liposomal formulation, Doxil, on CT26 and BxPC3. By using the BxPC3 model, they are able to show that the TGF-β inhibitor can improve 2 MDa dextran and Doxil penetration, leading to enhanced tumor inhibition (Fig. 3). This is due to the fact that low-dose TGF-β inhibitor can block pericyte proliferation without affecting the function of endothelial cells and tumor cells. Consistent with this finding, various types of TGFβ inhibitors, including small kinase inhibitor and its nano-formulation (Fig. 4), siRNA and antibodies have been shown to decrease pericyte coverage. These inhibitors increase vessel leakiness and improve the intratumoral penetration of sub-100 nm NP, including PEI-PEG-coated MSNP, liposome, and polymeric micelle in other high pericyte coverage tumors, such as diffuse-type gastric cancer and 4T1 breast cancer models [82–86]. This finding enabled delivery

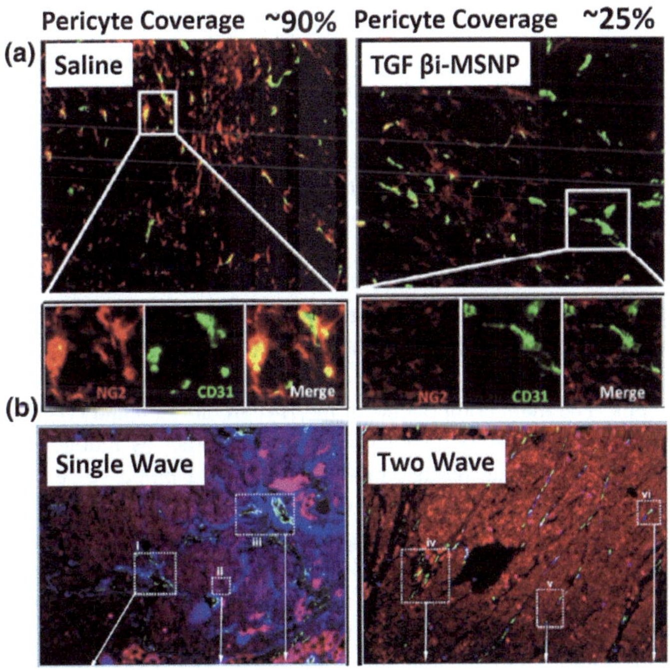

Fig. 4 Two wave nanotherapy was used to treat BxPC3 xenografts with high pericyte coverage. The first wave was MSNP loaded TGFβ inhibitor (TGFβi-MSNPs and the second was MSNP loaded gemcitabine. **a** Double staining of endothelial cells (CD31, *green*) and pericytes (NG2, *red*) in two treatment groups. The first wave TGFβi-MSNPs can significantly decrease the coverage of pericytes (*red*, NG2) on the endothelial wall (*green*, CD31). **b** Fluorescent images of tumor sections to show that TGFβi-MSNPs improve the extent of liposome intratumoral distribution in the BxPC3 xenografts. In **b**, liposomes were labeled with Texas-*red*, blood vessels were stained for CD31 (*green*) and pericyte was stained for NG2 (*blue*) (Reproduced from Meng et al. 2013, copyright of ACS publishing group)

optimization of various contrasting media. Two recent studies have indicated that TGF-β knockdown can improve MRI contrast [87, 88]. In contrast to the finding in the BxPC3 model, TGF-β inhibitors cannot increase particle penetration and improve therapeutic outcome in CT26, since the pericyte coverage was too low to achieve any additional effect. Interestingly, the VEGF inhibitor, Sorafenib, increased extravasation of 2 MDa dextran in the CT26 model (Fig. 3). Inhibition of VEGF-A can efficiently diminish nonfunctional microvessels with low-pericyte coverage while increase pericyte coverage in the functional microvessels. Thus, the tumor vasculature is "normalized." The increased pericyte coverage and tumor vasculature normalization were also observed by McDonald's group [89]. Recent research by Jain's group indicated that normalization of tumor blood vessels not only improves small molecule-based chemotherapy, but also facilitates the delivery of nanomedicine with smaller sizes [90]. Therefore, tumor vessel normalization is one explanation for enhanced NP penetration after treatment with a VEGF inhibitor in a CT26 tumor model. The combination of a VEGF inhibitor with Doxil can synergistically inhibit CT26 tumor growth. Thus, the relationship between pericyte coverage and NP extravasation varies with regard to the original pericyte coverage, blood vessel stabilization, and extracellular content. Pericyte coverage is an indispensable factor for vessel stabilization and maturation. Neither leaky, unmatured blood vessels with little coverage, nor over-matured vessels with abundant pericyte coverage are suitable for NP delivery. Moreover, pericyte coverage is just one of many factors that influence the intratumoral transport of NP. We need to take the other barriers into consideration when proposing strategies for the improvement of NP delivery.

2.3.3 Basement Membrane as Another Biophysical Barrier for Nanoparticle Extravasation into Interstitial Space

The BM is a specialized form of ECM that functions as a scaffold for endothelial and mural cells. The main components of the BM are laminin and type IV collagen, which form distinct sheet-like dispositions linked together by nidogen and heparin sulfates [56, 60]. In normal tissues, more than 99 % of blood vessels are covered with a thin layer of BM. It supports the architecture of the blood vessels and regulates vessel development through gradual secretion of pro-angiogenesis and proinflammatory cytokines, such as TGF-β and TNFα [36]. In contrast to normal blood vessel BM, the BM of tumor microvessels is primarily continuous but conspicuously abnormal [36]. Heterogeneous BM morphologies exist in different regions of the same tumor or different tumor types. The first type of BM is characterized by a loose association with endothelial cells. Spontaneous pancreatic islet cell tumors in RIP-Tag2 mice, MCa-IV breast carcinomas, and Lewis Lung carcinomas belong to this type of tumors [57]. Murine lung cancer 3LL and pancreatic cancer BxPC3 are marked by a second type of BM, with a distribution of brighter collagen nodules condensely overlapped with the capillary. The third type of BM can be observed in the 4T1 model. This model has a larger collagen content, which is completely dissociated from blood vessels. Different from the interstitial matrix, another category of ECM, BM does not induce elevated interstitial pressure, yet functions as a sieve to modulate extravasation of free drug and NP from capillaries into the TME.

Extravasation of 1 nm doxorubicin (DOX), 50 nm macromolecule FITC-tagged dextran and 80 nm PEGylated liposomes were evaluated on the BM/vessel overlapped 3LL model and the BM/vessel dissociated model 4T1. Results indicated that the extravasation pattern of small molecules (including DOX and dextran) were comparable, suggesting that vascular collagen could only modulate the transport of small molecules to a limited extent [57]. However, the extravasation of liposomes was significantly different between these two types of tumors. Extravasation only occurred from the vessels that were not tightly covered by collagen type IV. An in vitro collagen sleeve model was further developed to mimic collagen surrounding capillaries [57, 60]. Collagen sleeve thickness, which was modeled by changing the number of collagen fiber layers and the size of collagen mesh (with 50–200 nm openings) were evaluated to study their effects on passive diffusion behavior. Since the molecular size of DOX is substantially smaller than any opening in the mesh, both thickness and mesh size failed to provide any resistance. However, diffusion of particles with size larger than 100 nm could be severely impeded by mesh size and thickness. Therefore, the collagen type IV density, mesh size, number of layers, and association with vessels by itself could potentially be a biophysical barrier for limiting drug extravasation and therapeutic efficacy.

Angiogenesis of blood vessels requires degradation of collagen IV by recruiting metalloproteases MMP2 and MMP9, providing a transient niche with a leaky tumor vasculature, loose and thin BM, that is beneficial for NP delivery [91, 92]. However, this transient disruption of contact between endothelial cells and vessel BM leads to endothelial apoptosis and the formation of collagen fragments that antagonize angiogenesis. On the other hand, the residual nondegraded collagen IV accumulates during repeated remodeling, resulting in multiple distinct layers of BM. During the formation of multiple layers, Collagen density increases while mesh size decreases, generating a more resistant pattern for later NP penetration. Furthermore, the underlying cells can regenerate along the surviving BM that acts as a template or scaffold for generating axons, which renders NP diffusion even harder [36].

In conclusion, BM remodeling is a complicated and controversial procedure controlled by angiogenesis. In addition to digesting the BM via intravenous dosing of collagen IV degradation enzyme, closely monitoring angiogenesis process and dosing NP at the optimal interval is required to improve NP extravasation.

2.4 Extracellular Matrix Components Determine the Interstitial Transport of Nanoparticles and Macromolecules

Diffusion of NP across the thick interstitial matrix is the last step to approaching tumor cells. For tumors with a less interstitial matrix, such as melanoma and colorectal cancers, NP can easily diffuse across the interstitial barrier, access tumor cells and induce growth inhibition. However, for tumors with a thick interstitial matrix, this process is more intractable. In contrast to BM, whose major component is collagen IV in the form of sheet-like structure, the tumor interstitial matrix consists of a highly interconnected network of collagen fiber structures (mainly collagen I, II III) that interact with other molecules, such as proteoglycans and glycol aminoglycans [93–95].

Collagen content is the major determinant of interstitial transport [14]. Tumors rich in collagen inhibit diffusion to a greater extent than tumors with low collagen content. Recent studies showed that matrix modifiers such as bacterial collagenase, relaxin, and losartan (Fig. 6), an antifibrotic collagen I inhibitor, could modify the collagen network in tumors and improve the intratumoral spread of polystyrene NP, oncolytic virus HSV particles and Doxil [96–98]. Apart from collagen content, the orientation of the fiber net can also influence particle diffusion. During the development of tumors, collagen remodeling enzymes modify the architecture of the collagen scaffold from early, thin and relaxed collagens (curly fibrils) to thick, linearized, and aligned fibrils (Fig. 5a). Linearization of the collagen matrix stiffens the ECM, which thereafter not only elicits diverse effects on cellular differentiation and migration, but also narrows the interfiber spacing, reducing particle motility [41]. Alignment is one determinant of collagen crosslink and NP distribution. Stylianopoulos et al. established a mathematical model to evaluate particle diffusion across collagen fibers varying in degrees of alignments. This study indicates that the orientation of fibrils leads to diffusion anisotropy (Fig. 5b, c) [99]. An in vivo NP distribution study performed by Diop-Frimpong et al. further confirmed this concept [96]. In the MU89 tumor with a more aligned and organized collagen fiber network, penetration and diffusion of NP were more restricted to a limited direction and area, while in the HSTS26T tumor model, with a dense but more diffusive, less fiber-like collagen network, particle penetration was more scattered and diffusive (Fig. 5c, d).

The difference in the collagen crosslinking pattern causes disparate NP diffusion and leads to different therapeutic outcomes. Since collagen crosslinking is predominantly catalyzed by enzymes such as lysyl oxidase (LOX), regulated by fibronectin and organized by SPARC (secreted protein acidic and rich in cysteine), these molecules can be used as interesting target candidates to inhibit collagen crosslinking and fibril network organization [41]. For example, Kanapathipillai et al. designed a nanocarrier-based delivery system, a PLGA-conjugated LOX inhibitory antibody, to actively target to ECM. It decreased collagen crosslinking, inhibited tumor growth and metastasis, and improved therapy [6, 100].

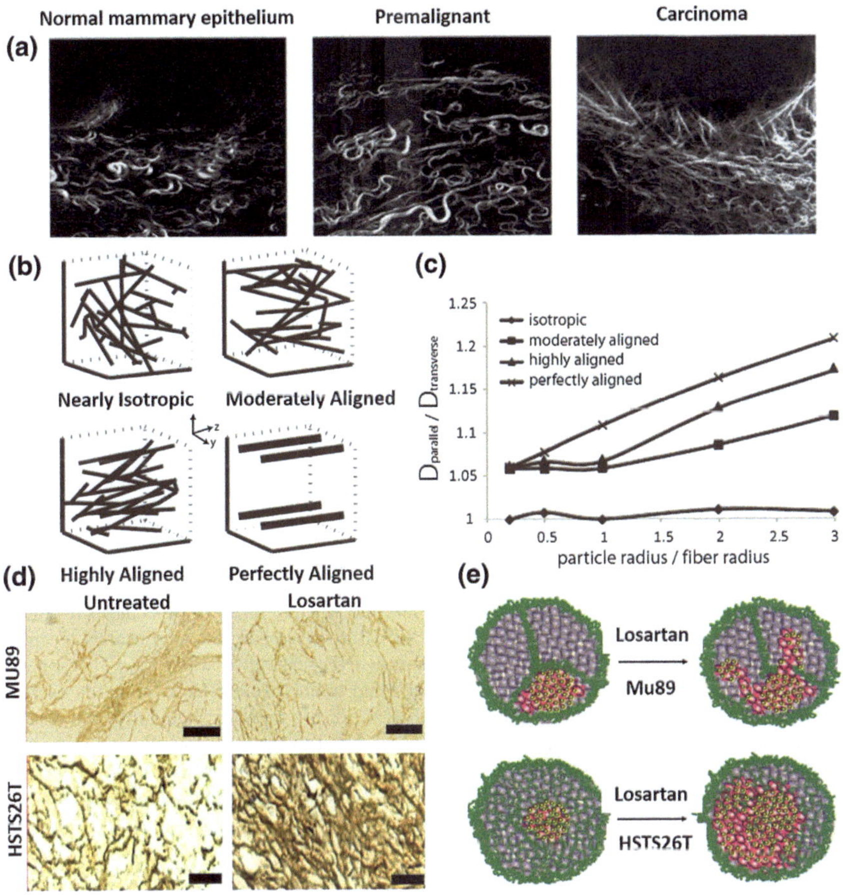

Fig. 5 The influence of collagen crosslinking and alignment on NP distribution. **a** The architecture of the collagen scaffold changes from early thin relaxed structures (curly fibrils) to thick, linearized, and highly aligned structures during malignancy development. (Reproduced from Egeblad et al. 2010, copyright of Elsevier) **b** Typical fiber structure established in vitro using a mathematical model. **c** Diffusion anisotropy as a function of the particle radius over the fiber radius for the fiber structures employed in the study. Results indicate that the more aligned one has a more strict distribution direction. (**b**, **c** reproduced from Stylianopoulos. 2010, copyright of Elsevier. Detailed description refers to the original manuscript.) **d** Immunostaining of collagen I in the two in vivo tumor models. Mu89 has highly aligned fibril structure that separate tumors into different compartments, while HSTS26T has dense collagen but less fibril structure. Losartan treatment can decrease the collagen content in both tumors. **e** A scheme hypothesis for losartan treatment in the two tumor models with a different collagen pattern. It indicates that the aligned and highly cross-linked tumors have limited NP perfusion after Losartan treatment, while the less fibril-like tumors have more scattered NP perfusion after Losartan treatment (**d**, **e** reproduced from Diop-Frimpong et al. 2011, copyright of National Academy of Sciences of the United States of America)

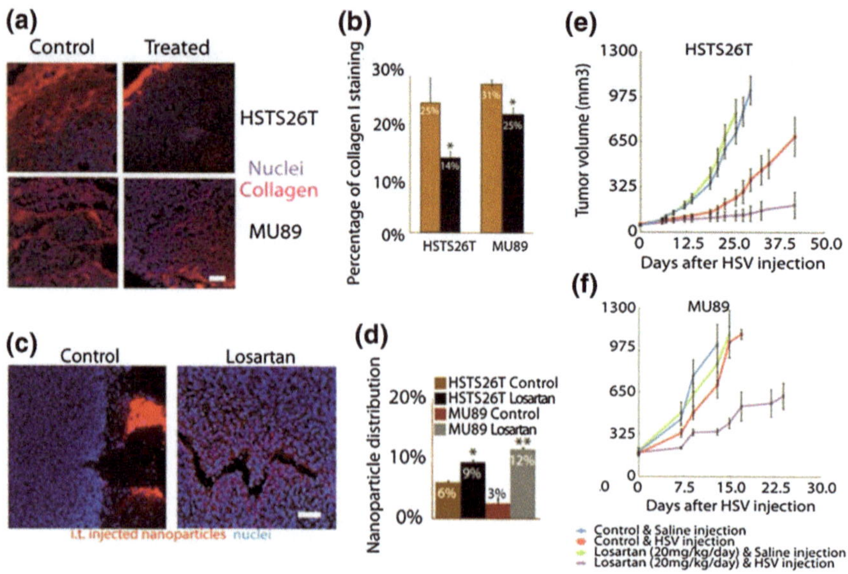

Fig. 6 Immunostaining of collagen I (*red*) (**a**) and quantitative analysis of collagen I content (**b**) in different treatment groups indicate that Losartan treatment can significantly decrease collagen content in both MU89 and HSTS26T tumors. NP penetration was further investigated. Fluorescence labeled NP was intravenously injected and tumors were then harvested to evaluate the NP penetration, **c** is the frozen slice to visualize NP distribution intratumorally, and **d** is the quantitative analysis of NP distribution. Results indicate that Losartan treatment can increase NP penetration, **e** and **f** are the growth inhibition curves of different treatments. Losartan can promote an antitumor effect in both MU89 and HSTS26T tumors (Reproduced from Diop-Frimpong et al. 2011, copyright of National Academy of Sciences of the United States of America)

Another determinant of interstitial transport is the glycosaminoglycan content. Hyaluronan (HA) is a nonsulfated glycosaminoglycan in the interstitial matrix. It consists of glucuronic acid disaccharide/N-acetyl glucosamine repeats of variable length and signals through CD44 to regulate receptor tyrosine kinase and small GTPase activity [101]. HA is implicated in the process of epithelial to mesenchymal transition, angiogenesis, and chemo resistance [102]. Anionic repeats of HA also capture mobile cations and solvate water, resulting in osmotic swelling and high interstitial pressure [103]. In pancreatic ductal adenocarcinoma (PDA) and the KPC pancreatic model, a clinically relevant genetically engineered mouse model (GEMMs) established by Tuveson et al., HA staining covered almost 100 % of the tumor sections and is predominantly associated with the desmoplastic stroma [101, 104]. HA depletion is reported to reverse the quiescent state of endothelium, induce fenestrae, and impair junctional integrity through disrupting CD44-dependent reorganization of endothelial actin cytoskeleton [101]. The ultrastructural changes and the vascular re-expansion lead to IFP reduction and have a multiplicative effect on intratumoral diffusion and convection [101]. Consistent with this finding, systemic administration of a PEGylated human recombinant PH20 hyaluronidase (PEGPH20), increases macromolecule permeability and augments chemotherapy responses in the KPC pancreatic model [98]. Intratumoral administration of bovine hyaluronidases also shows promise in several xenograft models [105, 106]. No recent study has shown effect of HA degradation on particle penetration, yet the specificity of this effect to the tumor suggests utility as a promising combinatory component to improve the delivery of agents with larger particle size [107] (Fig. 6).

Other than nonsulfated glycosaminoglycan, sulfated glycosaminoglycan can also affect interstitial transport. On one hand, these elongated and thin fibers increase the viscosity of the interstitial fluid; on the other hand, they also carry a highly negative charge, which can inhibit the transport of macromolecules or NP by forming aggregates [14]. For example, the electrostatic interaction between heparan sulfate and the diffusing NP decreases the diffusion coefficient of the NP by three orders of magnitude [108].

2.5 Strategies to Improve Therapy

From the aforementioned evidence, we conclude that insufficient transport of diagnostic and therapeutic NP in tumors results from the abnormal structure and function of tumor vessels and the dense ECM in the desmoplastic stroma. Therefore, therapeutic strategies to enhance drug delivery have focused on either remodeling the tumor vasculature to increase the function of the vascular network and decrease the interstitial pressure, or remodeling the tumor interstitial matrix so that NP can extravasate the capillary walls and penetrate faster and deeper inside the tumor.

2.5.1 Remodeling of Tumor Vasculature

The EPR effect improves NP accumulation in tumor microvessels, whereas the tortuous vessel structure, compressed diameter, deficient function, high interstitial pressure, abnormal pericytes, and BM coverage limit NP extravasation from blood vessels into the interstitial space. Therefore, tumor blood vessels are considered as potential targets to improve the therapeutic potential of nano-formulations. One strategy is to remodel tumor blood vessels to a leakier state by decreasing pericyte coverage, BM thickness, or by reducing interstitial pressure to facilitate convection. TGF-β receptor antagonists (including small molecule kinase inhibitors and siRNA) were the most frequently used therapeutic agents to inhibit pericyte recruitment and BM activation [109]. The combination of TGF-β antagonists with NP has shown enhanced particle diffusion and promising therapeutic outcome. TGF-β inhibitors were also found to lower interstitial hypertension. Decrease in IFP instantly increases vessel permeation. In addition to TGF-β inhibitors, VEGF inhibitors such as bevacizumab or anti-VEGF antibody can significantly decrease IFP. A monoclonal antibody against VEGF reduced glioblastoma IFP by more than 70 % [58, 110].

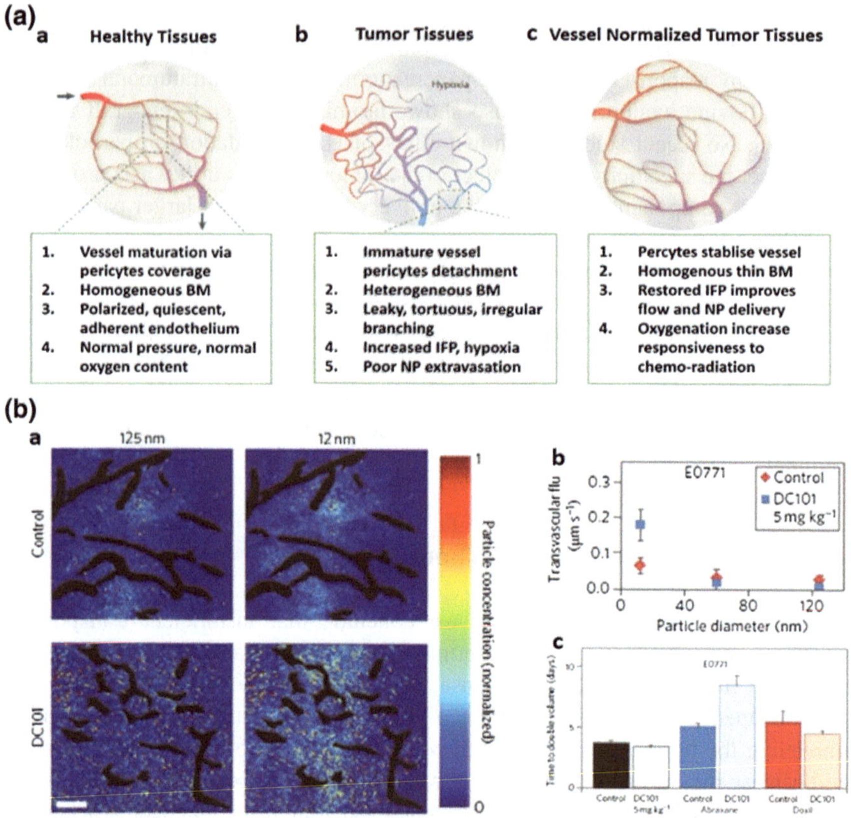

Fig. 7 Vessel normalization improves NP delivery. **a** Proposed role of vessel normalization in the response of tumors to anti-angiogenic therapy. *a* Normal vessel structure. *b* Tumor vasculature structure. Tumor vessel is structurally and functionally abnormal, providing resistance to the delivery of small molecules and NP. *c* Dynamic vascular normalization induced by VEGFR blockade. (Reproduced from Carmeliet et al. 2011, copyright of Nature publishing group) **b** Effects of vascular normalization on NP delivery and therapy in 4T1 and E0771 tumors. *a* NP penetration versus particle size in orthotopic 4T1 mammary tumors in response to normalizing therapy with DC101. NP concentrations (denoted by pseudocolor) are relative to initial intravascular levels, with vessels shown in *black*. *b* Penetration rates (transvascular flux) for NP in E0771 tumors in mice treated with DC 101, *a* and *b* indicates that normalization improves 12 nm NP penetration while not affecting 125 nm penetration. Scale bar, 100 μm. *c* Cytotoxic nanomedicine effectiveness by vascular normalization. Quantification of tumor growth rates based on the time to reach double the initial volume. Abraxane (10 nm) and Doxil (100 nm) monotherapy induce growth delays versus the control treatment. Normalization with DC101 enhances the effectiveness of the 10 nm Abraxane, but does not affect that of the 100 nm Doxil (Detailed description refers to the original manuscript, Chauhan et al. 2012, copyright of Nature publishing group)

Similar results have been demonstrated elsewhere in other types of cancer [111, 112]. Tumor IFP can also be lowered by using PDGF antagonists [58]. Tumors suitable for this type of treatment, such as pancreatic cancer (APC) and 3LL murine lung cancer, usually have hypovasculature, compressed vessels, extremely high IFP, thick pericyte coverage, and BM coating.

Another strategy is the so-called normalization of blood vessel (Fig. 7) [34, 71, 90]. This treatment type is suitable for tumors marked by hypervasculature with tortuous structure and less pericyte and ECM coverage. VEGF inhibitors decrease IFP and facilitate particle perfusion. Moreover, VEGF blockage (for e.g., using of VEGF receptor-2 blocking antibody DC101) prunes immature vessels, facilitates the recruitment of pericytes, decreases vessel density and diameter, and remodels the vasculature to more closely resemble the structure of normal vessels (Fig. 7a) [34, 90]. Agents with indirect anti-angiogenic effects, such as trastuzumab, can also lead to vascular normalization. A recent study indicated that transient vessel normalization can improve the performance of small anticancer molecule reagents [71]. Vessel normalization might compromise the transvascular transport of large NP (>100 nm) due to the decrease in pore size, but recent research indicates that it can improve the permeability of small hard NP (12 nm quantum dots) and soft NP (50 nm dextran) (Fig. 7b) [82, 90]. Strategies for blood vessel remodeling have to be adapted, based on tumor structure and particle size. In addition, radiotherapy and hyperthermia conditioning can also lead to transient leakiness of blood vessel and thus improve the intratumoral delivery of NP [113, 114].

2.5.2 Remodeling of Tumor Microenvironment

The ECM, particularly the collagen and glycosaminoglycan content, limits NP diffusion. To improve drug penetration, a common strategy is to degrade these components and increase the accessibility of the diffusing particles. In addition to hyaluronidase and collagenase mentioned in previous sections, matrix MMP-1 and MMP-8 are proteases frequently used to decrease the level of tumor glycoaminoglycans and improve convection [14, 115].

2.5.3 Design of Nanoparticle to Improve the Delivery

Besides remodeling of the TME, particle size also plays an important role to enable high-level NP penetration into tumor elements. The smaller the particles the better the transport. Notably, free drugs with smaller sizes can diffuse more rapidly than NP. However, small molecules not only distribute to normal tissue inducing adverse effects, but also fail to be trapped in the tumor tissue for optimized efficacy. Therefore, the size of NP needs to be optimized for each tumor and its metastasis sites Using dextran of various molecular weights in a FaDu tumor model, variable distribution relative to molecular weight has been demonstrated [73, 116]. In this study, 3.3 kDa dextran resembling small molecule drugs entered all tumor tissues quickly. 70 kDa dextran gradually extravasated the blood vessels into the ECM, while 2 MDa dextran remained in the vascular lumen. Polymeric micelles are one kind of NP used widely to delivery hydrophobic chemotherapy drugs. In another study, Cabral and Kataoka et al. prepared a series of micellar nanomedicines (micelle DACHPt) with a diameter ranging from 30 to 100 nm. They found that

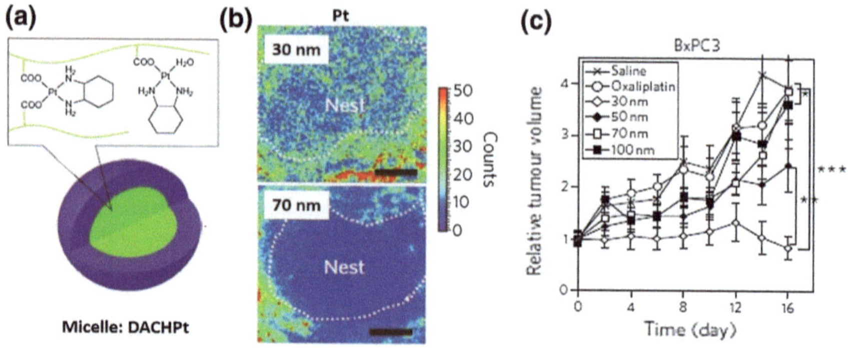

Fig. 8 **a** Structure of a DACHPt micelle. **b** Mapping of platinum atoms from DACHPt of varying sizes in BxPC3 xenografts by μ-SR-XRF 24 h after administration of micelles. Scale bars, 50 μm. This microdistribution figure indicates that small particles have better intratumoral distribution. **c** Antitumor activity of DACHpt micelles with different diameters. Smaller micelles have better antitumor activity. Detailed information refers to the original paper (Reproduced from Cabral et al. 2011, copyright of Nature publishing group)

penetration of NP decreased significantly upon increasing the particle size. Only small particles (30 nm) could penetrate the poorly permeable pancreatic cancer model, BxPC3, and caused promising therapeutic effect (Fig. 8) [85]. In addition, Pain's work using PEGylated quantum dots further inferred that diffusion of NP with smaller sizes (10–20 nm) can be increased after vasculature normalization, similar to free drugs. However, particles around 100 nm cannot achieve a similar effect [90]. These observations emphasize the importance of tailoring the diameter of NP products, even those with a diameter less than 100 nm.

In addition to particle size, the shape and surface charge of therapeutic NP also plays a key role in extravasation and interstitial transport. For example, cationic particles are more likely to target endothelial cells and exhibit a higher vascular permeability compared to its anionic and neutral counterpart [117]. While extravasated into the interstitial space, cationic particles can aggregate with negatively charged hyaluronan, and anionic particles can aggregate with positively charged collagen. Therefore, neutral particles diffuse faster and distributed more homogeneously inside the tumor interstitial place [118]. As for the influence of particle shape, it is reported that linear, semi-flexible macromolecules can diffuse more rapidly in the ECM than spherical particles with similar size [119]

Though condensed ECM functions as a barrier for NP diffusion, it can also be taken advantage of to improve the efficacy of nanomedicine. Based on the acidic pH, reduced oxygen pressure, and enzyme-rich properties of the TME, NP can be constructed to the advantage of these properties. For example, a multistage quantum dots embedded gelatin NP was engineered to degrade gradually and release 10 nm small quantum dots in response to MMPs, zinc-dependent endopeptidases that are abundant in the ECM [120]. Drug-polymer conjugates have also been designed with a cleavable linker, which is the substrate of MMP and fibroblast activation protein (FAP), a gelatinase that is expressed on TAFs [121–124]. Upon penetration of the ECM, free drug is released upon linker cleavage and diffuses more rapidly than the NP in the interstitial space for better therapeutic outcome. Based on this concept, pH sensitive particles have been designed to trigger the release of free drug from the cargo within tumor elements [125, 126]. In addition, external stimulants, such as electric pulses, magnetic field, ultrasound, heat, and light can also be used to improve NP penetration and free drug release [127–131].

3 The Relationship Between Nanoparticle Subtumoral Distribution and Tumor/Stroma Biological Interaction

In the previous section, we described the stroma as a physical barrier for NP extravasation from blood and diffusion into tumor cells. We then proposed methods to promote NP accumulation and improve diagnostic and therapeutic effects. Apart from being a physical barrier, the stroma, especially stroma cells, also support tumor growth through a direct cell adhesion interaction or in a paracrine manner mediated by secreted factors. Therefore, stroma cells and noncell components can be recognized as potential targets for antitumor therapy. In a preliminary experiment with a human bladder cancer model, we have quantified the intratumoral distribution of DiI-labeled liposomes and observed that around 20 % of liposomes were passively internalized by TAFs, one of the major stroma cells (Data not published). This raises the following questions: What is the amount of NP accumulated in the interstitial space that are actually internalized by tumor cells? What kind of stroma cells have taken up the accumulated NP? What is the response of stroma cells to the therapeutic NP? Will the stroma-tumor interactions be affected by NP assaulted stroma cells? These questions await answer in future experiments.

3.1 Nanoparticles that Target Endothelial Cells

NP resident within the tumor vasculature first encounter layers of endothelial cells and are ready to be internalized by them. Nontargeted PEGylated NP was internalized by endothelial cells through a low-density lipoprotein receptor-mediated pathway or other alternative pathways in vivo [132]. Since therapeutic strategies for regulating endothelial cells can result in tumor shrinkage via decreasing oxygen and nutrients supply, some nanomedicines are designed to actively target endothelial cells and increase cellular internalization. Targeting endothelial cells evades the stroma barriers and decreases the potential of drug-mediated resistance based on the genetic stability of endothelial cells. In addition, some endothelial cell markers also exist on the tumors, making these NP a dual targeting agent with a broad-spectrum of effects [65]. Integrin $\alpha v\beta 3$ is preferentially expressed on angiogenic endothelium in malignant tissue and widely used as an endothelial target [65]. Many recent studies have shown that functional therapeutic NP (loaded with doxorubicin, or antiangiogenesis agent) modified by integrin targeting peptide or cyclic or linear derivatives of RGD oligopeptide ligands can result in a strong inhibition of tumor growth [133–135]. RGD modification can also be used to improve vasculature imaging. For example, iRGD conjugated super-paramagnetic iron oxide NP (SPIONs) are able to image integrin αvβ3/β5-positive tumor neovasculature in vivo through MRI [136].

3.2 Nanoparticles that Target Macrophages

Preclinical studies indicate that TAMs represent an attractive target since they have been identified as an independent poor prognostic factor in several tumors types [47]. Antibodies, such as anti-CSF-1R, have been used to target TAMs and showed promise [47]. Since TAMs have a high expression of mannose receptor, mannose has been used as a targeting ligand for NP-based TAMs delivery [137, 138]. Many NP have also been engineered to image TAMs for diagnostic purposes [139]. The recognition that MRI-compatible nanomaterials can label TAMs dates back to the mid-1990s and has recently found renewed interest. Macrophage-specific PET imaging agents are also being developed [139]. TAMs were previously viewed as agents dispatched by the immune system to attack and eliminate tumors (M2 macrophage). However, extensive research over the past decade implicates that a sub-group of TAMs, known as M1 macrophages, has antitumorigenic properties [47]. Therefore, current focus has been shifted from exclusively depleting and imaging all TAMs to modulating the ratio of M1/M2 macrophages for improved therapy. Therapeutic NP should be able to target M2 macrophages, inhibiting M2 function or converting them into an antitumorigenic M1 subtype.

3.3 Nanoparticles that Target TAFs

TAFs are mesenchymal-like cells playing key roles in transformation, proliferation, and invasion of tumors [33, 140]. The majority of TAFs originate from trans-differentiation of resident fibroblasts, pericytes, or adipocytes in response to tumor secreted growth factors such as TGF-β, endothelin-1, and fibroblast growth factor 2 (FGF2). Alternatively, TAFs can also derive from distant sources such as bone marrow-derived mesenchymal stem cells (MSCs) [33]. Homing of MSCs to neoplastic sites induces their trans-differentiation into more aggressive α-SMA, fibroblast activation protein (FAP), tenascin-C and thrombospondin-1 expressing TAFs, and pericytes. In addition, TAFs can stem from epithelial cells following the initiation of an epithelial–mesenchymal transition, or from endothelial cells undergoing endothelial to mesenchymal transition (EndMT) [44, 46, 141]. TAFs synthesize and secrete ECM and regulate the release of degrading enzyme and growth factors. TAFs can activate angiogenesis through TGF-β mediated secretion of VEGF-A, or through recruiting of circulating endothelial progenitor cells [33]. TAFs can also secrete cytokines, such as CXCL12, to direct tumor lung metastasis or tumor immune invasion through binding with CXCR4 on remote premetastatic niche [142].

Recent research indicates that several passively diffused therapeutic NP can specifically distribute to TAFs and induce cell death. Cellex, a docetaxel-conjugate NP developed by Murakami et al., is a good example of this [143]. Cellex is a 120 nm NP that can reduce α-SMA content by 82 and 70 % in the 4T1 breast cancer model and the MDA-MB-231 model, respectively, native docetaxel and Abraxane exert no significant antistromal activity. Recently in our own lab, lipid-coated calcium phosphate NP encapsulating chemodrugs and siRNAs (LCP), and lipid-coated cisplatin NP (LPC) with cisplatin as both carrier and anticancer agents have been synthesized [15, 144, 145]. Both LCP and LPC NP can penetrate the TME barrier and distribute to TAFs. In a recent study, Zhang et al. indicated that a combination of gemcitabine LCP NP and cisplatin LPC NP can target TAFs and block α-SMA positive fibroblast recruitment by more than 87 % after multiple injections in a stroma-rich bladder cancer model (Fig. 9) [53]. In another study, cisplatin was also reported to deplete TAFs when co-delivered with an mTOR inhibitor in PLGA NP (Fig. 10) [146]. Transient depletion of TAFs increased tumor permeability, suppressed IFP, increased NP accumulation, and inhibited tumor metastasis [50, 95, 147, 148]. In both Zhang and Murakami's work, naïve TAFs are very sensitive to docetaxel and cisplatin, and show significant stromal depletion post single injection (Fig. 9). The mechanism of NP passively diffused to TAFs is not discussed in detail in the two aforementioned manuscripts. One possible reason may be that the majority of TAFs, especially α-SMA positive pericytes were localized around endothelial cells. When NPs are extravasated from the capillary wall, they immediately encounter these TAFs, which lead to their preferential internalization. Suitable particle size and materials with a high TAFs affinity can also explain the TAFs distribution. A significant depletion of α-SMA positive cells at the initial dose may also result from different responses of TAFs and tumor cells to therapeutic NP. That is to say, NP can be internalized by both tumor cells and TAFs. However, the latter has a lesser proliferating rate is more sensitive to chemotherapy and is less likely to induce resistance.

Due to the significant role of TAFs in mediating ECM formation and tumor cell progression, therapeutic NP that are designed to target fibroblasts within the tumor-stroma offer another treatment option. However, a lack of specific and unique surface targets limits the clinical application of this strategy. Recently, the identification of fibroblast activation protein (FAP) α as a target selectively expressed on TAFs has led to intensive efforts to exploit this novel cellular target for clinical benefit [149]. FAP is a membrane-bound serine protease of the prolyl oligopeptidase family with unique post-prolyl endopeptidase activity. Monoclonal antibody derivatives against FAP, prodrug, and drug-polymer conjugates with FAP cleavage bonds, and a DNA vaccine targeting FAP have been developed to improve the target therapies targeting TAFs [50, 149].

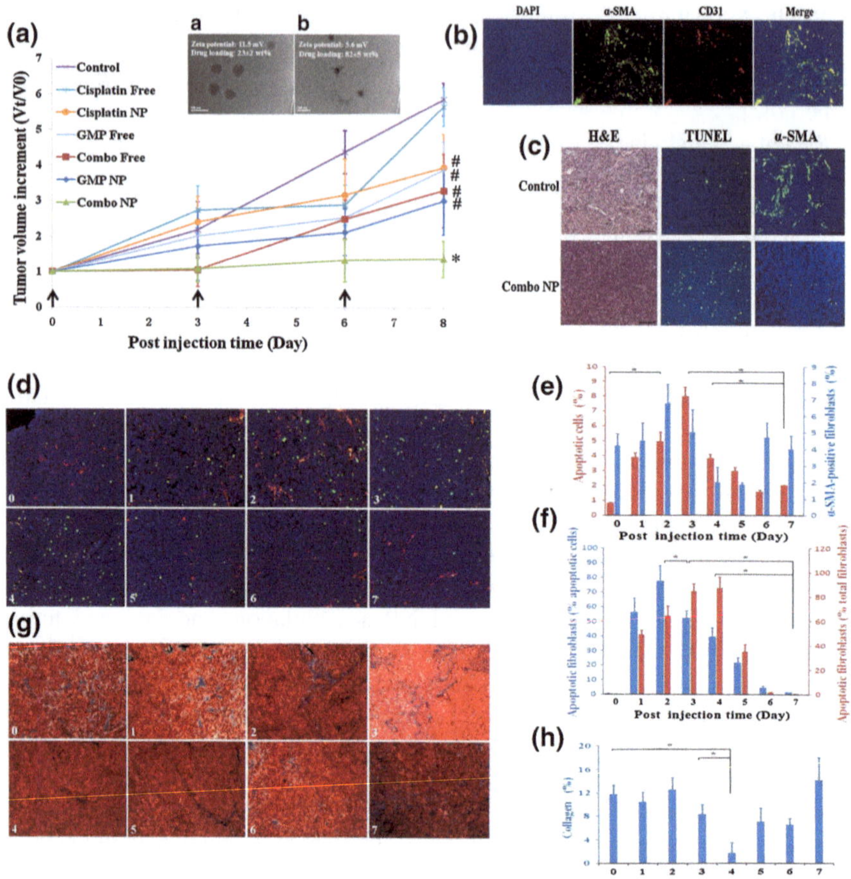

Fig. 9 Combination of GMP LCP NP and cisplatin LPC (Combo NP) target α-SMA positive TAFs and suppress tumor growth in a stroma-rich bladder cancer model. **a** Tumor growth inhibition of different formulations on stroma-rich tumor bearing mice. Combo NP showed the most significant antitumor effect. *a* TEM of GMP LCP NP. *b* TEM of cisplatin LPC NP. **b** The distribution of TAFs (α-SMA, *green*) and blood vessels (CD31, *red*) in the stroma-rich model. **c** Effect of Combo NP on the induction of apoptosis and inhibition of TAFs. Then, tumor bearing mice were further treated with a single injection of combo NP and tissues were collected and analyzed every day post injection. (From **d** to **h**). **d** Double staining for SMA positive TAFs (*red*), TUNEL (*green*) and apoptotic fibroblast (*yellow*). **e** Quantitative results for TUNEL-positive cells and α-SMA positive fibroblasts. **f** Quantitative results for apoptotic fibroblasts expressed as the percentage of total apoptotic cells and fibroblasts. **f** Masson's trichrome stain for collagen (*blue*). **g** Quantitative results for collagen expressed by the area (%). α-SMA positive TAFs. Collagen decreased significantly on the fourth day post single injection (Reproduced from Zhang et al. 2014, copyright of Elsevier)

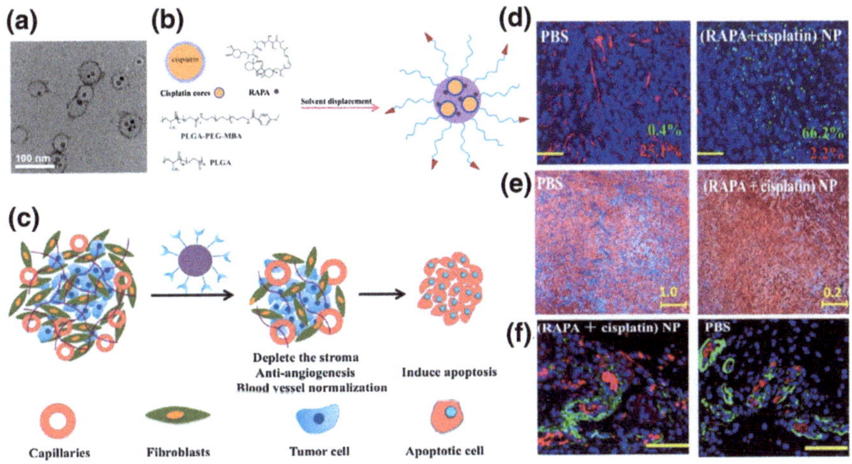

Fig. 10 **a**, **b** Preparation and TEM image of PLGA NP co-encapsulated with cisplatin cores and an mTOR inhibitor rapamycin (RAPA). **d** Double staining of TUNEL (*green*) for apoptosis and α-SMA (*red*) for TAFs. **e** Masson trichrome staining for collagen, **d** and **e** indicate that combination therapy can induce cell apoptosis, deplete α-SMA positive fibroblast and inhibit collagen synthesis in nude mice bearing A375 luc melanoma. **f** Shows that combinatory NP improve the penetration of DiI PLGA NP (*red*) in an A375 luc xenograft. The blood vessels were stained with CD31 (*green*). **c** Hypothesis: RAPA and cisplatin combination treatment remodels the tumor microenvironment. Combinatory PLGA NPs exhibited considerable antiangiogenesis effect and blood vessel normalization while also depleting the stroma (Reproduced from Guo et al. 2014, Copyright of ACS nano)

3.4 Paradoxical Outcome of Targeting and Depleting Stroma Cells

Stroma cells support tumor progress and migration. They can also form an innate niche that promotes resistance toward small molecule or NP-based chemotherapy. For example, fibroblast-secreted heparin growth factor (HGF) regulates MAPK and AKT signaling pathway, resulting in resistant to vemurafenib, a mutant Braf inhibitor, in the treatment of BrafV600E mutated melanoma [43, 46]. Another example is the inhibitory immune microenvironment caused by regulatory T cells and M2 macrophages that limits the efficacy of cancer vaccines [150]. However, stroma cell depletion acts as a double-edged sword. Feig et al. indicated FAP positive TAFs, secrete CXCL12 and direct tumor immune evasion in a model of pancreatic ductal adenocarcinoma (PDA) [142]. Paradoxically, Ozdemir et al. and Rhim et al. demonstrate that stroma targeted depletion results in undifferentiated and aggressive pancreatic cancer, uncovering a protective role of stroma in this cancer [48, 151, 152]. TAFs function differently with regards to cancer models. Targeted depletion of TAFs should be approached with caution when dealing with different tumor types. Moreover, chronic inhibition of stroma cells can lead to acquired resistance. In a recent study by Sun et al., it was observed that treatment-

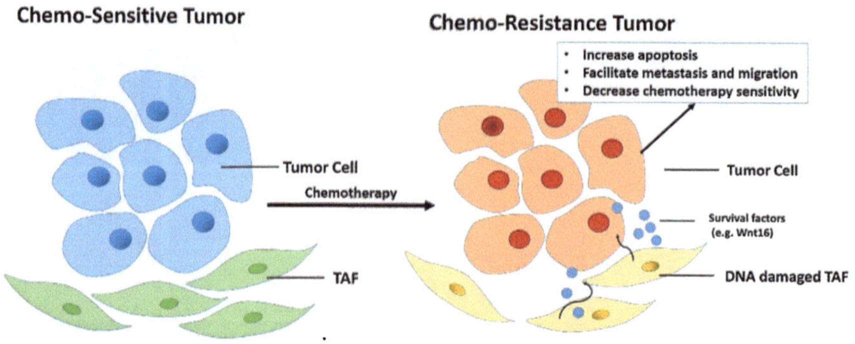

Fig. 11 Mechanism of tumor microenvironment and stroma cell induced acquired resistance

induced DNA damage in the neighboring benign stroma cells promotes prostate, breast, and ovarian cancer therapy resistance through paracrine secretion of Wnt16 [49]. Consistent with this finding, Krtolica found that senescent fibroblasts can promote epithelial growth [153]. These findings underline the acquired resistance elicited by TAFs following a chronic chemotherapy assault. Our own preliminary data indicate that chronic exposure of TAFs to cisplatin-containing NP can lead to the resistance of neighboring tumor cells along with the TAFs through paracrine signaling. Resistant TAFs secreted more extracellular molecules to stiffen the TME, promoting tumor growth while inhibiting NP penetration (Fig. 11). In order to overcome this stroma-induced resistance, combination strategies should be considered to deplete tumor cells and TAFs, and to inhibit the prosurvival crosstalk between these two types of cells. Nanotechnology provides the ability to co-load multiple modalities with different functions and targets. It is a preferred approach for targeting stroma cells while also inhibiting tumor-stroma crosstalk.

4 Conclusions and Future Perspectives

The dense ECM structure and aberrant tumor vasculature blocks NP penetration of tumor cells and results in limited therapeutic outcome. However, penetration is not the only standard to evaluate therapeutic response. Heterogeneous distribution of NP in the interstitial space and disparate internalization of NP to stroma cells may cause acquired resistance from TME and eventually lead to the treatment failure. Presently, the challenge is to design NP with multifunctional modalities to target both tumor and stroma cells, block the resistant tumor-stroma crosstalk, and uniformly deliver the designed NP homogenously across the tumor. In addition to delivering therapeutic and diagnostic NP to solid tumors, another critical task requiring further investigation is targeted delivery of NP to metastatic sites and inhibiting the formation of a stroma-rich metastatic niche. Recently, Swami et al. approached this challenge by engineering nanomedicine to target myeloma and the bone metastatic microenvironment [154]. Moreover, a recent discovery on the effect of melanoma-derived exosomes in inducing vascular leakiness at premetastatic sites may provide a means of passively targeting NP to metastatic sites [155]. Clinically relevant models should be established to prove the concept. In TME research, several mathematical in vitro models and clinically relevant in vivo models have been developed. However, up to now, the existing models are not able to sufficiently depict complicated interactions between NP and the TME. More sophisticated model systems together with more effective nanomaterials need to be developed to more adequately explore the TME.

Acknowledgments This work was supported by NIH grant support: CA149363, CA151652, CA149387 and DK100664. The authors thank Andrew Mackenzie Blair for his assistance in the chapter preparation.

References

1. Wang Y, Zhang L, Guo S, Hatefi A, Huang L (2013) Incorporation of histone derived recombinant protein for enhanced disassembly of core-membrane structured liposomal nanoparticles for efficient siRNA delivery. J Controlled Release Off J Controlled Release Soc 172(1):179–189. doi:10.1016/j.jconrel.2013.08.015
2. Chang HI, Yeh MK (2012) Clinical development of liposome-based drugs: formulation, characterization, and therapeutic efficacy. Int J Nanomed 7:49–60. doi:10.2147/IJN.S26766
3. Savla R, Taratula O, Garbuzenko O, Minko T (2011) Tumor targeted quantum dot-mucin 1 aptamer-doxorubicin conjugate for imaging and treatment of cancer. J Controlled Release Off J Controlled Release Soc 153(1):16–22. doi:10.1016/j.jconrel.2011.02.015
4. Allen PM, Liu W, Chauhan VP, Lee J, Ting AY, Fukumura D, Jain RK, Bawendi MG (2010) InAs(ZnCdS) quantum dots optimized for biological imaging in the near-infrared. J Am Chem Soc 132(2):470–471. doi:10.1021/ja908250r
5. Fang J, Nakamura H, Maeda H (2011) The EPR effect: unique features of tumor blood vessels for drug delivery, factors involved, and limitations and augmentation of the effect. Adv Drug Deliv Rev 63(3):136–151. doi:10.1016/j.addr.2010.04.009
6. Kanapathipillai M, Brock A, Ingber DE (2014) Nanoparticle targeting of anti-cancer drugs that alter intracellular signaling or influence the tumor microenvironment. Adv Drug Deliv Rev. doi:10.1016/j.addr.2014.05.005
7. Nishihara H (2014) Human pathological basis of blood vessels and stromal tissue for nanotechnology. Adv Drug Deliv Rev 74C:19–27. doi:10.1016/j.addr.2014.01.005
8. Kakkar V, Singh S, Singla D, Kaur IP (2011) Exploring solid lipid nanoparticles to enhance the oral bioavailability of curcumin. Mol Nutr Food Res 55(3):495–503. doi:10.1002/mnfr.201000310
9. Knezevic NZ, Trewyn BG, Lin VS (2011) Functionalized mesoporous silica nanoparticle-based visible light responsive controlled release delivery system. Chem Commun 47 (10):2817–2819. doi:10.1039/c0cc04424e
10. Elzoghby AO, Samy WM, Elgindy NA (2012) Albumin-based nanoparticles as potential controlled release drug delivery systems. J Controlled Release Off J Controlled Release Soc 157(2):168–182. doi:10.1016/j.jconrel.2011.07.031
11. Klibanov AL, Maruyama K, Torchilin VP, Huang L (1990) Amphipathic polyethyleneglycols effectively prolong the circulation time of liposomes. FEBS Lett 268 (1):235–237
12. Ruiz A, Hernandez Y, Cabal C, Gonzalez E, Veintemillas-Verdaguer S, Martinez E, Morales MP (2013) Biodistribution and pharmacokinetics of uniform magnetite nanoparticles

chemically modified with polyethylene glycol. Nanoscale 5(23):11400–11408. doi:10.1039/c3nr01412f
13. Bibby DC, Talmadge JE, Dalal MK, Kurz SG, Chytil KM, Barry SE, Shand DG, Steiert M (2005) Pharmacokinetics and biodistribution of RGD-targeted doxorubicin-loaded nanoparticles in tumor-bearing mice. Int J Pharm 293(1–2):281–290. doi:10.1016/j.ijpharm.2004.12.021
14. Jain RK, Stylianopoulos T (2010) Delivering nanomedicine to solid tumors. Nat Rev Clin Oncol 7(11):653–664. doi:10.1038/nrclinonc.2010.139
15. Guo S, Lin CM, Xu Z, Miao L, Wang Y, Huang L (2014) Co-delivery of cisplatin and rapamycin for enhanced anticancer therapy through synergistic effects and microenvironment modulation. ACS Nano. doi:10.1021/nn5010815
16. Xu Z, Ramishetti S, Tseng YC, Guo S, Wang Y, Huang L (2013) Multifunctional nanoparticles co-delivering Trp2 peptide and CpG adjuvant induce potent cytotoxic T-lymphocyte response against melanoma and its lung metastasis. J Control Release 172 (1):259–265. doi:10.1016/j.jconrel.2013.08.021
17. Zhang Y, Peng L, Mumper RJ, Huang L (2013) Combinational delivery of c-myc siRNA and nucleoside analogs in a single, synthetic nanocarrier for targeted cancer therapy. Biomaterials 34(33):8459–8468. doi:10.1016/j.biomaterials.2013.07.050
18. Ashley CE, Carnes EC, Phillips GK, Padilla D, Durfee PN, Brown PA, Hanna TN, Liu J, Phillips B, Carter MB, Carroll NJ, Jiang X, Dunphy DR, Willman CL, Petsev DN, Evans DG, Parikh AN, Chackerian B, Wharton W, Peabody DS, Brinker CJ (2011) The targeted delivery of multicomponent cargos to cancer cells by nanoporous particle-supported lipid bilayers. Nat Mater 10(5):389–397. doi:10.1038/nmat2992
19. Aryal S, Hu CM, Zhang L (2011) Polymeric nanoparticles with precise ratiometric control over drug loading for combination therapy. Mol Pharm 8(4):1401–1407. doi:10.1021/mp200243k
20. Kolishetti N, Dhar S, Valencia PM, Lin LQ, Karnik R, Lippard SJ, Langer R, Farokhzad OC (2010) Engineering of self-assembled nanoparticle platform for precisely controlled combination drug therapy. Proc Natl Acad Sci USA 107(42):17939–17944. doi:10.1073/pnas.1011368107
21. Vivero-Escoto JL, Taylor-Pashow KM, Huxford RC, Della Rocca J, Okoruwa C, An H, Lin W, Lin W (2011) Multifunctional mesoporous silica nanospheres with cleavable Gd(III) chelates as MRI contrast agents: synthesis, characterization, target-specificity, and renal clearance. Small 7(24):3519–3528. doi:10.1002/smll.201100521
22. Hu SH, Gao X (2010) Nanocomposites with spatially separated functionalities for combined imaging and magnetolytic therapy. J Am Chem Soc 132(21):7234–7237. doi:10.1021/ja102489q
23. Nasongkla N, Bey E, Ren J, Ai H, Khemtong C, Guthi JS, Chin SF, Sherry AD, Boothman DA, Gao J (2006) Multifunctional polymeric micelles as cancer-targeted, MRI-ultrasensitive drug delivery systems. Nano Lett 6(11):2427–2430. doi:10.1021/nl061412u
24. Nasongkla N, Shuai X, Ai H, Weinberg BD, Pink J, Boothman DA, Gao J (2004) cRGD-functionalized polymer micelles for targeted doxorubicin delivery. Angew Chem Int Ed Engl 43(46):6323–6327. doi:10.1002/anie.200460800
25. Banerjee R, Tyagi P, Li S, Huang L (2004) Anisamide-targeted stealth liposomes: a potent carrier for targeting doxorubicin to human prostate cancer cells. Int J Cancer 112(4):693–700. doi:10.1002/ijc.20452
26. Hu CM, Zhang L (2009) Therapeutic nanoparticles to combat cancer drug resistance. Curr Drug Metab 10(8):836–841
27. Hu CM, Zhang L (2012) Nanoparticle-based combination therapy toward overcoming drug resistance in cancer. Biochem Pharmacol 83(8):1104–1111. doi:10.1016/j.bcp.2012.01.008
28. Galluzzi L, Senovilla L, Vitale I, Michels J, Martins I, Kepp O, Castedo M, Kroemer G (2012) Molecular mechanisms of cisplatin resistance. Oncogene 31(15):1869–1883. doi:10.1038/onc.2011.384

29. Hamelers IH, Staffhorst RW, Voortman J, de Kruijff B, Reedijk J, van Bergen en Henegouwen PM, de Kroon AI (2009) High cytotoxicity of cisplatin nanocapsules in ovarian carcinoma cells depends on uptake by caveolae-mediated endocytosis. Clinical Cancer Res Off J Am Assoc Cancer Res 15(4):1259–1268. doi:10.1158/1078-0432.CCR-08-1702
30. Steichen SD, Caldorera-Moore M, Peppas NA (2012) A review of current nanoparticle and targeting moieties for the delivery of cancer therapeutics. Eur J Pharm Sci Off J Eur Fed Pharm Sci 48(3):416–427. doi:10.1016/j.ejps.2012.12.006
31. Davis ME, Chen ZG, Shin DM (2008) Nanoparticle therapeutics: an emerging treatment modality for cancer. Nat Rev Drug Discov 7(9):771–782. doi:10.1038/nrd2614
32. Rink JS, Plebanek MP, Tripathy S, Thaxton CS (2013) Update on current and potential nanoparticle cancer therapies. Curr Opin Oncol 25(6):646–651. doi:10.1097/CCO.0000000000000012
33. Orimo A, Gupta PB, Sgroi DC, Arenzana-Seisdedos F, Delaunay T, Naeem R, Carey VJ, Richardson AL, Weinberg RA (2005) Stromal fibroblasts present in invasive human breast carcinomas promote tumor growth and angiogenesis through elevated SDF-1/CXCL12 secretion. Cell 121(3):335–348. doi:10.1016/j.cell.2005.02.034
34. Carmeliet P, Jain RK (2011) Principles and mechanisms of vessel normalization for cancer and other angiogenic diseases. Nat Rev Drug Discov 10(6):417–427. doi:10.1038/nrd3455
35. Ellem SJ, De-Juan-Pardo EM, Risbridger GP (2014) In vitro modeling of the prostate cancer microenvironment. Adv Drug Deliv Rev. doi:10.1016/j.addr.2014.04.008
36. Baluk P, Morikawa S, Haskell A, Mancuso M, McDonald DM (2003) Abnormalities of basement membrane on blood vessels and endothelial sprouts in tumors. Am J Pathol 163 (5):1801–1815. doi:10.1016/S0002-9440(10)63540-7
37. Cao Y, Zhang ZL, Zhou M, Elson P, Rini B, Aydin H, Feenstra K, Tan MH, Berghuis B, Tabbey R, Resau JH, Zhou FJ, Teh BT, Qian CN (2013) Pericyte coverage of differentiated vessels inside tumor vasculature is an independent unfavorable prognostic factor for patients with clear cell renal cell carcinoma. Cancer 119(2):313–324. doi:10.1002/cncr.27746
38. Lokody I (2014) Microenvironment: tumour-promoting tissue mechanics. Nat Rev Cancer 14 (5):296. doi:10.1038/nrc3727
39. Duyverman AM, Steller EJ, Fukumura D, Jain RK, Duda DG (2012) Studying primary tumor-associated fibroblast involvement in cancer metastasis in mice. Nat Protoc 7(4):756–762. doi:10.1038/nprot.2012.031
40. Correia AL, Bissell MJ (2012) The tumor microenvironment is a dominant force in multidrug resistance. Drug Resist Updates Reviews Comment Antimicrob Anticancer Chemother 15(1–2):39–49. doi:10.1016/j.drup.2012.01.006
41. Egeblad M, Rasch MG, Weaver VM (2010) Dynamic interplay between the collagen scaffold and tumor evolution. Curr Opin Cell Biol 22(5):697–706. doi:10.1016/j.ceb.2010.08.015
42. Singleton PA, Mirzapoiazova T, Guo Y, Sammani S, Mambetsariev N, Lennon FE, Moreno-Vinasco L, Garcia JG (2010) High-molecular-weight hyaluronan is a novel inhibitor of pulmonary vascular leakiness. Am J Physiol Lung Cell Mol Physiol 299(5):L639–L651. doi:10.1152/ajplung.00405.2009
43. Straussman R, Morikawa T, Shee K, Barzily-Rokni M, Qian ZR, Du J, Davis A, Mongare MM, Gould J, Frederick DT, Cooper ZA, Chapman PB, Solit DB, Ribas A, Lo RS, Flaherty KT, Ogino S, Wargo JA, Golub TR (2012) Tumour micro-environment elicits innate resistance to RAF inhibitors through HGF secretion. Nature 487(7408):500–504. doi:10.1038/nature11183
44. Li H, Fan X, Houghton J (2007) Tumor microenvironment: the role of the tumor stroma in cancer. J Cell Biochem 101(4):805–815. doi:10.1002/jcb.21159
45. Yokoi K, Godin B, Oborn CJ, Alexander JF, Liu X, Fidler IJ, Ferrari M (2013) Porous silicon nanocarriers for dual targeting tumor associated endothelial cells and macrophages in stroma of orthotopic human pancreatic cancers. Cancer Lett 334(2):319–327. doi:10.1016/j.canlet.2012.09.001

46. Paraiso KH, Smalley KS (2013) Fibroblast-mediated drug resistance in cancer. Biochem Pharmacol 85(8):1033–1041. doi:10.1016/j.bcp.2013.01.018
47. Ries CH, Cannarile MA, Hoves S, Benz J, Wartha K, Runza V, Rey-Giraud F, Pradel LP, Feuerhake F, Klaman I, Jones T, Jucknischke U, Scheiblich S, Kaluza K, Gorr IH, Walz A, Abiraj K, Cassier PA, Sica A, Gomez-Roca C, de Visser KE, Italiano A, Le Tourneau C, Delord JP, Levitsky H, Blay JY, Ruttinger D (2014) Targeting tumor-associated macrophages with anti-CSF-1R antibody reveals a strategy for cancer therapy. Cancer Cell 25(6):846–859. doi:10.1016/j.ccr.2014.05.016
48. Ozdemir BC, Pentcheva-Hoang T, Carstens JL, Zheng X, Wu CC, Simpson TR, Laklai H, Sugimoto H, Kahlert C, Novitskiy SV, De Jesus-Acosta A, Sharma P, Heidari P, Mahmood U, Chin L, Moses HL, Weaver VM, Maitra A, Allison JP, LeBleu VS, Kalluri R (2014) Depletion of carcinoma-associated fibroblasts and fibrosis induces immunosuppression and accelerates pancreas cancer with reduced survival. Cancer Cell 25(6):719–734. doi:10.1016/j.ccr.2014.04.005
49. Sun Y, Campisi J, Higano C, Beer TM, Porter P, Coleman I, True L, Nelson PS (2012) Treatment-induced damage to the tumor microenvironment promotes prostate cancer therapy resistance through WNT16B. Nat Med 18(9):1359–1368. doi:10.1038/nm.2890
50. Loeffler M, Kruger JA, Niethammer AG, Reisfeld RA (2006) Targeting tumor-associated fibroblasts improves cancer chemotherapy by increasing intratumoral drug uptake. J Clin Investig 116(7):1955–1962. doi:10.1172/JCI26532
51. Guo S, Wang Y, Miao L, Xu Z, Lin CM, Zhang Y, Huang L (2013) Lipid-coated cisplatin nanoparticles induce neighboring effect and exhibit enhanced anticancer efficacy. ACS Nano 7(11):9896–9904. doi:10.1021/nn403606m
52. Min KH, Lee HJ, Kim K, Kwon IC, Jeong SY, Lee SC (2012) The tumor accumulation and therapeutic efficacy of doxorubicin carried in calcium phosphate-reinforced polymer nanoparticles. Biomaterials 33(23):5788–5797. doi:10.1016/j.biomaterials.2012.04.057
53. Zhang J, Miao L, Guo S, Zhang Y, Zhang L, Satterlee A, Kim WY, Huang L (2014) Synergistic anti-tumor effects of combined gemcitabine and cisplatin nanoparticles in a stroma-rich bladder carcinoma model. J Controlled Release Off J Controlled Release Soc 182:90–96. doi:10.1016/j.jconrel.2014.03.016
54. Armulik A, Genove G, Betsholtz C (2011) Pericytes: developmental, physiological, and pathological perspectives, problems, and promises. Dev Cell 21(2):193–215. doi:10.1016/j.devcel.2011.07.001
55. Sims DE (1986) The pericyte—a review. Tissue Cell 18(2):153–174
56. Inoue S (1989) Ultrastructure of basement membranes. Int Rev Cytol 117:57–98
57. Yokoi K, Kojic M, Milosevic M, Tanei T, Ferrari M, Ziemys A (2014) Capillary-wall collagen as a biophysical marker of nanotherapeutic permeability into the tumor microenvironment. Cancer Res 74(16):4239–4246. doi:10.1158/0008-5472.CAN-13-3494
58. Danquah MK, Zhang XA, Mahato RI (2011) Extravasation of polymeric nanomedicines across tumor vasculature. Adv Drug Deliv Rev 63(8):623–639. doi:10.1016/j.addr.2010.11.005
59. McDonald DM, Thurston G, Baluk P (1999) Endothelial gaps as sites for plasma leakage in inflammation. Microcirculation 6(1):7–22
60. Yurchenco PD, Ruben GC (1987) Basement membrane structure in situ: evidence for lateral associations in the type IV collagen network. J Cell Biol 105(6 Pt 1):2559–2568
61. Accardo A, Salsano G, Morisco A, Aurilio M, Parisi A, Maione F, Cicala C, Tesauro D, Aloj L, De Rosa G, Morelli G (2012) Peptide-modified liposomes for selective targeting of bombesin receptors overexpressed by cancer cells: a potential theranostic agent. Int J Nanomed 7:2007–2017. doi:10.2147/IJN.S29242
62. Lu Y, Low PS (2002) Folate-mediated delivery of macromolecular anticancer therapeutic agents. Adv Drug Deliv Rev 54(5):675–693

63. Zhang L, Gu FX, Chan JM, Wang AZ, Langer RS, Farokhzad OC (2008) Nanoparticles in medicine: therapeutic applications and developments. Clin Pharmacol Ther 83(5):761–769. doi:10.1038/sj.clpt.6100400
64. Su EP, Housman LR, Masonis JL, Noble JW Jr, Engh CA (2014) Five year results of the first US FDA-approved hip resurfacing device. J Arthroplast 29(8):1571–1575. doi:10.1016/j.arth.2014.03.021
65. Tianjiao Ji YZ, Ding Y, Nie G (2013) Using functional nanomaterials to target and regulate the tumor microenvironment: diagnostic and therapeutic applications. Adv Mater 25 (26):3508–3525. doi:10.1002/adma.201300299
66. Jain RK (1994) Barriers to drug delivery in solid tumors. Sci Am 271(1):58–65
67. Boucher Y, Baxter LT, Jain RK (1990) Interstitial pressure gradients in tissue-isolated and subcutaneous tumors: implications for therapy. Cancer Res 50(15):4478–4484
68. Swartz MA, Lund AW (2012) Lymphatic and interstitial flow in the tumour microenvironment: linking mechanobiology with immunity. Nat Rev Cancer 12(3):210–219. doi:10.1038/nrc3186
69. Boucher Y, Jain RK (1992) Microvascular pressure is the principal driving force for interstitial hypertension in solid tumors: implications for vascular collapse. Cancer Res 52 (18):5110–5114
70. Padera TP, Stoll BR, Tooredman JB, Capen D, di Tomaso E, Jain RK (2004) Pathology: cancer cells compress intratumour vessels. Nature 427(6976):695. doi:10.1038/427695a
71. Jain RK (2005) Normalization of tumor vasculature: an emerging concept in antiangiogenic therapy. Science 307(5706):58–62. doi:10.1126/science.1104819
72. Hanahan D, Weinberg RA (2011) Hallmarks of cancer: the next generation. Cell 144 (5):646–674. doi:10.1016/j.cell.2011.02.013
73. Kano MR (2014) Nanotechnology and tumor microcirculation. Adv Drug Deliv Rev 74C:2–11. doi:10.1016/j.addr.2013.08.010
74. Lindblom P, Gerhardt H, Liebner S, Abramsson A, Enge M, Hellstrom M, Backstrom G, Fredriksson S, Landegren U, Nystrom HC, Bergstrom G, Dejana E, Ostman A, Lindahl P, Betsholtz C (2003) Endothelial PDGF-B retention is required for proper investment of pericytes in the microvessel wall. Genes Dev 17(15):1835–1840. doi:10.1101/gad.266803
75. Song N, Huang Y, Shi H, Yuan S, Ding Y, Song X, Fu Y, Luo Y (2009) Overexpression of platelet-derived growth factor-BB increases tumor pericyte content via stromal-derived factor-1alpha/CXCR4 axis. Cancer Res 69(15):6057–6064. doi:10.1158/0008-5472.CAN-08-2007
76. Yu X, Radulescu A, Chen CL, James IO, Besner GE (2012) Heparin-binding EGF-like growth factor protects pericytes from injury. J Surg Res 172(1):165–176. doi:10.1016/j.jss.2010.07.058
77. Abramsson A, Lindblom P, Betsholtz C (2003) Endothelial and nonendothelial sources of PDGF-B regulate pericyte recruitment and influence vascular pattern formation in tumors. J Clin Investig 112(8):1142–1151. doi:10.1172/JCI18549
78. Greenberg JI, Shields DJ, Barillas SG, Acevedo LM, Murphy E, Huang J, Scheppke L, Stockmann C, Johnson RS, Angle N, Cheresh DA (2008) A role for VEGF as a negative regulator of pericyte function and vessel maturation. Nature 456(7223):809–813. doi:10.1038/nature07424
79. Furuhashi M, Sjoblom T, Abramsson A, Ellingsen J, Micke P, Li H, Bergsten-Folestad E, Eriksson U, Heuchel R, Betsholtz C, Heldin CH, Ostman A (2004) Platelet-derived growth factor production by B16 melanoma cells leads to increased pericyte abundance in tumors and an associated increase in tumor growth rate. Cancer Res 64(8):2725–2733
80. Hosaka K, Yang Y, Seki T, Nakamura M, Andersson P, Rouhi P, Yang X, Jensen L, Lim S, Feng N, Xue Y, Li X, Larsson O, Ohhashi T, Cao Y (2013) Tumour PDGF-BB expression levels determine dual effects of anti-PDGF drugs on vascular remodelling and metastasis. Nat Commun 4:2129. doi:10.1038/ncomms3129

81. Zhang L, Nishihara H, Kano MR (2012) Pericyte-coverage of human tumor vasculature and nanoparticle permeability. Biol Pharm Bull 35(5):761–766
82. Kano MR, Komuta Y, Iwata C, Oka M, Shirai YT, Morishita Y, Ouchi Y, Kataoka K, Miyazono K (2009) Comparison of the effects of the kinase inhibitors imatinib, sorafenib, and transforming growth factor-beta receptor inhibitor on extravasation of nanoparticles from neovasculature. Cancer Sci 100(1):173–180. doi:10.1111/j.1349-7006.2008.01003.x
83. Kano MR, Bae Y, Iwata C, Morishita Y, Yashiro M, Oka M, Fujii T, Komuro A, Kiyono K, Kaminishi M, Hirakawa K, Ouchi Y, Nishiyama N, Kataoka K, Miyazono K (2007) Improvement of cancer-targeting therapy, using nanocarriers for intractable solid tumors by inhibition of TGF-beta signaling. Proc Natl Acad Sci USA 104(9):3460–3465. doi:10.1073/pnas.0611660104
84. Liu J, Liao S, Diop-Frimpong B, Chen W, Goel S, Naxerova K, Ancukiewicz M, Boucher Y, Jain RK, Xu L (2012) TGF-beta blockade improves the distribution and efficacy of therapeutics in breast carcinoma by normalizing the tumor stroma. Proc Natl Acad Sci USA 109(41):16618–16623. doi:10.1073/pnas.1117610109
85. Cabral H, Matsumoto Y, Mizuno K, Chen Q, Murakami M, Kimura M, Terada Y, Kano MR, Miyazono K, Uesaka M, Nishiyama N, Kataoka K (2011) Accumulation of sub-100 nm polymeric micelles in poorly permeable tumours depends on size. Nat Nanotechnol 6 (12):815–823. doi:10.1038/nnano.2011.166
86. Meng H, Zhao Y, Dong J, Xue M, Lin YS, Ji Z, Mai WX, Zhang H, Chang CH, Brinker CJ, Zink JI, Nel AE (2013) Two-wave nanotherapy to target the stroma and optimize gemcitabine delivery to a human pancreatic cancer model in mice. ACS Nano 7(11):10048–10065. doi:10.1021/nn404083m
87. Minowa T, Kawano K, Kuribayashi H, Shiraishi K, Sugino T, Hattori Y, Yokoyama M, Maitani Y (2009) Increase in tumour permeability following TGF-beta type I receptor-inhibitor treatment observed by dynamic contrast-enhanced MRI. Br J Cancer 101(11):1884–1890. doi:10.1038/sj.bjc.6605367
88. Kumagai M, Kano MR, Morishita Y, Ota M, Imai Y, Nishiyama N, Sekino M, Ueno S, Miyazono K, Kataoka K (2009) Enhanced magnetic resonance imaging of experimental pancreatic tumor in vivo by block copolymer-coated magnetite nanoparticles with TGF-beta inhibitor. J Controlled Release Off J Controlled Release Soc 140(3):306–311. doi:10.1016/j.jconrel.2009.06.002
89. Mancuso MR, Davis R, Norberg SM, O'Brien S, Sennino B, Nakahara T, Yao VJ, Inai T, Brooks P, Freimark B, Shalinsky DR, Hu-Lowe DD, McDonald DM (2006) Rapid vascular regrowth in tumors after reversal of VEGF inhibition. J Clin Investig 116(10):2610–2621. doi:10.1172/JCI24612
90. Chauhan VP, Stylianopoulos T, Martin JD, Popovic Z, Chen O, Kamoun WS, Bawendi MG, Fukumura D, Jain RK (2012) Normalization of tumour blood vessels improves the delivery of nanomedicines in a size-dependent manner. Nat Nanotechnol 7(6):383–388. doi:10.1038/nnano.2012.45
91. Bauvois B (2012) New facets of matrix metalloproteinases MMP-2 and MMP-9 as cell surface transducers: outside-in signaling and relationship to tumor progression. Biochim Biophys Acta 1825(1):29–36. doi:10.1016/j.bbcan.2011.10.001
92. Mammoto T, Jiang A, Jiang E, Panigrahy D, Kieran MW, Mammoto A (2013) Role of collagen matrix in tumor angiogenesis and glioblastoma multiforme progression. Am J Pathol 183(4):1293–1305. doi:10.1016/j.ajpath.2013.06.026
93. Danhier F, Feron O, Preat V (2010) To exploit the tumor microenvironment: passive and active tumor targeting of nanocarriers for anti-cancer drug delivery. J Controlled Release Off J Controlled Release Soc 148(2):135–146. doi:10.1016/j.jconrel.2010.08.027
94. Jain RK (2013) Normalizing tumor microenvironment to treat cancer: bench to bedside to biomarkers. J Clin Oncol Off J Am Soc Clin Oncol 31(17):2205–2218. doi:10.1200/JCO.2012.46.3653

95. Alderton GK (2014) Microenvironment: an exercise in restraint. Nat Rev Cancer 14(7):449. doi:10.1038/nrc3769
96. Diop-Frimpong B, Chauhan VP, Krane S, Boucher Y, Jain RK (2011) Losartan inhibits collagen I synthesis and improves the distribution and efficacy of nanotherapeutics in tumors. Proc Natl Acad Sci USA 108 (7):2909–2914. doi:10.1073/pnas.1018892108
97. Ganesh S, Gonzalez Edick M, Idamakanti N, Abramova M, Vanroey M, Robinson M, Yun CO, Jooss K (2007) Relaxin-expressing, fiber chimeric oncolytic adenovirus prolongs survival of tumor-bearing mice. Cancer Res 67(9):4399–4407. doi:10.1158/0008-5472.CAN-06-4260
98. McKee TD, Grandi P, Mok W, Alexandrakis G, Insin N, Zimmer JP, Bawendi MG, Boucher Y, Breakefield XO, Jain RK (2006) Degradation of fibrillar collagen in a human melanoma xenograft improves the efficacy of an oncolytic herpes simplex virus vector. Cancer Res 66 (5):2509–2513. doi:10.1158/0008-5472.CAN-05-2242
99. Stylianopoulos T, Diop-Frimpong B, Munn LL, Jain RK (2010) Diffusion anisotropy in collagen gels and tumors: the effect of fiber network orientation. Biophys J 99(10):3119–3128. doi:10.1016/j.bpj.2010.08.065
100. Kanapathipillai M, Mammoto A, Mammoto T, Kang JH, Jiang E, Ghosh K, Korin N, Gibbs A, Mannix R, Ingber DE (2012) Inhibition of mammary tumor growth using lysyl oxidase-targeting nanoparticles to modify extracellular matrix. Nano Lett 12(6):3213–3217. doi:10.1021/nl301206p
101. Jacobetz MA, Chan DS, Neesse A, Bapiro TE, Cook N, Frese KK, Feig C, Nakagawa T, Caldwell ME, Zecchini HI, Lolkema MP, Jiang P, Kultti A, Thompson CB, Maneval DC, Jodrell DI, Frost GI, Shepard HM, Skepper JN, Tuveson DA (2013) Hyaluronan impairs vascular function and drug delivery in a mouse model of pancreatic cancer. Gut 62(1):112–120. doi:10.1136/gutjnl-2012-302529
102. Provenzano PP, Hingorani SR (2013) Hyaluronan, fluid pressure, and stromal resistance in pancreas cancer. Br J Cancer 108(1):1–8. doi:10.1038/bjc.2012.569
103. Chahine NO, Chen FH, Hung CT, Ateshian GA (2005) Direct measurement of osmotic pressure of glycosaminoglycan solutions by membrane osmometry at room temperature. Biophys J 89(3):1543–1550. doi:10.1529/biophysj.104.057315
104. Olive KP, Jacobetz MA, Davidson CJ, Gopinathan A, McIntyre D, Honess D, Madhu B, Goldgraben MA, Caldwell ME, Allard D, Frese KK, Denicola G, Feig C, Combs C, Winter SP, Ireland-Zecchini H, Reichelt S, Howat WJ, Chang A, Dhara M, Wang L, Ruckert F, Grutzmann R, Pilarsky C, Izeradjene K, Hingorani SR, Huang P, Davies SE, Plunkett W, Egorin M, Hruban RH, Whitebread N, McGovern K, Adams J, Iacobuzio-Donahue C, Griffiths J, Tuveson DA (2009) Inhibition of Hedgehog signaling enhances delivery of chemotherapy in a mouse model of pancreatic cancer. Science 324(5933):1457–1461. doi:10.1126/science.1171362
105. Beckenlehner K, Bannke S, Spruss T, Bernhardt G, Schonenberg H, Schiess W (1992) Hyaluronidase enhances the activity of adriamycin in breast cancer models in vitro and in vivo. J Cancer Res Clin Oncol 118(8):591–596
106. Brekken C, de Lange Davies C (1998) Hyaluronidase reduces the interstitial fluid pressure in solid tumours in a non-linear concentration-dependent manner. Cancer Lett 131(1):65–70
107. Neesse A, Michl P, Frese KK, Feig C, Cook N, Jacobetz MA, Lolkema MP, Buchholz M, Olive KP, Gress TM, Tuveson DA (2011) Stromal biology and therapy in pancreatic cancer. Gut 60(6):861–868. doi:10.1136/gut.2010.226092
108. Thorne RG, Lakkaraju A, Rodriguez-Boulan E, Nicholson C (2008) In vivo diffusion of lactoferrin in brain extracellular space is regulated by interactions with heparan sulfate. Proc Natl Acad Sci USA 105(24):8416–8421. doi:10.1073/pnas.0711345105
109. Yingling JM, Blanchard KL, Sawyer JS (2004) Development of TGF-beta signalling inhibitors for cancer therapy. Nat Rev Drug Discov 3(12):1011–1022. doi:10.1038/nrd1580
110. Lee CG, Heijn M, di Tomaso E, Griffon-Etienne G, Ancukiewicz M, Koike C, Park KR, Ferrara N, Jain RK, Suit HD, Boucher Y (2000) Anti-Vascular endothelial growth factor

treatment augments tumor radiation response under normoxic or hypoxic conditions. Cancer Res 60(19):5565–5570

111. Willett CG, Boucher Y, di Tomaso E, Duda DG, Munn LL, Tong RT, Chung DC, Sahani DV, Kalva SP, Kozin SV, Mino M, Cohen KS, Scadden DT, Hartford AC, Fischman AJ, Clark JW, Ryan DP, Zhu AX, Blaszkowsky LS, Chen HX, Shellito PC, Lauwers GY, Jain RK (2004) Direct evidence that the VEGF-specific antibody bevacizumab has antivascular effects in human rectal cancer. Nat Med 10(2):145–147. doi:10.1038/nm988
112. Hurwitz HI, Fehrenbacher L, Hainsworth JD, Heim W, Berlin J, Holmgren E, Hambleton J, Novotny WF, Kabbinavar F (2005) Bevacizumab in combination with fluorouracil and leucovorin: an active regimen for first-line metastatic colorectal cancer. J Clin Oncol Off J Am Soc Clin Oncol 23(15):3502–3508. doi:10.1200/JCO.2005.10.017
113. Kong G, Braun RD, Dewhirst MW (2000) Hyperthermia enables tumor-specific nanoparticle delivery: effect of particle size. Cancer Res 60(16):4440–4445
114. Hainfeld JF, Dilmanian FA, Slatkin DN, Smilowitz HM (2008) Radiotherapy enhancement with gold nanoparticles. J Pharm Pharmacol 60(8):977–985. doi:10.1211/jpp.60.8.0005
115. Mok W, Boucher Y, Jain RK (2007) Matrix metalloproteinases-1 and -8 improve the distribution and efficacy of an oncolytic virus. Cancer Res 67(22):10664–10668. doi:10.1158/0008-5472.CAN-07-3107
116. Dreher MR, Liu W, Michelich CR, Dewhirst MW, Yuan F, Chilkoti A (2006) Tumor vascular permeability, accumulation, and penetration of macromolecular drug carriers. J Natl Cancer Inst 98(5):335–344. doi:10.1093/jnci/djj070
117. Campbell RB, Fukumura D, Brown EB, Mazzola LM, Izumi Y, Jain RK, Torchilin VP, Munn LL (2002) Cationic charge determines the distribution of liposomes between the vascular and extravascular compartments of tumors. Cancer Res 62(23):6831–6836
118. Stylianopoulos T, Poh MZ, Insin N, Bawendi MG, Fukumura D, Munn LL, Jain RK (2010) Diffusion of particles in the extracellular matrix: the effect of repulsive electrostatic interactions. Biophys J 99(5):1342–1349. doi:10.1016/j.bpj.2010.06.016
119. Pluen A, Netti PA, Jain RK, Berk DA (1999) Diffusion of macromolecules in agarose gels: comparison of linear and globular configurations. Biophys J 77(1):542–552. doi:10.1016/S0006-3495(99)76911-0
120. Wong C, Stylianopoulos T, Cui J, Martin J, Chauhan VP, Jiang W, Popovic Z, Jain RK, Bawendi MG, Fukumura D (2011) Multistage nanoparticle delivery system for deep penetration into tumor tissue. Proc Natl Acad Sci USA 108(6):2426–2431. doi:10.1073/pnas.1018382108
121. Huang S, Fang R, Xu J, Qiu S, Zhang H, Du J, Cai S (2011) Evaluation of the tumor targeting of a FAPalpha-based doxorubicin prodrug. J Drug Target 19(7):487–496. doi:10.3109/1061186X.2010.511225
122. Aggarwal S, Brennen WN, Kole TP, Schneider E, Topaloglu O, Yates M, Cotter RJ, Denmeade SR (2008) Fibroblast activation protein peptide substrates identified from human collagen I derived gelatin cleavage sites. Biochemistry 47(3):1076–1086. doi:10.1021/bi701921b
123. Hatakeyama H, Akita H, Harashima H (2011) A multifunctional envelope type nano device (MEND) for gene delivery to tumours based on the EPR effect: a strategy for overcoming the PEG dilemma. Adv Drug Deliv Rev 63(3):152–160. doi:10.1016/j.addr.2010.09.001
124. LeBeau AM, Brennen WN, Aggarwal S, Denmeade SR (2009) Targeting the cancer stroma with a fibroblast activation protein-activated promelittin protoxin. Mol Cancer Ther 8(5):1378–1386. doi:10.1158/1535-7163.MCT-08-1170
125. Lim EK, Huh YM, Yang J, Lee K, Suh JS, Haam S (2011) pH-triggered drug-releasing magnetic nanoparticles for cancer therapy guided by molecular imaging by MRI. Adv Mater 23(21):2436–2442. doi:10.1002/adma.201100351
126. Chen W, Meng F, Cheng R, Zhong Z (2010) pH-Sensitive degradable polymersomes for triggered release of anticancer drugs: a comparative study with micelles. J Controlled Release Off J Controlled Release Soc 142(1):40–46. doi:10.1016/j.jconrel.2009.09.023

127. Ge J, Neofytou E, Cahill TJ 3rd, Beygui RE, Zare RN (2012) Drug release from electric-field-responsive nanoparticles. ACS Nano 6(1):227–233. doi:10.1021/nn203430m
128. Oliveira H, Perez-Andres E, Thevenot J, Sandre O, Berra E, Lecommandoux S (2013) Magnetic field triggered drug release from polymersomes for cancer therapeutics. J Control Release Off J Controlled Release Soc 169(3):165–170. doi:10.1016/j.jconrel.2013.01.013
129. Zderic V (2008) Ultrasound-enhanced drug and gene delivery: a review. In: conference proceedings: annual international conference of the IEEE engineering in medicine and biology society IEEE engineering in medicine and biology society annual conference, p 4472. doi:10.1109/IEMBS.2008.4650205
130. Huschka R, Zuloaga J, Knight MW, Brown LV, Nordlander P, Halas NJ (2011) Light-induced release of DNA from gold nanoparticles: nanoshells and nanorods. J Am Chem Soc 133(31):12247–12255. doi:10.1021/ja204578e
131. Manzoor AA, Lindner LH, Landon CD, Park JY, Simnick AJ, Dreher MR, Das S, Hanna G, Park W, Chilkoti A, Koning GA, ten Hagen TL, Needham D, Dewhirst MW (2012) Overcoming limitations in nanoparticle drug delivery: triggered, intravascular release to improve drug penetration into tumors. Cancer Res 72(21):5566–5575. doi:10.1158/0008-5472.CAN-12-1683
132. Kim HR, Gil S, Andrieux K, Nicolas V, Appel M, Chacun H, Desmaele D, Taran F, Georgin D, Couvreur P (2007) Low-density lipoprotein receptor-mediated endocytosis of PEGylated nanoparticles in rat brain endothelial cells. Cell Mol Life Sci CMLS 64(3):356–364. doi:10.1007/s00018-007-6390-x
133. Agemy L, Friedmann-Morvinski D, Kotamraju VR, Roth L, Sugahara KN, Girard OM, Mattrey RF, Verma IM, Ruoslahti E (2011) Targeted nanoparticle enhanced proapoptotic peptide as potential therapy for glioblastoma. Proc Natl Acad Sci USA 108(42):17450–17455. doi:10.1073/pnas.1114518108
134. Murphy EA, Majeti BK, Barnes LA, Makale M, Weis SM, Lutu-Fuga K, Wrasidlo W, Cheresh DA (2008) Nanoparticle-mediated drug delivery to tumor vasculature suppresses metastasis. Proc Natl Acad Sci USA 105(27):9343–9348. doi:10.1073/pnas.0803728105
135. Sugahara KN, Teesalu T, Karmali PP, Kotamraju VR, Agemy L, Girard OM, Hanahan D, Mattrey RF, Ruoslahti E (2009) Tissue-penetrating delivery of compounds and nanoparticles into tumors. Cancer Cell 16(6):510–520. doi:10.1016/j.ccr.2009.10.013
136. Lee HY, Li Z, Chen K, Hsu AR, Xu C, Xie J, Sun S, Chen X (2008) PET/MRI dual-modality tumor imaging using arginine-glycine-aspartic (RGD)-conjugated radiolabeled iron oxide nanoparticles. J Nucl Med Off Publ Soc Nucl Med 49(8):1371–1379. doi:10.2967/jnumed.108.051243
137. Cui Z, Hsu CH, Mumper RJ (2003) Physical characterization and macrophage cell uptake of mannan-coated nanoparticles. Drug Dev Ind Pharm 29(6):689–700. doi:10.1081/DDC-120021318
138. Zhu S, Niu M, O'Mary H, Cui Z (2013) Targeting of tumor-associated macrophages made possible by PEG-sheddable, mannose-modified nanoparticles. Mol Pharm 10(9):3525–3530. doi:10.1021/mp400216r
139. Weissleder R, Nahrendorf M, Pittet MJ (2014) Imaging macrophages with nanoparticles. Nat Mater 13(2):125–138. doi:10.1038/nmat3780
140. Yamashita M, Ogawa T, Zhang X, Hanamura N, Kashikura Y, Takamura M, Yoneda M, Shiraishi T (2012) Role of stromal myofibroblasts in invasive breast cancer: stromal expression of alpha-smooth muscle actin correlates with worse clinical outcome. Breast Cancer 19(2):170–176. doi:10.1007/s12282-010-0234-5
141. Siemann DW (2011) Tumor microenvironment. Wiley, Chichester
142. Feig C, Jones JO, Kraman M, Wells RJ, Deonarine A, Chan DS, Connell CM, Roberts EW, Zhao Q, Caballero OL, Teichmann SA, Janowitz T, Jodrell DI, Tuveson DA, Fearon DT (2013) Targeting CXCL12 from FAP-expressing carcinoma-associated fibroblasts synergizes with anti-PD-L1 immunotherapy in pancreatic cancer. Proc Natl Acad Sci USA 110(50):20212–20217. doi:10.1073/pnas.1320318110

143. Murakami M, Ernsting MJ, Undzys E, Holwell N, Foltz WD, Li SD (2013) Docetaxel conjugate nanoparticles that target alpha-smooth muscle actin-expressing stromal cells suppress breast cancer metastasis. Cancer Res 73(15):4862–4871. doi:10.1158/0008-5472.CAN-13-0062
144. Guo S, Miao L, Wang Y, Huang L (2014) Unmodified drug used as a material to construct nanoparticles: delivery of cisplatin for enhanced anti-cancer therapy. J Control Release 174:137–142. doi:10.1016/j.jconrel.2013.11.019
145. Li J, Yang Y, Huang L (2012) Calcium phosphate nanoparticles with an asymmetric lipid bilayer coating for siRNA delivery to the tumor. J Control Release Off J Control Release Soc 158(1):108–114. doi:10.1016/j.jconrel.2011.10.020
146. Guo S, Lin CM, Xu Z, Miao L, Wang Y, Huang L (2014) Co-delivery of cisplatin and rapamycin for enhanced anticancer therapy through synergistic effects and microenvironment modulation. ACS Nano 8(5):4996–5009. doi:10.1021/nn5010815
147. Mao Y, Keller ET, Garfield DH, Shen K, Wang J (2013) Stromal cells in tumor microenvironment and breast cancer. Cancer metastasis rev 32(1–2):303–315. doi:10.1007/s10555-012-9415-3
148. Hwang RF, Moore T, Arumugam T, Ramachandran V, Amos KD, Rivera A, Ji B, Evans DB, Logsdon CD (2008) Cancer-associated stromal fibroblasts promote pancreatic tumor progression. Cancer Res 68(3):918–926. doi:10.1158/0008-5472.CAN-07-5714
149. Brennen WN, Isaacs JT, Denmeade SR (2012) Rationale behind targeting fibroblast activation protein-expressing carcinoma-associated fibroblasts as a novel chemotherapeutic strategy. Mol Cancer Ther 11(2):257–266. doi:10.1158/1535-7163.MCT-11-0340
150. Xu Z, Wang Y, Zhang L, Huang L (2014) Nanoparticle-delivered transforming growth factor-beta siRNA enhances vaccination against advanced melanoma by modifying tumor microenvironment. ACS Nano 8(4):3636–3645. doi:10.1021/nn500216y
151. Gore J, Korc M (2014) Pancreatic cancer stroma: friend or foe? Cancer Cell 25(6):711–712. doi:10.1016/j.ccr.2014.05.026
152. Rhim AD, Oberstein PE, Thomas DH, Mirek ET, Palermo CF, Sastra SA, Dekleva EN, Saunders T, Becerra CP, Tattersall IW, Westphalen CB, Kitajewski J, Fernandez-Barrena MG, Fernandez-Zapico ME, Iacobuzio-Donahue C, Olive KP, Stanger BZ (2014) Stromal elements act to restrain, rather than support, pancreatic ductal adenocarcinoma. Cancer Cell 25(6):735–747. doi:10.1016/j.ccr.2014.04.021
153. Krtolica A, Parrinello S, Lockett S, Desprez PY, Campisi J (2001) Senescent fibroblasts promote epithelial cell growth and tumorigenesis: a link between cancer and aging. Proc Natl Acad Sci USA 98(21):12072–12077. doi:10.1073/pnas.211053698
154. Swami A, Reagan MR, Basto P, Mishima Y, Kamaly N, Glavey S, Zhang S, Moschetta M, Seevaratnam D, Zhang Y, Liu J, Memarzadeh M, Wu J, Manier S, Shi J, Bertrand N, Lu ZN, Nagano K, Baron R, Sacco A, Roccaro AM, Farokhzad OC, Ghobrial IM (2014) Engineered nanomedicine for myeloma and bone microenvironment targeting. Proc Natl Acad Sci USA 111(28):10287–10292. doi:10.1073/pnas.1401337111
155. Peinado H, Aleckovic M, Lavotshkin S, Matei I, Costa-Silva B, Moreno-Bueno G, Hergueta-Redondo M, Williams C, Garcia-Santos G, Ghajar C, Nitadori-Hoshino A, Hoffman C, Badal K, Garcia BA, Callahan MK, Yuan J, Martins VR, Skog J, Kaplan RN, Brady MS, Wolchok JD, Chapman PB, Kang Y, Bromberg J, Lyden D (2012) Melanoma exosomes educate bone marrow progenitor cells toward a pro-metastatic phenotype through MET. Nat Med 18 (6):883–891. doi:10.1038/nm.2753

How Nanoparticles Interact with Cancer Cells

Abdullah Syed and Warren C.W. Chan

Abstract

There are currently no nanoparticle formulations that optimally target diseased cells in the body. A small percentage of nanoparticles reach these cells and most accumulate in cells of the mononuclear phagocytic system. This chapter explores the interactions between nanoparticles and cells that may explain the causes for off-target accumulation of nanoparticles. A greater understanding of the nanoparticle-cellular interactions will lead to improvements in particle design for improved therapeutic outcome.

Keywords

Nanomedicine · Nano-Biointeractions · Nanoparticle Cell Interaction · Cancer Targeting

Contents

A. Syed · W.C.W. Chan (✉)
Institute of Biomaterials and Biomedical Engineering,
Terrence Donnelly Centre for Cellular and Biomolecular Research,
University of Toronto, 164 College St., 407, Toronto, ON M5S 3G9, Canada
e-mail: Warren.chan@utoronto.ca

C.A. Mirkin et al. (eds.), *Nanotechnology-Based Precision Tools for the Detection and Treatment of Cancer*, Cancer Treatment and Research 166,
DOI 10.1007/978-3-319-16555-4_10

1 Introduction

The in vivo accumulation of nanoparticles in target and off-target tissues depends on the interactions of the nanoparticles with different types of cells (i.e. diseased cells and reticuloendothelial cells). By manipulating these interactions, it may be possible to improve the accumulation of therapeutics at target tissues and reduce off-target exposure. More precise targeting can improve the therapeutic indices of chemotherapeutic drugs because it can reduce the toxicity that results from drugs interacting with off-target organs. Currently, the realization of this prospect is hampered by the fact that most nanoparticles do not reach target sites, and technologies based upon them have demonstrated only modest efficacy in clinical trials [1, 2]. Overcoming this challenge requires an improved understanding of the processes that determine the biodistribution and pharmacokinetics of nanoparticles.

When nanoparticles are injected in a live animal, several key steps occur before the nanoparticle encounters cancer cells. First, when exposed to blood or serum, proteins adsorb on the nanoparticle surface forming a layer known as the "protein corona" (see Sect. 3). The types of proteins bound to the surface and the kinetics of aggregation are dependent on the initial geometry, surface chemistry, and material composition of the nanoparticle as well the composition of the biological environment. Once bound, these proteins may cause the nanoparticles to aggregate or direct them to bind to phagocytic cells located in the organs of the mononuclear phagocyte system which will remove the nanoparticles from circulation before they can reach cancer cells.

It has been proposed that the nanoparticles remaining in circulation encounter the tumor microenvironment by diffusing through leaky vasculature [3]. Decorating the nanoparticle surface with molecules that bind cancer cells is presumed to enhance the rate of endocytosis into the cells. Subsequently, nanoparticles may escape endosomes and enter the cytoplasm in order to exert a therapeutic effect. Alternatively, nanoparticles may manipulate cell function through surface receptors that control major signaling pathways without undergoing endocytosis. These potential therapeutic effects of nanoparticles can only be realized if we can transport the nanoparticle to the diseased site.

The complexity of how nanoparticles reach the targeted site is fascinating, but would be difficult to fully explore in this book chapter. We recommend the following excellent review articles on the topic [2, 4–7]. This book chapter focuses on what happens to nanoparticles once they reach the cancer cells. This chapter begins with a general introduction of the mechanisms of endocytosis that are relevant to understanding nanoparticle-cell interactions and the relative rates of uptake of nanoparticles by cancer cells and phagocytes (see Sect. 2). The remaining sections follow the chronological process that nanoparticles undergo in vivo. This process begins with the nanoparticle's exposure to blood serum (see Sect. 3) and from there proceeds to their binding and/or uptake by phagocytes and cancer cells (see Sect. 4). This step is followed by the nanoparticle's endosomal escape (see Sect. 5) and exocytosis (see Sect. 6), which are also important factors that determine nanoparticle accumulation. A full understanding of these processes should allow the extraction of principles that can be used to design nanoparticles with deliberately programmed behavior.

2 Uptake of Nanoparticles Proceeds Through Distinct Mechanisms in Both Cancer Cells and Professional Phagocytes

The mechanisms of endocytosis employed by cells depend on the characteristics of the nanoparticles themselves, including the presence of ligands that bind and activate specific receptors on the cell surface [8–10]. The cell membrane is generally impermeable to most macromolecules and nanoparticles, so the transport of both nutrients and nanoparticles proceeds mostly through energy-dependent mechanisms of uptake. This has been confirmed with the observation that uptake of nanoparticles is almost completely inhibited when such energy-dependent mechanisms are blocked [10]. Some of these endocytotic mechanisms are shown in Fig. 1. Since endocytotic mechanisms are critical components of cell signaling pathways there may be unique mechanisms of uptake by diseased cells. Currently we know that phagocytosis is associated with off-target accumulation and it has been claimed that

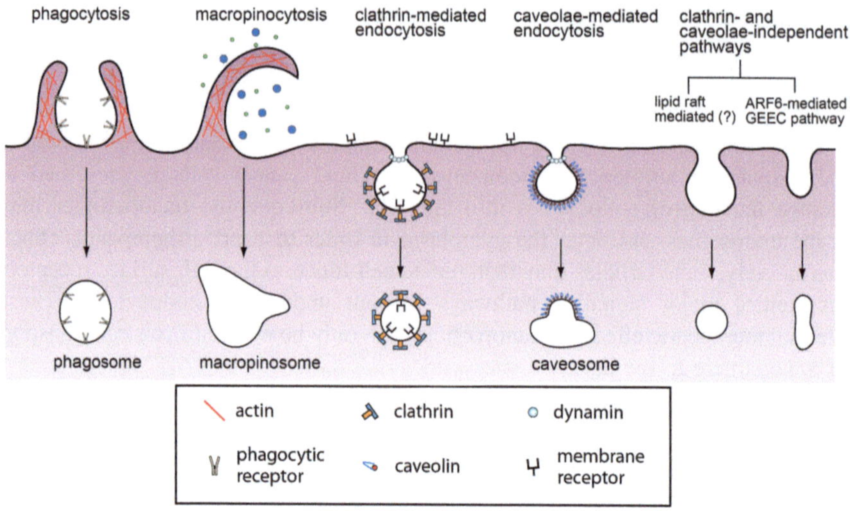

Fig. 1 Schematic of some of the mechanisms of endocytosis that have so far been elucidated. The focus in this chapter will be on clathrin-mediated, caveolar-type, macropinocytosis, and phagocytosis mechanisms. Adapted from [68] with permission of The Royal Society of Chemistry

caveolin-mediated endocytosis is associated with improved transfection efficiency but it is unclear if any known endocytotic mechanism behaves differently in a broad range of cancer cells [11]. More research on mechanisms of endocytosis may uncover further patterns that differentiate cancer cells from normal and off-target cells. This may lead to new methods to redirect nanoparticles towards particular endocytotic mechanisms as an alternate means to improve nanoparticle specificity. To this end, further studies are needed to determine which mechanisms of uptake are commonly employed by cancer cells compared to non-target cells as well as studies that relate nanoparticle uptake through particular endocytotic pathways.

2.1 Clathrin-Mediated Endocytosis Is Normally Involved in Receptor-Mediated Uptake by Cancer Cells

The most well-known and studied uptake mechanism is clathrin-mediated endocytosis. This mechanism is responsible for the endocytosis of iron-saturated transferrin, low-density lipoproteins, the influenza virus and many types of nanoparticles in a variety of cell types including cancer cells [12]. It is also presumed to be responsible for the uptake of nanoparticles when their surface is coated with certain targeting ligands [13]. The initiation of clathrin-mediated endocytosis is hypothesized to be spontaneous, and therefore non-specific, beginning with the assembly of FCHO and AP2 (and several other proteins) on the cell membrane [14, 15]. The binding of a molecule or nanoparticle onto multiple membrane receptors can also trigger the process in a ligand specific manner by creating membrane curvature, which initiates the assembly of AP2 [14]. Following initiation, a clathrin coat forms underneath the cell membrane, and the membrane is pulled into the cell to form a pit [14, 15]. The pit and the clathrin coat can be visualized unambiguously using TEM [16]. At this point, the involvement of a cargo on a receptor is required for maturation of the pit into an endocytic vesicle [17]. Once the clathrin assembly matures, dynamin catalyzes membrane scission and separates the newly formed vesicle from the cell surface [17]. Clathrin vesicles have a minimal diameter of 31 nm as determined by the cryo-TEM of vesicles, but the upper limit may be much higher [18]. Tsuji et al. confirmed the clathrin-mediated uptake of 523 nm diameter particles by directly visualizing the clathrin coat using TEM [16]. Veiga et al. and others have reported the clathrin-dependent uptake of particles as large as 1 μm in diameter through fluorescence co-localization and siRNA knockdown techniques [16, 19].

2.2 Clathrin-Independent Endocytic Mechanisms Are Prevalent in Many Cell Types, but Are Numerous and More Difficult to Study

Nanoparticle uptake through clathrin-independent endocytosis pathways has been observed [20]. However, multiple pathways are involved and their characteristics are not completely understood because it is difficult to distinguish between them. [21]. The most commonly cited clathrin-independent pathway is caveolin-mediated uptake. Both spherical nucleic acids-gold nanoparticles and polyethylenimine—DNA complexes have demonstrated cellular uptake through this pathway [11, 22]. Caveolin is a membrane protein that is enriched in flask-shaped invaginations on the cell membrane. It was assumed that enclosed intracellular structures positive for caveolin1 as observed by TEM were endocytic vessels formed in caveolae; however, Sandvig et al. have recently argued that these structures are an artifact of sectioning and do not get internalized [21]. Nevertheless, caveolin1 is strongly associated with endocytosis perhaps through another pathway [11, 21, 23].

Unfortunately, our understanding of other clathrin-independent mechanisms is even more limited [21]. It is often difficult to distinguish these mechanisms from each other and from caveolin-mediated endocytosis [24]. One way to distinguish them is to determine their dependence on dynamin (which can be inhibited by dynasore and other agents), cholesterol (which is inhibited by methyl beta cyclodextrin and others) and other molecules essential to endocytosis. Inhibition of these pathways by small molecule inhibitors is simple to implement, but suffers from drawbacks due to the lack of their specificity for particular endocytic pathways [25]. An alternative approach is to measure the co-localization of nanoparticles with labeled proteins specific to each endocytic pathway using confocal microscopy [19]. This procedure is more expensive and time-consuming to implement. As a result, it is not commonly used for studying nanoparticle-cellular interactions.

2.3 Larger Particles Are Normally Taken up by Micropinocytosis in Most Cancer Cells

Although the association of defined particle sizes with specific uptake mechanisms has not been determined and is under heavy investigation, macropinocytosis is presumed to be responsible for the uptake of particles larger than 200 nm diameter in many cases. Unlike the mechanisms discussed earlier, the uptake of large volumes is usually actin-dependent since large changes to the cell membrane are required in order to engulf such particles. Macropinocytosis proceeds through the formation of extensive membrane ruffles, which non-specifically engulf nanoparticles and other entities [26]. This process requires both actin and cholesterol [12] and is distinctly observed using electron microscopy. It was shown by Herd et al. to be involved (along with phagocytosis) in the uptake of large silica particles (>200 nm diameter) and also by Iverson et al. to be involved in the uptake of ricinB-coated quantum dots (diameter of only 20 nm). These studies highlight the lack of consensus in establishing strict size limits for endocytic mechanisms [27]. It is likely that size is only one factor in determining the uptake mechanism of nanoparticles and the impact of other properties is still being evaluated.

2.4 Off-Target Accumulation of Nanoparticles Is Presumed to Be Due to Phagocytosis

Phagocytosis is thought to be responsible for the receptor-mediated uptake of particles larger than 500 nm in diameter, although strict size limits continue to be debated. Physiologically, this mechanism is used predominantly by professional phagocytes, which include monocytes, macrophages, neutrophils, dendritic cells, and mast cells. These cells are responsible for the clearance of most types of nanoparticles [28]. This mechanism is strictly receptor-mediated, but can be activated by opsonins bound to the nanoparticle surface that cause the phagocytes to engulf diverse cargo. The normal mechanism is initiated by the clustering of receptors induced by the binding of cargo to the cell surface. Endocytosis then proceeds by engulfment mediated by the assembly of an actin skeleton [12]. Since phagocytosis can operate on micron-sized particles, this process can be observed by light microscopy on live cells. Visualizing cell and particle interaction directly, Champion et al. showed that phagocytosis is sensitive to the curvature of the particle at the point of first contact and that beyond a contact angle of 45°, it is completely inhibited [8]. Phagocytosis can also be prevented by mimicking the molecules present on normal cells. Since professional phagocytes avoid ingesting healthy cells by recognizing molecular markers on their surface, such as CD47, Rodriguez et al. synthesized a peptide to mimic this marker. This "self peptide", when functionalized on polystyrene nanoparticles improved their blood circulation half-life two-fold compared to nanoparticles functionalized with scrambled peptides (30 min compared to 15 min) [29]. Strategies that reduce the phagocytic uptake of nanoparticles may be key to reducing off-target accumulation of nanoparticles.

3 Contact with Blood/Serum

The rapid adsorption of proteins on the surface of nanoparticles occurs immediately following their exposure to biological media [30]. This layer is referred to as the protein corona and it can change the size, shape, surface characteristics, aggregation state, and subsequent biological behaviors of the nanoparticles [31–33]. It is important to understand the binding and activation of opsonins as these steps are required for the receptor-mediated phagocytosis of nanoparticles by professional phagocytes as discussed earlier. This binding and activation affects nanoparticle fate in vivo. The influence of these proteins on mechanisms of off-target and on-target uptake is explored in this section.

3.1 Adsorption of Serum Proteins Changes Fundamental Nanoparticle Properties

Protein adsorption changes a broad range of fundamental nanoparticle properties. For example, protein adsorption reduced the cytotoxicity of carbon nanotubes and protected PEGylated quantum dots from the fluorescence quenching that normally occurs in physiological buffers [34, 35]. Typically, the formation of the protein corona increases the hydrodynamic diameter of nanoparticles by 3–35 nm and may cause their aggregation into larger clusters (as shown in Fig. 2) [33, 36, 37].

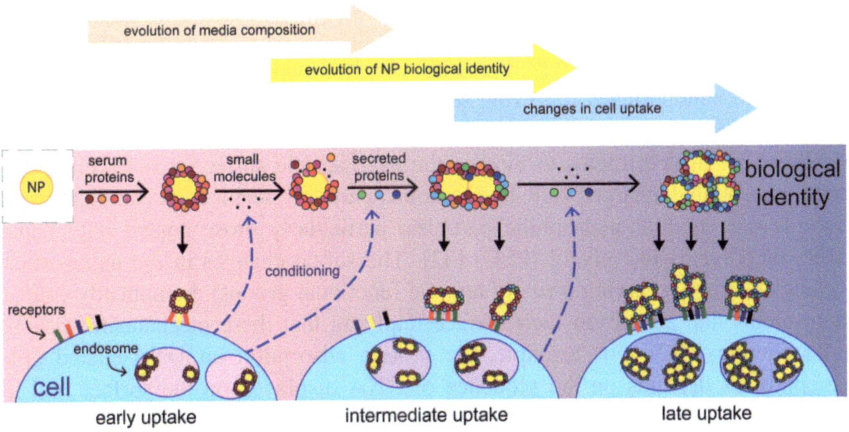

Fig. 2 Since the adsorption of serum proteins occurs on the nanoparticle surface, all subsequent surface-mediated interactions can be affected by the presence of these proteins. Reprinted with permission from [36]. Copyright 2014 American Chemical Society

In addition, they often display a slight negative charge (zeta potential between −7 and −25 mV), regardless of their original charge prior to their interaction with serum [30]. Since the size, shape, and charge of nanoparticles are known to affect the in vitro behavior of nanoparticles, the changes to these properties can drastically change observed uptake patterns.

3.2 Serum Protein Adsorption on Nanoparticle Surface Redirects Their Fate In Vitro

The clearest observed consequence of serum protein adsorption is the decreased uptake of highly charged particles that is observed after exposure to serum or other proteins [32, 33]. Cationic nanoparticles have been observed to have higher uptake than anionic particles in general, and this effect is retained (although less prominent) after their exposure to serum, even if both particle types have the same zeta potentials. This suggests that the difference in uptake is mediated by the proteins present on the surfaces of the nanoparticles [30]. Additionally, nanoparticles with targeting ligands on their surface may lose their binding specificity when exposed to serum, presumably due to competing mechanisms of uptake mediated by these serum proteins [38]. In fact, so many properties are modified by serum protein adsorption that the protein corona has been shown to be more predictive of the observed cell uptake of nanoparticles than any other characteristic of the construct [32].

3.3 The Mechanisms Employed by Adsorbed Proteins to Bind to the NP Surface and Redirect Their Fate Are not Fully Understood

The presumed mechanism of off-target accumulation of nanoparticles is the coating of their surface with opsonins, which mediate the uptake of these nanoparticles by professional phagocytes. Over 100 different proteins have been identified that adsorb onto nanoparticles following serum exposure, including well-known opsonins such as complement factors and immunoglobulins [39]. The mechanism employed by these proteins to bind to the nanoparticle surface is not well-understood except that charged and hydrophobic particles seem to accumulate more proteins than neutral hydrophilic particles, particularly those coated with dense layers of polyethylene glycol (PEG) [32]. The subtle changes in the nanoparticle surface chemistry (varying types of anionic functional groups) dramatically change the proteins that adsorb on their surface implying that these mechanisms may be quite complex [32]. For example, the protein adsorption on DNA-coated gold nanoparticles depends on the sequence of DNA displayed on the surface and its secondary structure [40]. Given, this apparent selectivity and complexity, we need to understand which proteins are involved in this process and how their binding can be modulated to provide greater control of the interactions of nanoparticles with cells after serum exposure.

3.4 Surface Passivation Is the Predominant Strategy to Reduce the Influence of Serum Protein Adsorption

Since the mechanisms of serum protein-mediated nanoparticle-cell interactions are not well-understood, the predominant strategy for reducing these undesirable outcomes has been to passivate the surface of nanoparticles with polyethylene glycol (PEG) moieties [41]. Certain zwitterionic surface ligands are also purported to reduce the binding of serum proteins, but more detailed experiments are required to determine how this strategy compares to PEGylation [42]. Surface passivation reduces both macrophage uptake and increases blood circulation half-life for a variety of nanoparticle types [43–45]. In addition, the loss of specificity due to serum protein adsorption and nanoparticle aggregation is also drastically reduced with appropriate passivation of the surface with PEG [37, 46]. However, proteins continue to adsorb on the nanoparticle surface even when the nanoparticle surface is decorated with a very dense layer of PEG [43]. Further, these proteins significantly affect the cell uptake of these particles [32]. Other limitations of PEG include the difficulty of obtaining a high density of PEG on many particle types as well as the fact that PEG increases the hydrodynamic size of nanoparticles which may make the nanoparticle too large for some applications. Alternatives to PEG, including the zwitterionic ligands discussed above, continue to be investigated for this purpose.

3.5 Serum Protein Adsorption Can Be Exploited to Create Novel NPs

Despite the daunting complexity of the interactions of serum proteins with nanoparticles, some efforts have been made to exploit these interactions in order to simplify the design of nanoparticles. Since the adsorption of proteins is essentially inevitable, Prapainop et al. used a small molecule to induce the binding and misfolding of the protein apolipoprotein B to trigger the specific receptor-mediated uptake of these particles in a macrophage cell line [47]. Although nanoparticles are efficiently taken up by macrophages in the absence of specific targeting ligands, this strategy has the potential to be a new direction for creating particles that are easier to synthesize (due to the greater stability of small molecules compared to protein-based targeting ligands) and that display specific targeting in the presence of serum proteins. Hamad-Schifferli's group has controlled the deposition of serum proteins on gold nanorods to create agglomerate protein corona nanoparticles with high payload capacities [48]. These particles were capable of encapsulating both small molecules (Doxorubicin) and DNA that slowly leach over multiple days; the release of such cargo can also be triggered using the near-IR excitation of the gold nanorods. Unfortunately, both approaches are still preliminary, and it remains to be seen if either strategy can be used for directing the specific uptake of nanoparticles in tumors. This is likely to be the focus of future research.

3.6 The Impact of Serum Protein Adsorption Needs Be Acknowledged in Future Studies About Nanoparticle-Cell Interactions

Despite the increased popularity of research on nanoparticle serum protein adsorption in recent years, it remains unclear how this issue should be addressed. New strategies, in addition to PEGylation, are needed to mitigate the effects of serum protein adsorption on nanoparticle behavior. Nevertheless, the research presented here can be applied in many studies involving nanoparticles. To begin with, researchers should study the effects of nanoparticle-cell interactions in the presence of serum since nanoparticle behavior is known to be dramatically altered by serum protein adsorption [49]. Secondly, nanoparticles that undergo aggregation or large changes in size and shape due to protein adsorption should be avoided in place of more stable particles to improve the reproducibility of observed interactions with cells. Lastly, a greater focus on exploiting the unique behavior of protein adsorption and assembly on nanoparticle surfaces should be pursued since these proteins likely mediate many of the interactions that nanoparticles experience.

4 Endocytosis Is Dependent on Nanoparticle Properties

After entering blood circulation and interacting with serum proteins nanoparticles are subject to two competing processes: clearance through renal or phagocytic pathways and accumulation into the tumor microenvironment. Both processes occur simultaneously and the overall biodistribution and pharmacokinetics observed for nanoparticles is dependent on their relative rates. Although systemic factors, such as the perfusion rate of the liver and kidneys compared to that of the tumors and the leakiness of the tumor vasculature, play a major role in dictating these rates, the focus of this chapter will be on cellular factors, specifically the rates of cell uptake in cancer cells and macrophages. The dimensions of nanoparticles, in particular , play a major role in determining their rate of uptake into cells as illustrated by Fig. 3.

4.1 Endocytosis of Nanoparticles into a Variety of Cancer Cells Is Sensitive to Nanoparticle Size

Nanoparticle size determines the curvature of the contact surface between the nanoparticle and the cell membrane and thus dictates the nature of the interaction between these two species. Since membrane curvature has been shown to initiate clathrin-mediated endocytosis as well as phagocytosis it is expected that cell uptake would be dramatically affected by nanoparticle geometry [12, 14]. Indeed, for polymer nanoparticles (PEG-acrylate derivatives) Gratton et al. showed that the efficiency of uptake by HeLa cells (cervical adenocarcinoma) decreased as the

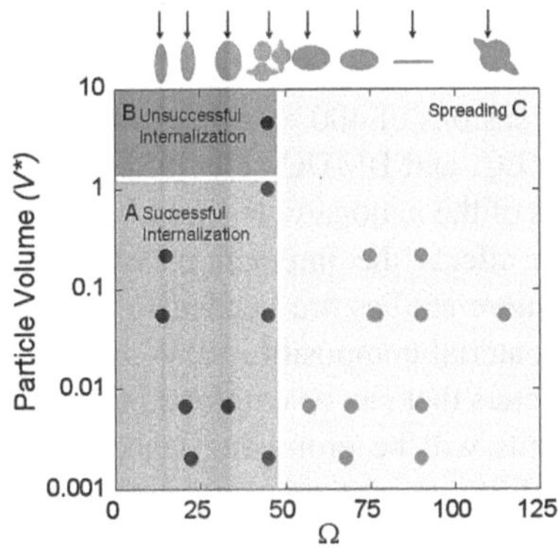

Fig. 3 Effects of geometric parameters on the phagocytosis of polystyrene particles. Reproduced with permission from [8]. Copyright (2006) National Academy of Sciences, U.S.A

particle diameters increased from 200 nm to 5 μm among particles with aspect ratios of 1 [50]. Several groups also explored the uptake of spherical nanoparticles smaller than 100 nm in diameter. Chithrani et al. showed that gold nanoparticles ~50 nm in diameter exhibited greater levels of uptake in HeLa cells than other nanoparticles between 15 and 100 nm [51]. Lu et al. and Varela et al. also obtained similar results using mesoporous silica nanoparticles (maximum uptake at 50 nm diameter) and polystyrene nanoparticles (maximum uptake at 40 nm diameter) despite employing varying cell types [52, 53]. Recently, some of these results have been criticized because increasing particle size also favors sedimentation over diffusion thereby increasing the uptake of gold particles that are 50 nm in diameter or larger [54]. Sedimentation also affects polymer particles (PEG-diacrylate) but becomes significant only for particles with diameters greater than 325 nm (disc shaped with height of 100 nm) due to their reduced density compared to gold nanoparticles [55]. Presumably sedimentation applies to many other particles and may complicate the interpretation of current results but this has not been characterized for most particle types. However, the overall trends observed across a variety of particle types still suggest that particles with diameters around 40–50 nm exhibit more rapid endocytosis than particles of other sizes.

4.2 Aspect Ratio Is also Important in Determining the Rate of Endocytosis

The aspect ratios of nanoparticles also affect the contact areas between the particles and the cells and therefore affect endocytic rates across a variety of particle types. This was illustrated by Gratton et al., who observed higher uptake of polymer nanorods with dimensions of 450 × 150 × 150 nm compared to particles with dimensions of 200 × 200 × 200 nm in HeLa cells [50]. However, Agarwal et al.

recently reported that polymer nanodiscs (PEG-diacrylate particles) [dimensions of 220 × 220 × 100 nm and 325 × 325 × 100 nm] were taken up faster than nanorods of the same material [dimensions of 400 × 100 × 100 nm and 800 × 100 × 100 nm] by HeLa, HEK293, HUVEC, and BMDC cells [55]. One difference between these studies is that the surfaces of the nanoparticles used by Gratton et al. were positively charged and charge also affects the interactions of nanoparticles with cells. To address these concerns, more studies are needed to decouple the effects of aspect ratio, particle size, core material composition, and surface capping ligands in order to establish design parameters that can be employed to control the endocytic rates of nanoparticles. These results will be promising since aspect ratio allows a second degree of freedom (in addition to particle size) as a means to design nanoparticles and manipulate their behavior in vivo.

4.3 Decorating the Nanoparticle Surface with Targeting Ligands Improves Endocytic Specificity by Increasing the Binding and Uptake of Nanoparticles into Target Cells

Another major challenge to nanoparticle design for in vivo applications is to manipulate the relative endocytic rates between cell types such that nanoparticle accumulation occurs more rapidly in target cell types and more slowly in off-target cell types. By decorating the nanoparticle surface with peptides, proteins, aptamers, or small molecules that specifically bind target cell types, the binding affinity and kinetics of binding of nanoparticles to the cell surface can be increased. Often, the binding affinity of the nanoparticle exceeds that of the free ligand. Wiley et al. modified the surface of gold nanoparticles with varying densities of transferrin and observed that the dissociation constant (K_d) of the nanoparticle from Neuro2A cells was between 4.9 and 0.014 nM, indicating that the nanoparticles bound these cells much more strongly than free transferrin (K_d of 144 nM) or nanoparticles without transferrin (K_d too large to measure reliably) [56]. Because this enhanced binding is due to the multivalent interaction to multiple cellular ligands, such functionalization may also directly trigger endocytosis by crosslinking surface ligands and initiating the assembly of a clathrin pit [57, 58].

This increase in endocytic rates in target cells has also been shown to improve the specificity of uptake. Gao et al. observed that the uptake of nanoparticles in U87 gliomablastoma cells doubled when PEG-polycaprolactone nanoparticles were modified with IL-13 peptides (which are specific to U87 cells) while uptake in RAW 264.7 (mouse macrophage) cells remained constant [13].

The binding to nanoparticles to cells through multiple ligands on the nanoparticle surface can also affect the physiological function of the receptors and provide a secondary means to achieve a therapeutic effect. In the case of ErbB2-positive cancer cells, the function of the ErbB2 receptor is to dimerize and provide growth signals that the cancer cell needs to survive and proliferate [59]. By decorating the surface of gold nanoparticles with Herceptin, Jiang et al. found that they could down-regulate the expression of ErbB2 and trigger apoptosis in SK-BR-3 (breast adenocarcinoma) cells [60]. This effect was greater for functionalized nanoparticles than for free Herceptin and strongest for 40–50 nm diameter particles because of the faster internalization of particles in that size range [60].

Although advantageous in the context of cancer therapy, targeting increases the complexity of nanoparticle design and targeting specificity may be lost in physiological environments. When exposed to serum, nanoparticles are non-specifically coated with many serum proteins which may block the function of targeting ligands [61]. As a result, Salvati et al. observed that nanoparticles capable of specific targeting in serum-free conditions using transferrin did not display targeting specificity after being incubated in serum [61]. This effect can be countered by appropriate surface modification schemes, but at the expense of further complicating the design of the nanoparticle [46].

5 Endosomal Escape Is a Limiting Factor for Delivering Cargo to Specific Cell Compartments Using Nanoparticles

Following endocytosis, many nanoparticles have been observed to accumulate in endosomal compartments, where the nanoparticle and its contents may be degraded before they can exert their therapeutic effect. In order to exert a therapeutic effect, the nanoparticle or its cargo must enter other cellular compartments. This problem also affects pathogens, such as the influenza virus, which have responded by employing specific peptides that disrupt or fuse with the endosomal membrane and deliver their cargo into the cellular cytoplasm. A schematic of how this may be achieved with a nanoparticle system is shown in Fig. 4. Inspired by this design, Plank et al. incorporated INF3D1, a peptide dimer construct from the influenza virus, into a polylysine DNA nanoparticle and were able to increase transfection efficiency of a plasmid construct by 5000 fold compared to bare constructs [62]. The improved transfection efficiency implied that the nanoparticle or its cargo (in this case DNA sequences) was able to escape the endosomal compartment and therefore could potentially reach other sites inside the cell. The endosome was disrupted because this peptide dimer assembles into an elongated alpha-helical construct in the acidic compartment of an endosome [62]. This assembly can be achieved with purely synthetic peptides, which are also capable of disrupting endosomes [63]. Hatakeyama et al. employed a synthetic peptide, known as “GALA” due to the repeating unit of its protein sequence, to improve the transfection efficiency of siRNA-containing liposomes [64]. In a manner similar to the influenza peptide mentioned above, this peptide undergoes a conformational transition at pH 5.0 (the typical pH in late endosomes and lysosomes) to form an amphipathic alpha helix that disrupts lipid bilayers and causes endosomes to leak. As a result,

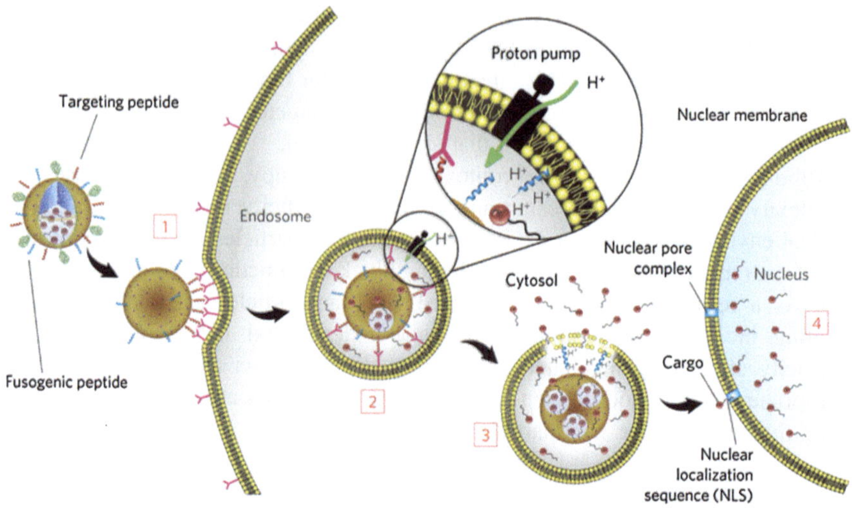

Fig. 4 Schematic of a nanoparticle system escaping the endosomal compartment (step 3) using a fusogenic peptide before delivering cargo to the nucleus. Delivery of therapeutics to cell compartments requires nanoparticles to first escape from endosomal processing before they can exert other effects. Adapted by permission from Macmillan Publishers Ltd: Nature Materials [69], copyright (2014)

Hatakeyama et al. observed a 100-fold enhancement in the transfection efficiency (quantified in this case as the expression of luciferase) of their liposomal constructs after modifying them with GALA [64]. If similar techniques are employed for other types of nanoparticles, dramatic improvements in their efficacy may also be possible.

6 Exocytosis of Particles Affects Rates of Accumulation

Once nanoparticles are endocytosed into cancer cells or phagocytic cells, they can release their cargo to exert a therapeutic effect. However, the strength of this effect depends not only on the rate of endocytosis but also on the accumulation and residence time of the nanoparticles inside cells. The accumulation of nanoparticles inside cells at a given time depends on the relative rates of endocytosis and exocytosis and therefore manipulating the rates of exocytosis offers another route for optimizing the therapeutic effect of nanoparticles. For example, some particles are efficiently endocytosed but they show poor accumulation inside cells due to rapid exocytosis. Yanes et al. observed that 95 % of mesoporous silica nanoparticles were exocytosed within 48 h from A549 (lung carcinoma) cells [65]. In addition the cytotoxicity of these particles when loaded with camptothecin was improved when exocytosis was inhibited suggesting that inhibition of exocytosis may be a viable route for improving therapeutic response in vivo. Other cancer cell types had much lower rates of exocytosis, for instance only 35 % of particles were exocytosed in 24 h for MDA-MB-435 (melanoma) compared to 95 % for A549 cells which may explain the variable responses to nanoparticle therapy observed in mouse models using different cell lines [65].

The rates of exocytosis are also dependent on particle size and surface chemistry. Chithrani et al. observed that nanoparticle exocytosis reached 40 % for 14 nm diameter transferrin coated-gold nanoparticles but only 10 % for 100 nm particles with the same surface chemistry in HeLa cells (cervical adenocarcinoma) [66]. The surface chemistry of nanoparticles also influenced their rate of exocytosis. Using gold nanoparticles with anionic, cationic, zwitterionic, and PEG surface modifications, Park et al. observed that exocytosis rates were much lower for cationic nanoparticles than PEG nanoparticles in human macrophages (obtained by differentiating U937 cells) [37]. However, even PEGylated cationic nanoparticles demonstrated different behavior depending on the type of cationic functional group used. Kim et al. showed that particles with an aromatic surface functionality at the end of a cationic PEG modification were exocytosed to a greater extent than nanoparticles with an aliphatic surface functionality at the end of the same cationic PEG structure [67]. This observation that minor changes in surface chemistry alter the rates of exocytosis is reminiscent of the observations of the sensitivity of protein adsorption to similarly subtle changes in the nanoparticle surface. These similarities may be due to the effects of proteins in mediating exocytosis either through manipulating cell response directly or inducing varying levels of aggregation that radically alter the processing nanoparticles. As a result, a greater understanding of how protein corona formation occurs may also provide a route to manipulating exocytosis of nanoparticles and thereby improve therapeutic efficacy.

7 Conclusion

Due to the popularity of nanoparticle research, a variety of avenues have been explored to improve the therapeutic effects of nanoparticles. This chapter describes the lifecycle of an administered nanoparticle from the first serum interaction to cell receptor binding, uptake, and exocytosis. Each step provides engineering opportunities and limitations for designing nanoparticles to target diseased cells. Strategies are available that can improve nanoparticle targeting, but the strategies usually require careful modification of the nanoparticle either by changing its size and shape or by conjugating its surface with targeting ligands, endosomolytic agents or other functional groups. However, this requires an increase in the design complexity of the nanoparticles. The focus of future research should be to develop more sophisticated particles that can incorporate a variety of design traits in order to achieve dramatic improvements in nanoparticle targeting and therapeutics.

References

1. Park K (2013) Facing the truth about nanotechnology in drug delivery. ACS Nano 7(9):7442–7447
2. Jain RK, Stylianopoulos T (2010) Delivering nanomedicine to solid tumors. Nat Rev Clin Oncol 7(11):653–664
3. Hobbs SK et al (1998) Regulation of transport pathways in tumor vessels: role of tumor type and microenvironment. Proc Natl Acad Sci USA 95(8):4607–4612
4. Stylianopoulos T, Jain RK (2013) Combining two strategies to improve perfusion and drug delivery in solid tumors. Proc Natl Acad Sci USA 110(46):18632–18637
5. Chauhan VP, Jain RK (2013) Strategies for advancing cancer nanomedicine. Nat Mater 12(11):958–962
6. Jain RK (2013) Normalizing tumor microenvironment to treat cancer: bench to bedside to biomarkers. J Clin Oncol 31(17):2205–2218
7. Choi HS et al (2007) Renal clearance of quantum dots. Nat Biotechnol 25(10):1165–1170
8. Champion JA, Mitragotri S (2006) Role of target geometry in phagocytosis. Proc Natl Acad Sci USA 103(13):4930–4934
9. Iversen TG, Frerker N, Sandvig K (2012) Uptake of ricinB-quantum dot nanoparticles by a macropinocytosis-like mechanism. J Nanobiotechnol 10:33
10. Rejman J et al (2004) Size-dependent internalization of particles via the pathways of clathrin- and caveolae-mediated endocytosis. Biochem J 377(Pt 1):159–169
11. Gabrielson NP, Pack DW (2009) Efficient polyethylenimine-mediated gene delivery proceeds via a caveolar pathway in HeLa cells. J Control Release 136(1):54–61
12. Doherty GJ, McMahon HT (2009) Mechanisms of endocytosis. Annu Rev Biochem 78:857–902
13. Gao H et al (2013) Ligand modified nanoparticles increases cell uptake, alters endocytosis and elevates glioma distribution and internalization. Sci Rep 3:2534
14. Henne WM et al (2010) FCHo proteins are nucleators of clathrin-mediated endocytosis. Science 328(5983):1281–1284
15. Honing S et al (2005) Phosphatidylinositol-(4,5)-bisphosphate regulates sorting signal recognition by the clathrin-associated adaptor complex AP2. Mol Cell 18(5):519–531
16. Tsuji T, Yoshitomi H, Usukura J (2013) Endocytic mechanism of transferrin-conjugated nanoparticles and the effects of their size and ligand number on the efficiency of drug delivery. Microscopy (Oxford) 62(3):341–352
17. Ehrlich M et al (2004) Endocytosis by random initiation and stabilization of clathrin-coated pits. Cell 118(5):591–605
18. Cheng Y et al (2007) Cryo-electron tomography of clathrin-coated vesicles: structural implications for coat assembly. J Mol Biol 365(3):892–899
19. Veiga E, Cossart P (2005) Listeria hijacks the clathrin-dependent endocytic machinery to invade mammalian cells. Nat Cell Biol 7(9):894–900
20. Georgieva JV et al (2011) Surface characteristics of nanoparticles determine their intracellular fate in and processing by human blood-brain barrier endothelial cells in vitro. Mol Ther 19 (2):318–325
21. Sandvig K et al (2008) Clathrin-independent endocytosis: from nonexisting to an extreme degree of complexity. Histochem Cell Biol 129(3):267–276
22. Choi CH et al (2013) Mechanism for the endocytosis of spherical nucleic acid nanoparticle conjugates. Proc Natl Acad Sci USA 110(19):7625–7630
23. Nevins AK, Thurmond DC (2006) Caveolin-1 functions as a novel Cdc42 guanine nucleotide dissociation inhibitor in pancreatic beta-cells. J Biol Chem 281(28):18961–18972
24. Bae YM et al (2012) Endocytosis, intracellular transport, and exocytosis of lanthanide-doped upconverting nanoparticles in single living cells. Biomaterials 33(35):9080–9086
25. Iversen TG, Skotland T, Sandvig K (2011) Endocytosis and intracellular transport of nanoparticles: Present knowledge and need for future studies. Nano Today 6(2):176–185

26. Grimmer S, van Deurs B, Sandvig K (2002) Membrane ruffling and macropinocytosis in A431 cells require cholesterol. J Cell Sci 115(Pt 14):2953–2962
27. Herd H et al (2013) Nanoparticle geometry and surface orientation influence mode of cellular uptake. ACS Nano 7(3):1961–1973
28. Owens DE 3rd, Peppas NA (2006) Opsonization, biodistribution, and pharmacokinetics of polymeric nanoparticles. Int J Pharm 307(1):93–102
29. Rodriguez PL et al (2013) Minimal "Self" peptides that inhibit phagocytic clearance and enhance delivery of nanoparticles. Science 339(6122):971–975
30. Tenzer S et al (2013) Rapid formation of plasma protein corona critically affects nanoparticle pathophysiology. Nat Nanotechnol 8(10):772–781
31. Tenzer S et al (2011) Nanoparticle size is a critical physicochemical determinant of the human blood plasma corona: a comprehensive quantitative proteomic analysis. ACS Nano 5(9):7155–7167
32. Walkey CD et al (2014) Protein corona fingerprinting predicts the cellular interaction of gold and silver nanoparticles. ACS Nano 8(3):2439–2455
33. Lesniak A et al (2012) Effects of the presence or absence of a protein corona on silica nanoparticle uptake and impact on cells. ACS Nano 6(7):5845–5857
34. Ge C et al (2011) Binding of blood proteins to carbon nanotubes reduces cytotoxicity. Proc Natl Acad Sci USA 108(41):16968–16973
35. Sahoo B et al (2007) Spontaneous formation of a protein corona prevents the loss of quantum dot fluorescence in physiological buffers. Chem Phys Lett 445(4–6):217–220
36. Albanese A et al (2014) Secreted biomolecules alter the biological identity and cellular interactions of nanoparticles. ACS Nano 8(6):5515–5526
37. Oh N, Park JH (2014) Surface chemistry of gold nanoparticles mediates their exocytosis in macrophages. ACS Nano 8(6):6232–6241
38. Salvati A et al (2013) Transferrin-functionalized nanoparticles lose their targeting capabilities when a biomolecule corona adsorbs on the surface. Nat Nanotechnol 8:137–143
39. Walkey CD, Chan WC (2012) Understanding and controlling the interaction of nanomaterials with proteins in a physiological environment. Chem Soc Rev 41(7):2780–2799
40. Chinen AB, Guan CM, Mirkin CA (2014) Spherical nucleic acid nanoparticle conjugates enhance g-quadruplex formation and increase serum protein interactions. Angew Chem Int Ed 54(2):527–531
41. Kamaly N et al (2012) Targeted polymeric therapeutic nanoparticles: design, development and clinical translation. Chem Soc Rev 41(7):2971–3010
42. García KP et al (2014) Zwitterionic coatings: zwitterionic-coated "stealth" nanoparticles for biomedical applications: recent advances in countering biomolecular corona formation and uptake by the mononuclear phagocyte system (small 13/2014). Small 10(13):2505
43. Walkey CD et al (2012) Nanoparticle size and surface chemistry determine serum protein adsorption and macrophage uptake. J Am Chem Soc 134(4):2139–2147
44. Zhang Y, Kohler N, Zhang M (2002) Surface modification of superparamagnetic magnetite nanoparticles and their intracellular uptake. Biomaterials 23(7):1553–1561
45. Gref R et al (2000) 'Stealth' corona-core nanoparticles surface modified by polyethylene glycol (PEG): influences of the corona (PEG chain length and surface density) and of the core composition on phagocytic uptake and plasma protein adsorption. Colloids Surf B 18(3–4): 301–313
46. Dai Q, Walkey C, Chan WC (2014) Polyethylene glycol backfilling mitigates the negative impact of the protein corona on nanoparticle cell targeting. Angew Chem Int Ed Engl 53 (20):5093–5096
47. Prapainop K, Witter DP, Wentworth P Jr (2012) A chemical approach for cell-specific targeting of nanomaterials: small-molecule-initiated misfolding of nanoparticle corona proteins. J Am Chem Soc 134(9):4100–4103
48. Kah JC et al (2012) Exploiting the protein corona around gold nanorods for loading and triggered release. ACS Nano 6(8):6730–6740

49. Monopoli MP et al (2012) Biomolecular coronas provide the biological identity of nanosized materials. Nat Nanotechnol 7(12):779–786
50. Gratton SE et al (2008) Microfabricated particles for engineered drug therapies: elucidation into the mechanisms of cellular internalization of PRINT particles. Pharm Res 25 (12):2845–2852
51. Chithrani BD, Ghazani AA, Chan WC (2006) Determining the size and shape dependence of gold nanoparticle uptake into mammalian cells. Nano Lett 6(4):662–668
52. Lu F et al (2009) Size effect on cell uptake in well-suspended, uniform mesoporous silica nanoparticles. Small 5(12):1408–1413
53. Varela JA et al (2012) Quantifying size-dependent interactions between fluorescently labeled polystyrene nanoparticles and mammalian cells. J Nanobiotechnol 10:39
54. Cho EC, Zhang Q, Xia Y (2011) The effect of sedimentation and diffusion on cellular uptake of gold nanoparticles. Nat Nanotechnol 6(6):385–391
55. Agarwal R et al (2013) Mammalian cells preferentially internalize hydrogel nanodiscs over nanorods and use shape-specific uptake mechanisms. Proc Natl Acad Sci USA 110 (43):17247–17252
56. Wiley DT et al (2013) Transcytosis and brain uptake of transferrin-containing nanoparticles by tuning avidity to transferrin receptor. Proc Natl Acad Sci USA 110(21):8662–8667
57. Gao H, Shi W, Freund LB (2005) Mechanics of receptor-mediated endocytosis. Proc Natl Acad Sci USA 102(27):9469–9474
58. Henne WM et al (2007) Structure and analysis of FCHo2 F-BAR domain: a dimerizing and membrane recruitment module that effects membrane curvature. Structure 15(7):839–852
59. Harari D, Yarden Y (2000) Molecular mechanisms underlying ErbB2/HER2 action in breast cancer. Oncogene 19(53):6102–6114
60. Jiang W et al (2008) Nanoparticle-mediated cellular response is size-dependent. Nat Nanotechnol 3(3):145–150
61. Salvati A et al (2013) Transferrin-functionalized nanoparticles lose their targeting capabilities when a biomolecule corona adsorbs on the surface. Nat Nanotechnol 8(2):137–143
62. Plank C et al (1994) The influence of endosome-disruptive peptides on gene transfer using synthetic virus-like gene transfer systems. J Biol Chem 269(17):12918–12924
63. Cho YW, Kim JD, Park K (2003) Polycation gene delivery systems: escape from endosomes to cytosol. J Pharm Pharmacol 55(6):721–734
64. Hatakeyama H et al (2009) A pH-sensitive fusogenic peptide facilitates endosomal escape and greatly enhances the gene silencing of siRNA-containing nanoparticles in vitro and in vivo. J Controlled Release 139(2):127–132
65. Yanes RE et al (2013) Involvement of lysosomal exocytosis in the excretion of mesoporous silica nanoparticles and enhancement of the drug delivery effect by exocytosis inhibition. Small 9(5):697–704
66. Chithrani BD, Chan WC (2007) Elucidating the mechanism of cellular uptake and removal of protein-coated gold nanoparticles of different sizes and shapes. Nano Lett 7(6):542–550
67. Kim CS et al (2014) The role of surface functionality in nanoparticle exocytosis. Adv Healthc Mater 3:5–7
68. Chou LY, Ming K, Chan WC (2011) Strategies for the intracellular delivery of nanoparticles. Chem Soc Rev 40(1):233–245
69. Ashley CE et al (2011) The targeted delivery of multicomponent cargos to cancer cells by nanoporous particle-supported lipid bilayers. Nat Mater 10(5):389–397

Engineering the Nanoparticle-Protein Interface for Cancer Therapeutics

Amir Ata Saie, Moumita Ray, Morteza Mahmoudi and Vincent M. Rotello

Abstract

Intracellular delivery of functional proteins using nanoparticles can be a game-changing approach for cancer therapy. However, cytosolic release of functional protein is still a major challenge. In addition, formation of protein corona on the surface of the nanoparticles can also alter the behavior of the nanoparticles. Here, we will review recent strategies for protein delivery into the cell. Finally we will discuss the issue of protein corona formation in light of nanoparticle-protein interactions.

A.A. Saie
Department of Medical Biochemistry and Biophysics, Karolinska Institutet, Stockholm, Sweden

A.A. Saie · M. Mahmoudi
Faculty of Pharmacy, Nanotechnology Research Center and Department of Nanotechnology, Tehran University of Medical Sciences, Tehran, Iran

M. Mahmoudi
Division of Cardiovascular Medicine, Department of Medicine, Stanford University School of Medicine, Stanford, CA, USA

M. Mahmoudi
Division of Cardiology, Department of Pediatrics, School of Medicine, Stanford University, Stanford, CA, USA

M. Ray · V.M. Rotello (✉)
Department of Chemistry, University of Massachusetts, Amherst, 710 North Pleasant Street, Amherst, MA 01003, USA
e-mail: rotello@chem.umass.edu

C.A. Mirkin et al. (eds.), *Nanotechnology-Based Precision Tools for the Detection and Treatment of Cancer*, Cancer Treatment and Research 166,
DOI 10.1007/978-3-319-16555-4_11

Contents

1 Introduction

The interaction of proteins with nanomaterials is a double-edged sword. On one hand, protein delivery is a potentially powerful strategy for the development of new therapeutics. Concurrently, protein corona formation significantly alters the behavior of nanomaterials in an often unpredictable fashion.

Protein delivery is a challenging goal that would open new pathways in cancer therapy. Both in vitro and in vivo delivery would have important uses. In vitro, the delivery of proteins into cells could be used as a potentially game-changing approach to stimulate effective immune cells in immunotherapy applications [1]. The development of efficient in vivo protein delivery vectors would provide therapeutic replacement tools to agonize or antagonize key intracellular pathways for of cancer. To date, around 100 proteins, targeting a wide range of disease states including cancer, have been found to be transported into cells in using various animal models. Some of these protein systems have made it to clinical trials [2, 3].

Two key challenges exist in the delivery of proteins into cells. First, transporting the protein into the cell in its active form presents a significant hurdle. This issue can be addressed through protein conjugation and modification (e.g., with cell-penetrating peptides). These strategies allow proteins to take advantage of native cellular uptake processes, such as phagocytosis and endocytosis [4]. Transport into the cell, however, is actually the lesser of the two challenges. Once inside the cell, proteins are generally sequestered in vesicular entities (e.g., endosomes) after internalization, preventing them from accessing the cytosol, where they can be most effective [5]. Substantial progress has been made on both fronts; however, the delivery of active proteins to the cytosol remains a major challenge.

Protein corona formation is the second area where understanding of nanoparticle-protein interactions is essential. An accurate understanding of the structural and dynamic properties of nanoparticle-protein interactions will allow us to better engineer systems that behave as desired in delivery processes. In particular, the use of guided corona generation provides the potential to develop new targeting strategies.

Taken together, it is essential to understand protein-nanoparticle interactions in order to progress in the development of therapeutic systems. In this chapter, we will start off by discussing the basics of nanoparticle-protein interactions, and then use this understanding as a foundation for discussing recent research in this area.

2 Nanoparticle-Protein Interactions

Nanoparticles (NPs) have unique properties that can enable applications in diverse fields, such as bio-imaging [6], delivery [7], and sensing [8]. The biological recognition properties of NPs are dependent on how they interface with biomolecules (such as proteins) and their surroundings. Therefore, a basic knowledge of the biological behavior of NPs is a prerequisite for their use. The tunability of the core size and tailorability of the NP surface can facilitate the engineering of protein interactions. NPs can display characteristic opto-electronic and magnetic properties, whereas proteins have catalytic activity, recognition and inhibition [9].

The interactions between NPs and proteins are dependent on multiple parameters, such as the sizes, shapes, charges, and chemical functionalities of these moieties. These parameters regulate noncovalent interactions, such as van der Waals (vdw) interactions, electrostatic interactions, hydrogen bonding (H-bond), hydrophobic interactions, salt bridges, and π–π stacking interactions that lead to NP-protein assembly [10]. The characteristics of these interactions are summarized in Table 1.

Clearly, it is important to develop a fundamental understanding of the thermodynamic parameters behind NP-protein interactions. Isothermal titration calorimetry (ITC) is a powerful tool for investigating the fundamental thermodynamics at the

Table 1 Characteristics of the various chemical interactions between NPs and proteins (reused with permission from Ref. [10])

Forces	Strength	Range (nm)*	Specificity	Main factors
vdw forces	Weak	0–10	No	Interface complementarity
H-bond	Moderate	<0	Partial	Hydrogen donor/acceptor at interface
Electrostatic forces	Moderate	0–10	No	Charge state, ionic strength
Hydrophobic interaction	Strong	0–10	Partial	Hydrophobic surface
π–π stacking	Strong	0–5	Yes	Aromatic ring orientation
Salt bridge	Strong	<1	Yes	Multiple recognition

*0 indicates direct atom contact (based on vdw radius)

NP-protein interface. A model based on ITC results [11] describes the overall complexation process between NPs and proteins using the following equations:

$$\text{Protein} + \text{NP} \rightleftharpoons \text{Protein} - \text{NP} \tag{1}$$

$$\text{xH}_2\text{Op} + \text{yH}_2\text{On} \rightleftharpoons (\text{x} + \text{y} - \text{z})\text{H}_2\text{Op} - \text{n} + \text{zH}_2\text{O} \tag{2}$$

$$\text{Protein} \cdot \text{xH}_2\text{Op} + \text{NP} \cdot \text{yH}_2\text{On} \rightleftharpoons \text{Protein} - \text{NP} \cdot (\text{x} + \text{y} - \text{z})\text{H}_2\text{Op} - \text{n} + \text{zH}_2\text{O} \tag{3}$$

Here, H_2Op, H_2On, and H_2Op-n denote water molecules associated with the protein, NP, and NP-protein complex, respectively. Equation 1 describes NP-protein complex formation, where the changes in enthalpy (ΔH) and entropy (ΔS) are negative. Equation 2 refers to the solvent reorganization process involved in NP-protein complexation, where $\Delta H > 0$ and $\Delta S > 0$. Depending on the contribution of these two processes, the overall complexation (Eq. 3) of NP and protein can be either endothermic ($\Delta H > 0$ and $\Delta S > 0$) or exothermic ($\Delta H < 0$ and $\Delta S < 0$). Therefore, these results suggest that for NP-protein complexation, enthalpy and entropy can be balanced to get a favorable free energy change ($\Delta G < 0$). In the following sections, we will discuss how NP-protein interactions can be tuned by changing the chemical functionalities at the NP surface and how this affects the structure and function of the proteins.

2.1 Reversible *versus* Irreversible Interactions

Proteins can either bind to NPs reversibly or irreversibly. While both modes of interaction can alter protein structure and behavior, irreversible binding presents by far a greater challenge in understanding and in functional applications of nanomaterials. In order to be functional, proteins must retain their structures and activities. The denaturation of proteins on particle surfaces gives rise to a "hard" protein corona that can dramatically alter particle behavior in vivo.

In early studies, anionic NPs were used to inhibit the enzymatic activity of proteins, such as α-chymotrypsin (ChT) [12]. These NPs interacted electrostatically through cationic patches at the active site of this protein. Enzymatic inhibition was followed by slow irreversible denaturation of the secondary structure of the protein. The structural denaturation of this protein was due to the interaction of the nonpolar interior of the NP with the hydrophobic residues of ChT. Most applications of protein-NP conjugates require retention of protein activity, making retention of structure an important priority. The surfaces of NPs can be appropriately tailored to prevent the structural denaturation of proteins. For instance, introduction of oligo (ethylene glycol) (OEG) [13] functionality on the NP surface drastically reduces the rate of denaturation of ChT.

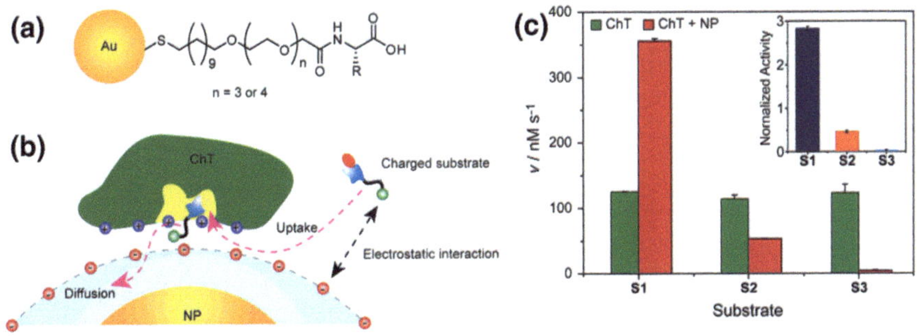

Fig. 1 **a** The chemical structure of anionic amino acid-functionalized gold nanoparticles. **b** Schematic representation of the interaction between negatively charged NPs and ChT, illustrating the effect of the charge not only on the binding, but also on the regulation of the enzymatic activity. **c** Generation rates for each substrate (S1 cationic, S2 neutral, and S3 anionic) as evidence that the binding selectivity is due to electrostatic interactions (Reproduced with permission from reference [12] © 2006 American Chemical Society)

An interesting area of research in bioconjugate chemistry is the generation of ensembles, where the collective properties of the conjugate structure differ from the properties of the individual precursor components. Enzymatic inhibition of ChT by anionic NPs arises not only from the blocking of the ChT active site by the NPs, but also depends on the charge state of the substrate (Fig. 1). The interaction of ChT with anionic NPs leads to the three-fold increase [12] in the catalytic activity of ChT for cationic substrates, while reducing its activity by 50 and 95 % for neutral and anionic substrates, respectively. This phenomenon of substrate selectivity can be attributed to the unique electrostatic environment presented by the protein-NP conjugate.

2.2 Structural Implications of Protein-Nanoparticle Interactions. Nanoparticles Can Stabilize or Destabilize Bound Proteins

The stability of NP-ChT complexes has been studied in greater detail using amino acid-functionalized gold nanoparticles (AuNPs). The functionalization of amino acids on AuNP surfaces provides direct access to biomimetic structural diversity. To probe the interaction between ChT and amino acid functionalized AuNPs, ChT catalyzed hydrolysis of *N*-succinyl-*L*-phenylalanine *p*-nitroanilide (SPNA) [14] was first examined in the presence of different concentrations of NPs. All anionic amino acid-functionalized AuNPs showed enzymatic inhibition (Fig. 2a), indicating that complementary electrostatic interactions are essential for ChT-AuNP complex formation. AuNPs with polar side chains, such as aspartic acid (Asp), glutamic acid (Glu), and asparagine (Asn), inhibited ChT activity by ~80 %, while NPs with hydrophobic side chains inhibited ChT activity by ~60 %. Presumably, the

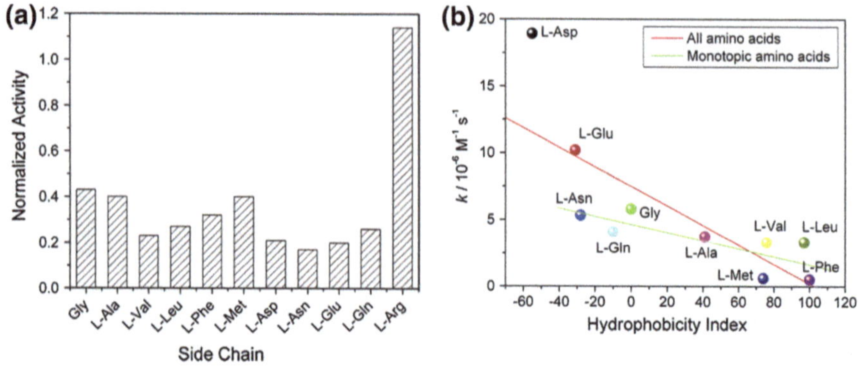

Fig. 2 **a** Normalized activity of ChT (3.2 μM) with nanoparticles (0.8 μM) bearing various amino acid side chains. **b** Correlation between the denaturation rate constants of ChT and the hydrophobicity index of amino acid side chains in nanoparticles (Reproduced with permission from reference [14] © 2006 American Chemical Society)

hydrophobic active site of ChT was more accessible to SPNA when the hydrophobic amino acids were present on the AuNP surface in comparison to when it was surrounded by NPs bearing hydrophilic side chains. The binding constants of the ChT-NP complexes were also dependent on the surface functionalization of the NPs. NPs with hydrophilic side chains, such as Asp-NP, Glu-NP, Asn-NP, and Gln-NP, showed a binding constant of $3 \times 10^6\ M^{-1}$, whereas Met-NP had a binding constant of $1.3 \times 10^7\ M^{-1}$, proving that hydrophobic interactions assisted in ChT-NP complex formation.

Another important concern that needs to be addressed is the influence of amino acid-functionalized NPs on protein conformation. Circular dichroism (CD) and fluorescence studies showed that NPs with hydrophobic side chains such as Phe-NP, Leu-NP, Met-NP, Val-NP, and Ala-NP, had very little effect on the native structure of ChT. On the other hand, NPs with hydrophilic side chains such as Gln-NP, Asn-NP, Asp-NP, and Glu-NP, showed drastic change in the conformation of ChT. As expected, Arg-NP had no effect on the native structure of ChT. These observations strongly suggest that the carboxylate functional groups facilitated the denaturation process. Competitive H-bonding interactions might destabilize the α-helical structure of ChT leading to the disruption of its secondary structure. Moreover, the above observation could be attributed to the fact that the hydroxide ions in the dianionic side chains were involved in the breakage of the salt bridge between the N-terminus of Ile16 and the Asp194 side chain. A plot of the denaturation rate constants of ChT with various NPs *versus* the hydrophobicity index of amino acid side chains (Fig. 2b) showed that the dianionic side chains strongly increased the rate of denaturation. The other side chains showed lesser, but still evident, correlation between hydrophobicity and denaturation rate, indicating that the hydrophobicity of the amino acid side chains on the NPs played a role in ChT denaturation/stabilization.

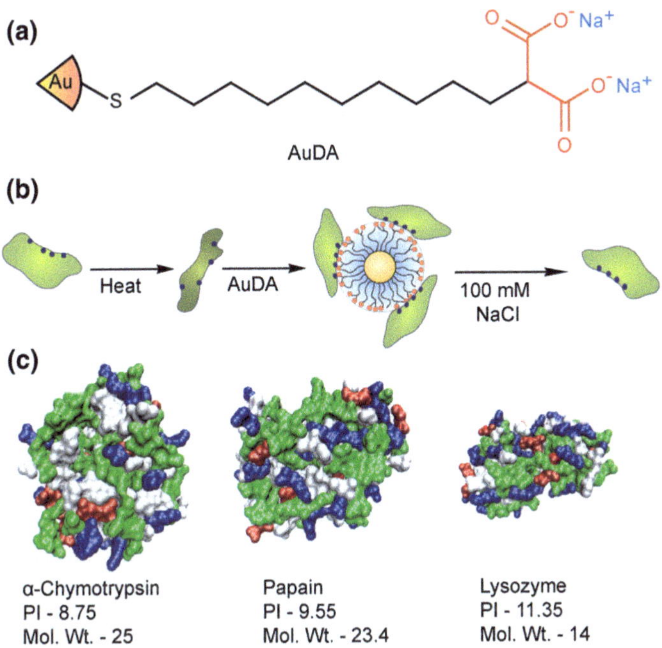

Fig. 3 **a** Schematic representation of the structure of AuDA and **b** the NP-mediated refolding of proteins after thermal denaturation. **c** Space filling models showing the surface structures of the three positively charged proteins used in the refolding study (Reproduced with permission from reference [15] © 2008 The Royal Society of Chemistry)

On the contrary, functionalized NPs can also act as chaperones to refold denatured proteins. Rotello et al. have used highly charged 2-(10-mercaptodecyl)malonic acid functionalized AuNPs (AuDA) (Fig. 3a) [15] to fold thermally denatured cationic proteins, such as ChT, papain and lysozyme. Thermal denaturation exposes the hydrophobic internal core of these proteins. Upon application of AuDAs, NP-protein complexes are formed via electrostatic interactions between the NP and the positive residues of the protein. The high negative charge of the AuDA prevents aggregation and thus facilitates refolding. After partial folding, proteins can be released from the NPs by increasing the ionic strength of the solution; they then self-fold into their native structures (Fig. 3b).

3 Delivery of Functional Proteins

Cells contain thousands of important proteins that participate in cellular functions [16]. The malfunctioning proteins can induce diseases [17]; thus, protein therapy is emerging as a promising approach to treat these illnesses. The success of this approach is, however, hindered by biological barriers, including inefficient delivery, endosomal entrapment, and proteolysis of proteins. One well-recognized strategy to overcome these challenges is through the use of nanocarrier-based drug/biomolecule delivery systems. Nanocarriers can protect the drug/biomolecules against degradation, control the pharmacokinetics/biodistribution of proteins for maximizing potency while minimizing attributed side effects, and enhance plasma half-life, reducing the frequency of drug administration to enable higher patient compliance [18]. An efficient nanocarrier for anticancer use has to pass several barriers, including the clearance by the reticuloendothelial system (RES) [19] and the crossing of the peritumoral endothelium [20] and the dense extracellular matrix (ECM) that often exists in solid tumors [21]. A primary driving force in nanoscale drug delivery is the enhanced permeability and retention (EPR) effect that relies on the porous architecture and poor lymphatic drainage of tumor vasculature [22]. Several years of research on nanoparticulate drug delivery systems have helped scientists to ascertain several design criteria that can be used to improve the functionality of NPs in vivo. Smaller NPs (>100 nm) are less likely to be engulfed by RES cells. Furthermore, hydrophilic polymers, such as poly(ethylene glycol) (PEG), when grafted to the surface of NPs, will increase circulation times leading to higher targeting capabilities of nanocarriers. To further increase the targeting capability of NPs, vector molecules (e.g., antibodies, aptamers, and small ligands) can be attached to their surface. It is noteworthy that tumors are composed of a heterogeneous population of multiple cancer cells [23]. Thus, multifunctional nanotherapeutics should be designed to target multiple cell lines in parallel.

Recent advances in peptide/protein drug development in cancer research include the introduction of antigens as cancer vaccines (e.g., PROVENGE is the first FDA-approved cancer vaccine for the prevention of prostate cancer) [24] and the application of cytokines, such as interferons (IFN)-β1a, IFN-α2a, granulocyte colony stimulating factor (GCSF), tumor necrosis factor(TNF)-α, and interleukins (ILs), along with small molecule drugs for the clinical treatment of cancer. Most importantly, monoclonal antibodies kill cancer cells by antibody-dependent, cell-mediated cytotoxicity and many are under consideration in clinical trials for the targeted therapy of breast, lung, colorectal, gastric, and brain solid tumors, metastatic events, and hematological malignancies. Indeed, several antibodies have already found their way into the clinic. Obviously, devising ways to specifically deliver such molecules to certain targets within the human body is an emerging area of research in NP-mediated drug delivery.

3.1 In Vitro

Specific nanomaterials, such as liposomes, polymeric systems, and even biomimetic particles have been proposed for the intracellular delivery of proteins for cancer therapeutics. However, the delivery of proteins using these vectors can be limited by instability and alteration of protein activity.

3.1.1 Inorganic NPs

Nanoparticles (NPs) have been widely considered for protein delivery applications. For instance, Ghosh et al. [25] have prepared gold NPs coated with short peptides, which effectively noncovalently complexed and delivered active anionic proteins (e.g., β-galactosidase) into a variety of cell lines (COS-1 (monkey kidney cells), MCF7 (human breast cancer cells), and even hard-to-transfect muscle cells (C2C12)). The engineered gold NPs were composed of interior alkyl chains for promoting core stability, a corona of tetraethylene glycol (TEG) for the prevention of denaturation and nonspecific interactions with biomolecules, and external arginine-rich peptide-tags working as a recognition unit. The authors hypothesized that the release of β-galactosidase from the particles was mediated by glutathione, which is present in the intracellular environment.

3.1.2 Nanocapsules

Hollow nanocarriers (e.g., nanocapsules and nanoliposomes) can provide more efficient systems for protein delivery applications due to their enhanced payload capacity. Besides protection against degradation by pH and light, hollow nanocarriers can also help reduce tissue irritation and provide more controllable targeting release of therapeutic agents. Tang et al. [7] fabricated nanocapsules for the delivery of functional proteins and enzymes to the cytoplasm. The functional proteins were incorporated in a capsule shell. The effectiveness of these capsules in the cytoplasmic delivery of proteins was confirmed by the delivery of fully functional caspase-3 to HeLa cells (Fig. 4) with concomitant apoptosis.

In another study, Yan et al. [26] produced nanocapsules composed of a protein core (e.g., green fluorescent protein (EGFP), horseradish peroxidase (HRP), bovine serum albumin (BSA), superoxide dismutase (SOD) and caspase-3), and a thin permeable polymeric shell anchored covalently to the protein core. The pH sensitive shell of the nanocapsules enabled controlled delivery of the proteins at specific pH values. The size, surface charge, and degradability of the nanocapsules could be tuned by varying the composition of the core, the nature of the crosslinkers as well as the type of employed monomers in construction of the shell. While non-degradable capsules showed long-term stability, the degradable ones released their ingredients inside the cells after breaking their shells. This system was shown to deliver multiple proteins to cells in a nontoxic and highly efficient manner, mainly through the caveolae-mediated endocytosis pathway. The cell transduction efficiency of nanocapsules containing EGFP in HeLa was 2–3 times higher than that of TAT–EGFP fusion proteins or antennapedia–EGFP conjugates. Thus, depending on the type of employed core proteins, nanocapsules may be suitable for a variety of biomedical applications, such as cellular imaging, cancer therapies, and anti-aging solutions. For example, the combination of indole-3-acetic acid (IAA) and HRP was recently proposed as a potential prodrug for the treatment of cancer [27]. This prodrug works via the HRP-mediated conversion of IAA into a free radical intermediate, which finally results in the induction of apoptosis in mammalian cells [28].

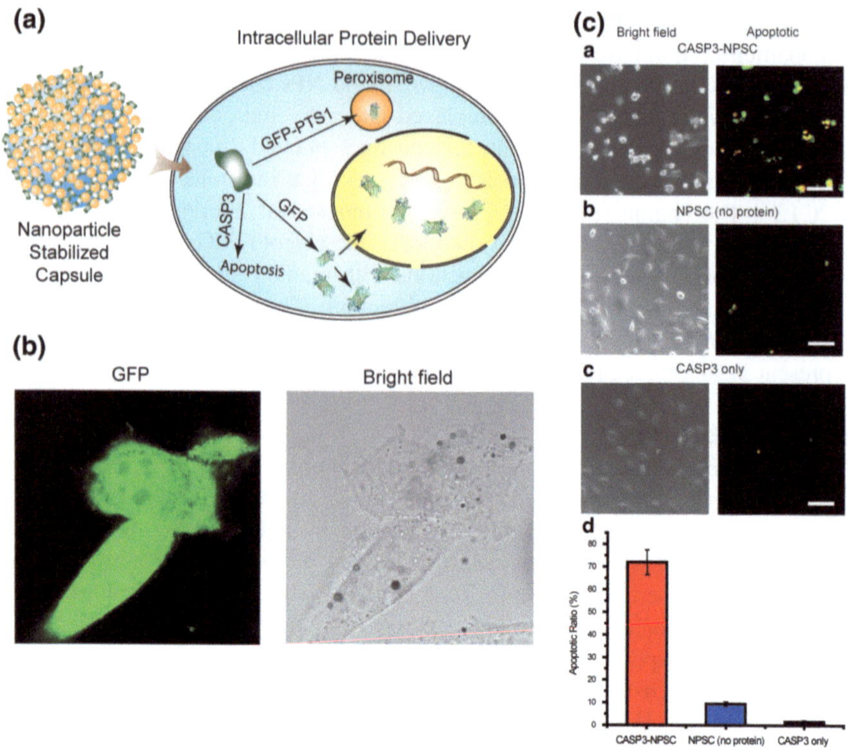

Fig. 4 **A** A schematic representation of the intracellular delivery of GFP and caspase-3 (CASP3) using nanoparticle-stabilized capsules (NPSCs). **B** Confocal images showing GFP delivery into HeLa cells by NPSCs (Scale bars: 20 μm). **C** Delivery of caspase-3 into HeLa cells. Cells were incubated for 1 h with (**a**) CASP3-NPSC, (**b**) NPSC without CASP3, and (**c**) only CASP3 without NPSC. Subsequently, cells were stained using Yopro-1 (*green* fluorescence) and 7-AAD (*red* fluorescence) for 30 min, and the overlapped images are presented as apoptotic. (**d**) Apoptosis ratios of the cells after CASP3 delivery (Scale bars: 100 μm). The error bars represent the standard deviations of three parallel measurements (Reproduced with permission from reference [7] © 2013 American Chemical Society)

Only nanoencapsulated HRP, and not the native HRP, could decrease cell viability with increasing IAA concentration.

Yeh et al. [29] synthesized multifunctional NP-stabilized capsules consisting of fluorescent polyhistidine-tagged proteins attached to CdSe/ZnS core–shell quantum dots through metal-affinity coordination. This negatively charged complex could interact with cationic gold NPs, which are anchored to the fatty acid core through guanidinium–carboxylate interactions. This lipophilic core functions as a reservoir for endosome-disrupting agents. The self-assembled system features stimuli-responsive delivery of multiple proteins into cell cytosol combined with FRET-based fluorescence tracking of cytosolic and vesicular distribution (see Fig. 5).

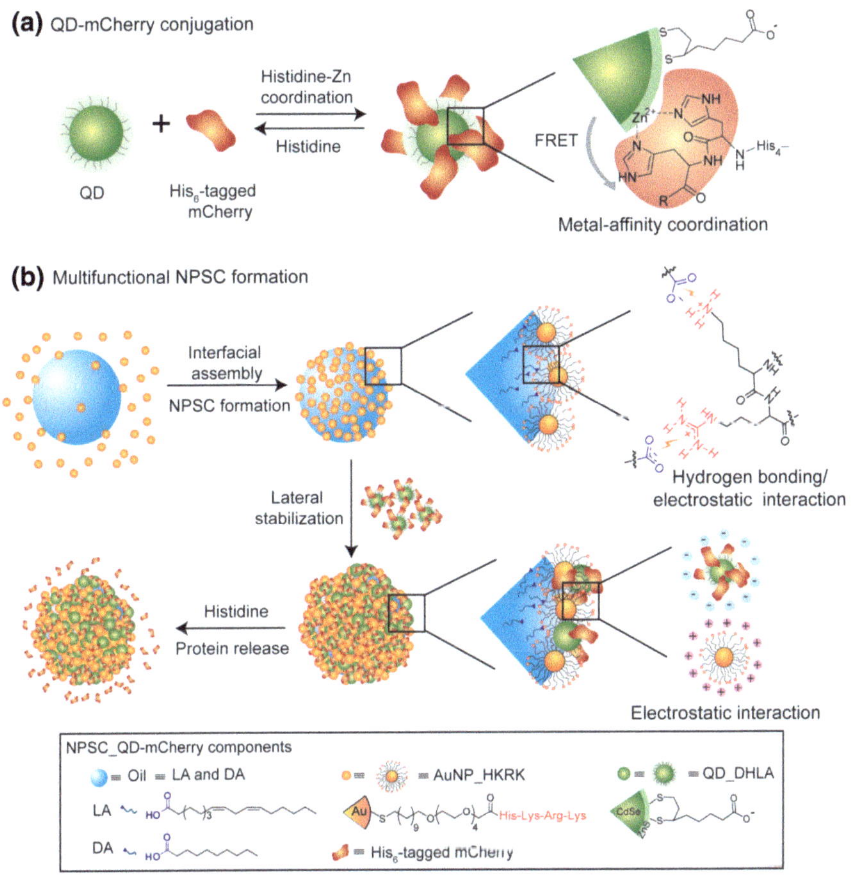

Fig. 5 **a** Dihydrolipoic acid-functionalized CdSe/ZnS core–shell quantum dots (QD_DHLA) and hexahistidine-tagged (His6-tagged) mCherry were conjugated by the reserved metal-affinity coordination using histidine as competitive molecules. **b** Fabrication of the template droplets by agitation of AuNP_HKRK (AuNPs functionalized with peptide (histidine-lysine-arginine-lysine)) and oil (linoleic acid (LA) and decanoic acid (DA)) in 5 mM phosphate buffer (pH 7.4). Then, the prepared template droplets were introduced into an AuNP_HKRK solution and crosslinked by anionic QD–mCherry conjugates to form NPSC_QD–mCherry (Reproduced with permission from reference [29] © 2014 Wiley-VCH Verlag GmbH & Co.)

3.1.3 Liposomes

Liposomes, another class of hollow nanocarriers, are vesicle-like structures that can retain therapeutics either in their hydrophilic core or their hydrophobic lipidic layers [30]. In the context of cancer (besides the well-known small molecule Doxil liposomal formulation), liposomes have been widely employed for delivery of various biomolecules, including IL-1, IL-2, IL-6, granulocyte-moncyte colony stimulating factor (GM-CSF), INF-γ [31], and granulocyte colony stimulating factor (GCSF) [32]. Liposomal encapsulation increases the protein half-life and enables the sustained release of encapsulants. For instance, insulin and peptides, such as leuprolide, enkephalin, and octreotide encapsulated in DepoFoam multivesicular liposomes were released at a controlled rate over a period from a few days to several weeks [33]. Liguori et al. [34] formulated pro-apoptotic membrane proteins (containing the voltage-dependent anion channel (VDAC) and pro-apoptotic Bak) into natural proteoliposomes to examine growth inhibitory effects in human colorectal carcinoma cells. The recombinant proteins were integrated into the lipidic bilayer of the liposomes. The induction of apoptosis was detected after a 24 h incubation period, which was accompanied by cytochrome c release and activation of caspases-3, -7, and -9 together with poly ADP ribose polymerase (PARP). The VDAC regulates the release of cytochrome c from mitochondria and Bak controls the permeability of mitochondrial membrane upon activation. The uptake of recombinant proteoliposomes by mammalian cells induces apoptosis by release of cytochrome c and activation of caspases. Gao et al. [35] formulated *Pseudomonas* exotoxin-based immunotoxins (PE38KDEL) into PEGylated anti-HER2 Fab′-functionalized liposomes, which showed promising results in the treatment of HER2-overexpressing breast cancers in vitro.

3.1.4 Polymeric Nanocarriers

Polymeric systems also have some applications in the field of drug delivery to cancer. 250 nm poly(γ-glutamic acid)-based NPs were used as protein carriers to deliver tumor vaccines, antigenic proteins, to antigen-presenting cells [36]. The ultimate goal of immunotherapy is to fight cancer through amplification of tumor immunity. With such a strategy, the immune system can distinguish the cells which express the tumor-associated antigens (TAAs) from normal cells. However, in this regard, TAA-specific cytotoxic T lymphocytes must be stimulated by the efficient delivery of TAA to antigen-presenting cells. Since antigen presenting cells only incorporate particles with a size range of 50–3 μm [37], nanoparticles can be useful for delivery of TAAs. The poly(γ-glutamic acid) NPs fabricated in this study had higher protein entrapment efficiency of 55–60 %. The NPs were shown to effectively deliver the encapsulated antigenic proteins inside the dendritic cells through cytosolic translocation from the endosomes and the ovalbumin (OVA) antigen was processed into epitope peptides and was subsequently presented via MHC molecules [36]. The authors investigated the inhibitory effects of NP/OVA in the lung metastasis of B16-OVA tumors expressing OVA as a TAA. Triplicate immunization with this formulation decreased the number of lung metastasis nodules in mice even more than Freund's complete adjuvant/OVA vaccine. Since poly(γ-glutamic acid) NPs are relatively stable in vivo, they could be used as controlled release systems of entrapped proteins. Authors also further analyzed the effect of entrapment process on the activity of catalase as a model protein. It was shown that entrapped catalase retains more than 90 % of its enzymatic activity [36].

3.1.5 Biomimetic Particles

Virus-like particles (VLPs) provide biogenic vectors for delivery. Kaczmarczyka et al. [38] used VLPs derived from an avian retrovirus for delivery of proteins to different target cells. The VLPs can be used to deliver proteins either as part of Gag fusion proteins (for intracellular delivery) or on the surface of VLPs. Since avian retroviruses do not replicate in human cells, the safety of this system for use in human applications is potentially high. VLPs containing Gag-Cre recombinase, Gag-Fcy:Fur, and Gag-human caspase-8 were synthesized. The authors successfully delivered caspase-8 to PC3 cells. Furthermore, in addition to the cellular delivery of proteins, in the same study, it was shown that murine IFN-γ and human TNF-related, apoptosis-inducing ligand (TRAIL) could be displayed on the surface of VLPs, and subsequently, these modified VLPs could activate the appropriate cellular receptors on the surface of cell membranes (in a mouse macrophage cell line developed from a C57BL/6 mouse that expresses IFN-γ receptor and PC3 cells, respectively).

3.2 In Vivo

Many proteins are believed to be promising candidates for cancer therapy; however, their clinical application is hampered by their removal through RES organs (spleen, lymph nodes, and liver) [19], their degradation by proteases [39], and their distribution and accumulation in non-target organs [40]. Furthermore, since high concentrations of particular proteins must be used due to the non-targeted nature of the therapy, the incidence of systemic side effects could be high. Nanoparticulate drug delivery systems could thus reduce the concentration of the drug required for clinical efficacy.

3.2.1 Inorganic NPs

TNF-α [41] was shown to cause hemorrhagic necrosis of solid tumors. Such a surprising finding led scientists to rapidly develop TNF-α for clinical testing; however, studies were discontinued due to some life-threatening toxicity observed in patients. Similar side effects were also later reported for other cytokines. Further attempts to localize the delivery of TNF to solid tumors, by surgical procedures, which could significantly improve its therapeutic index [42, 43], showed that by confining the therapy to the intended target, one may actually increase the efficacy and reduce the side effects associated with this therapy.

In later studies, Paciotti et al. [44], developed 26 nm colloidal gold NPs, on which CYT-6091, a multivalent drug, was mounted and meant to deliver TNF to its target. In this system, TNF was bound to PEGylated gold NPs. TNF not only has antitumor activity per se, but also functions as a targeting moiety and directs the NPs to MC-38 solid tumors (seven–ten-fold increase in targeting efficiency). A mere 7.5 mg of this formulation was shown to be as effective as 15 mg of native TNF. Furthermore, in mice receiving native TNF, the mortality rates were 33 and

15 % (with 15 and 7.5 mg, respectively); none of the animals receiving CYT-6091 died.

To enhance the antitumor efficacy of CYT-6091, which is a single agent therapy, the same authors later designed the CYT-21001 vector, which delivers TNF and paclitaxel simultaneously to the solid tumors [45]. The results of a phase I clinical trial of CYT-6091, consisting of TNF-α covalently linked to PEGylated colloidal gold NPs were published in 2009 [46]. The intravenously (IV) injected formulation has proven to be safe in 29 patients with solid tumors, and it showed a prolonged half-life compared to the native TNF-α and trafficking of the attached protein to the tumor, as assessed by electron microscopy in tumor biopsies.

Brinas et al. [47] have recently produced 3–5 nm gold NPs that were subsequently coated with two tumor-associated glycopeptide antigens (mucin (MUC4) and the Thomsen-Friedenreich antigen). The third coating agent was composed of a 28-residue peptide from the complement-derived protein C3d which functioned as a B-cell activating adjuvant. The immunization of mice with this vaccine led to the production of both IgM and IgG in the serum, showing the potential of this system in designing vaccines using TAAs.

3.2.2 Liposomes

Liposomes show high promise for in vivo applications. For instance, liposomes have been used for the cytosolic delivery of exogenous soluble protein antigens into antigen-presenting cells [48]. In this method, which adopts a cytosol-invading listerial for endosomal escape, listeriolysin O (which is a pore-forming protein that contributes to *Listeria monocytogenes* escape from the endosome into the cytosol) is co-encapsulated inside liposomes with OVA. Immunization of mice with this construct stimulated the activity of OVA-specific cytotoxic T lymphocytes and increased antigenic peptide-specific cytotoxic T lymphocyte precursor frequency. Overall, vaccination conferred protection to mice from lethal challenges with antigen-expressing tumor cells. The same group later demonstrated killing of B16 melanoma cells by listeriolysin O-liposome-mediated delivery of the protein toxin, gelonin. Gelonin is capable of enzymatically inactivating ribosomes and arresting protein synthesis in cancerous cells [49]. While the liposomal formulation had an IC_{50} of 0.1 nM with an incubation time of only 1 h, the free gelonin or listeriolysin-free formulation led to no detectable cytotoxicity. Direct intratumoral injection into mice bearing subcutaneous (SC) solid B16 melanoma showed that the listeriolysin O-liposomes were more effective than control formulations in restraining tumor growth.

Cruz et al. [50, 51] showed the enhanced survival of animals bearing PI534 tumors upon treatment with asparginase-encapsulated liposomes. Furthermore, a reduced occurrence of neutralizing antibodies was noted. Further studies by the same group demonstrated the superior efficacy of liposomal IL-2 over free IL-2 in inhibition of experimental M5076 metastases in mice [52]. In another study, liposomes harboring unmethylated cytosine-phosphorothioate-guanine containing oligodeoxynucleotides co-encapsulated with OVA were shown to inhibit the tumor growth and to completely cure ~50 % of OVA-expressing EG-7 tumor-bearing C57BL/6 mice upon intradermal inoculation [53].

A preclinical study demonstrated that VacciMax® (a vaccine composed of liposomes encapsulating human papilloma virus (HPV) 16 E7-derived cytotoxic T lymphocyte epitope fused to the T helper epitope PADRE and combined with CpG or lipopeptide adjuvant) could protect C57BL/6 mice against post-tumor challenge with HPV 16-expressing C3 tumor cells [54]. Neelapu et al. [55] incorporated a TAA, the idiotype of the Ig on B-cell malignancy in dimyristoylphosphatidylcholine lipid liposomes containing IL-2 as an adjuvant and the vaccine was shown to protect mice against lymphoma in preclinical studies. Later, the same group tested the safety and immunogenicity of the vaccine in 10 patients with advanced-stage follicular lymphoma [56]. The liposomal cancer vaccine proved to be safe and induced, sustained antitumor immune response through tumor-specific CD^{4+} and CD^{8+} T-cells in patients. After 50 months follow-up, 6 of the 10 patients were reported to remain in continuous first complete remission.

Broekhoven et al. [54] constructed B16-OVA-derived plasma membrane vesicles containing a new metal-chelating lipid, 3(nitrilotriacetic acid)-ditetradecylamine, which was used to engraft the recombinant single chain antibody fragments (ScFvs) to the dendritic cell surface molecules CD11c and DEC-205 onto the vesicle surface. The formulation was shown to stimulate strong B16-OVA-specific cytotoxic T lymphocyte responses in splenic T cells and to protect syngeneic mice against tumor growth.

3.2.3 Polymeric Nanocarriers

Polymeric systems are among the most extensively used platforms for protein therapy. For example, RGD peptide-loaded glycol chitosan NPs modified with 5β-cholanic acid were shown to possess antiangiogenic and antitumoral efficacy against solid tumors in mice upon IV injection or intratumoral administration [57].

Lim et al. [58] employed a formulation containing PEGylated heparin, TRAIL, and poly-L-lysine to increase its short biological half-life, improve its inherent instability, and eliminate its potential hepatotoxicity. IV injection of NPs in HCT-116 tumor-bearing BALB/c athymic mice efficiently suppressed tumor growth (>70 %) and induced significant tumor cell apoptosis without inducing liver toxicity.

PE38KDEL or fusion protein truncated *Pseudomonas* Exotoxin A has been used to prepare immunotoxin with monoclonal antibodies. Gao et al. [59] fabricated PE38KDEL-I-loaded poly(lactic-co-glycolic acid) (PLGA) NPs conjugated with F (ab') fragments of a humanized SM5-1 monoclonal antibody (PE-NP-S). After binding to SM5-1 binding protein-expressing hepatocellular carcinoma cell lines, PE-NP-S was internalized by these cells, and induced significant cytotoxicity. Further, in vivo studies on SM5-1 binding protein-overexpressing tumor xenograft model demonstrated that administration of the formulation significantly suppressed tumor development and even induced tumor regression. Due to the increased cancer therapeutic efficacy compared to older counterparts of the drug as well as reduced nonspecific toxicity and immunogenicity, this formulation might be a promising modality for cancer therapy.

Other studies have also revealed that poly(methoxypolyethyleneglycol cyanoacrylate–co-*n*-hexadecylcyanoacrylate) NPs could enhance the half-life of TNF-α in tumor bearing mice along with the targeting efficiency and antitumor potency [60]. The increased antitumor activity might arise from the higher accumulation of nanotherapeutics in tumor tissues and longer plasma circulation time.

4 The Protein Corona

As discussed in the introduction, protein-particle interactions that generate the protein corona are an important general issue in therapeutic delivery. Plasma proteins play crucial roles in blood, and they are found at high concentrations in the bloodstream. The layer of proteins that instantaneously covers NP surfaces when NPs enter biological fluids (e.g., blood, plasma, cell culture media, and intracellular environment) is called the protein corona. Even NPs functionalized with vector proteins (not considered as a corona) would pick up an additional layer of corona proteins upon entrance to a biological fluid. Protein coronas have the capability to significantly affect the biokinetics and in vivo fate of NPs. The adsorption of proteins lends a new bio-identity to NP surfaces, influencing their biological interactions. NP size [61], NP shape [62], and several surface characteristics, including the presence of surface functional groups [63], surface topography, uniformity, roughness [64], and surface hydrophobicity/hydrophilicity, [65–67] influence the nature of proteins adsorbed in the protein corona. Furthermore, interaction temperature [63, 68], NP dispersal environment [69], protein source [70], incubation time with protein source [71], concentration of protein source (proteome variability) [71–73], and gradient concentration [74] can also influence the corona composition. In turn, depending on the nature and amount of specific proteins adsorbed, NP biokinetics and their fate will be different. The adsorption of proteins to the surfaces mainly stems from the increase in the total entropy of the proteins on the surface as well as the nonspecific interactions between the proteins and NP surface [75]. While proteins with isoelectric points (pI) < 5.5 primarily bind to NPs with basic surfaces, those with pI > 5.5 are bound to NPs with acidic surfaces [76]. However, it is noteworthy that the adsorption of proteins to some surfaces might also be specific. For example, in the case of spherical nucleic acids, different sequences lead to formation of coronas with different compositions and this subsequently affects the level of cellular uptake [77]. The characterization of corona composition and its evolution in biological systems has been challenging due to the enormous complexity of the proteins and their interactions on NP surfaces [78].

4.1 "Hard" Versus "Soft" Corona

The protein corona is a protein layer on the surface of the NPs and can be differentiated into "soft" and "hard" varieties [78]. For "soft coronas", an exchange of biomacromolecules from the surrounding medium and the NP surface provides a dynamic structure. "Hard coronas" consist of macromolecules fixed to NP surfaces in a static fashion, generally involving extensive protein denaturation. The volatility of the proteins in the "soft corona" complicates detailed investigation, thus more studies have focused on materials with hard coronas [78]. Lundquist and co-workers [79] found that for a fixed material type, the biological activity of the proteins in the corona is strongly determined by the size as well as the zeta-potential of the NPs. Furthermore, the adsorption of blood serum proteins onto NPs is time dependent [80]: Proteins with the highest mobility are bound to the surface first and later they are replaced by less motile proteins, which show a higher affinity for the surface. It is believed that this process can take several hours as Casals et al. [71] confirmed. A "soft corona" loosely attached to the NP surface changes to an irreversibly attached "hard corona" over time. However, very recent reports reveal that "hard coronas" can form within less than a minute with no further compositional change [61]. It is noteworthy that differing reports might be due to the recently discovered "personalized protein corona" effect [81]. More specifically, changes in plasma composition in response to health conditions and diseases may lead to formation of different protein coronas on identical NPs in different individuals. It was shown that some proteins specifically appear or disappear in the hard protein coronas of identical NPs in different diseases. [81] Therefore, differences in proteome composition should also be taken into account in interpretation of results.

4.2 The Role of Protein Coronas in Delivery

The adsorption of proteins to NP surfaces has the ability to alter NP biokinetics and fate (Fig. 6). While the adsorption and enrichment of opsonin proteins, such as fibrinogen, immunoglobulins (Igs), and complement proteins, promotes NP phagocytosis by specialized immune cells (e.g., macrophages and phagocytic cells, such as hepatic Kupffer cells [82–85]), the adsorption of dysopsonin proteins, such as albumin and apolipoproteins (Apo), results in longer circulation times [86–89]. Fibrinogen, a glycoprotein involved in the coagulation process, binds to foreign surfaces and elicits the subsequent attachment of immune cells, such as monocytes, macrophages, and neutrophils. Through binding to CD11/CD18 on phagocytes, fibrinogen enhances phagocytosis [90, 91]. Furthermore, the complement C3 protein promotes the attachment of neutrophils to the NP surface [92]. Therefore, it might be hypothesized that, depending on the degree of binding, specific proteins might direct NPs to specific cells. For example, the binding of complement protein C3 and IgG to lecithin-coated polystyrene NPs influences NP uptake by murine Kupffer cells [82].

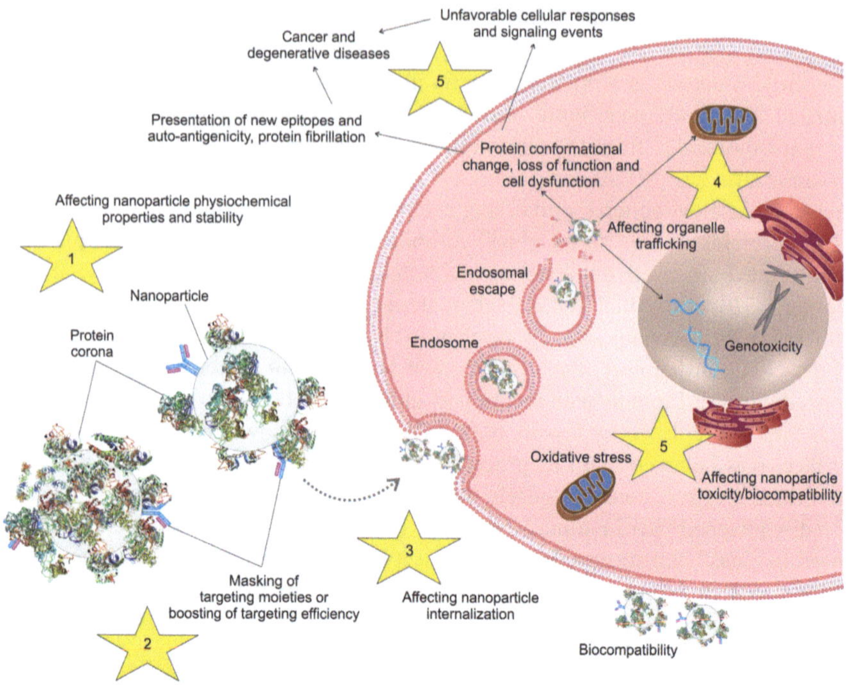

Fig. 6 The protein corona effects: *1* the physiochemical properties and stability of NPs, *2* NP targeting efficiency, *3* NP uptake and internalization, *4* NP distribution, trafficking and persistence within cells, *5* NP toxicity and biocompatibility, and *6* NP pharmacokinetics (not shown in this picture)

The formation of protein coronas has introduced complications in the field of drug delivery. Tandia et al. [93] have studied how human plasma proteins bind to cationic lipid DNA complexes and affect their transfection efficiencies. The adsorption of plasma proteins blocks the electrostatic interactions between 3-tetradecylamino-*N*-tertbutyl-*N'*-tetradecylpropionamidine (diC14-Amidine)/protamine/pCMV-luc complexes and the cell surface and brings about a decrease in cellular uptake and subsequent transfection efficiency. Apart from reduction of uptake rate, other corona-mediated phenomena are observed. For example, the protein corona formed on multifunctionalized colloidal mesoporous silica NPs hampers the release of camptothecin from the pores of NPs [94]. Moreover, the authors showed the possibility that certain biomolecules, which are present at the NP dispersal medium, can be loaded into the internal hydrophobic space in mesoporous silica NPs and, along with the NPs, be internalized into the living cells in an artificial manner [94].

4.3 Challenges Generated by Corona Proteins

The protein corona impacts the cellular interactions of NPs, especially their uptake rate. Various reports indicate that the cellular uptake of NPs is considerably higher in serum-free conditions compared to the cell media containing serum complement. More specifically, serum-derived coronas have been shown to reduce the uptake of a variety of NPs including those composed of gold [67], silica [95], FePt [96], carbon [97], Fe_3O_4 [98] and polystyrene [83, 99]. Cell-NP interactions are complicated, but relatively nonspecific phenomenon (i.e., uptake strongly depends on the amount of adsorbed proteins and their exposure sites rather than the exclusive presence of a certain protein within the corona). However, the adsorption of a specific protein, in some cases, can favor the uptake of a specific NP by cells through a certain route of entry. The adsorption of lung surfactant protein A on magnetite NPs enhances their uptake by macrophages compared to BSA-precoated NPs [100]. Furthermore, the presence of albumin on the surface of polystyrene NPs is required for caveolae-mediated endocytosis by living endothelial cells [101]. In one study, Lunov et al. [83] demonstrated that the uptake of NH_2-polystyrene NPs by macrophages changes from clathrin-mediated endocytosis to phagocytosis when protein-free medium is replaced with serum-enriched medium. In addition, DC-Chol–DOPE/DNA lipoplexes exhibit an unusual increase in plasmid transfection efficiency in serum [102]. The rich protein corona layer formed on lipoplexes in serum, lowers the interbilayer electrostatic repulsions between cationic lipoplexes, and promotes lipoplex aggregation. This increase in size switches the internalization mode from a clathrin-dependent to caveolae-mediated pathway, boosting transfection efficiency.

Drug delivery systems are often modified with functional moieties (small molecule ligands, such as folate, peptides, monoclonal antibodies, or antibody fragments) that are designed to recognize and target specific cells. However, protein coronas may significantly reduce the targeting efficiency of nanocarriers by blocking/masking the functional targeting moieties. Studies have shown that the targeting ability of functionalized NPs may be lost when they are exposed to a biological environment containing proteins. For example, Pitek et al. [103] discovered that a large decrease in the ability of carboxylated transferrin-coated NPs to detect transferrin occurs when they are incubated in plasma. Salvati et al. [104] attached transferrin molecules by a thiol–PEG linker to fluorescent PEG-modified silica NPs. In this study, more than 100 NPs, under various coupling conditions and with different PEG lengths, were tested to find the optimal PEG chain length for transferrin-mediated uptake of NPs. With an increase in the concentration of serum, NP-transferrin receptor interactions and NP uptake through the transferrin receptor decreased and eventually became negligible. Similar results also confirmed that loss of targeting happens in human serum. Using a click chemistry-based approach, Mirshafiee et al. [105] designed an experiment to demonstrate that the protein corona can limit the access of the targeting ligand on the NP surface (see Fig. 7). For this purpose, a copper-free click reaction was used as a model targeting reaction

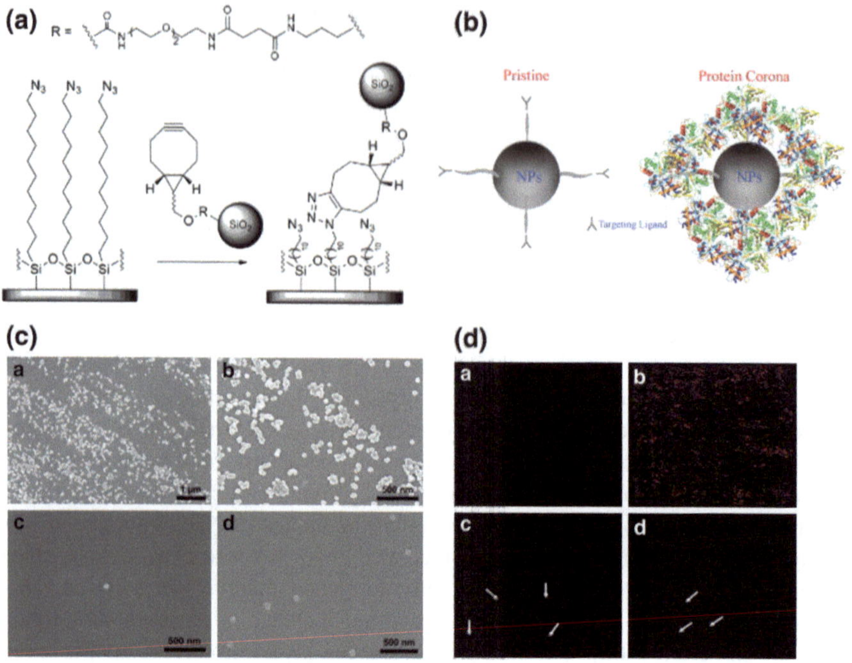

Fig. 7 **a** A copper-free click reaction between BCN and azide-functionalized groups at the surface of NPs and silicon substrates, respectively). **b** A simplified schematic showing the shielding effect of the protein corona. **c** SEM images of the azide-terminated silicon substrates after incubation with (*a* and *b*) pristine BCN-functionalized NPs, *c* 10 % serum corona BCN-NPs, and *d* 100 % serum corona BCN-NPs. **d** Fluorescence microscopy images of the incubated silicon substrates (5 mm by 5 mm) with pristine BCN-NPs and those coated with a protein corona (*a* pristine BCN-NPs to the azide-free substrate; *b* pristine BCN-NPs with the azide-functionalized substrate; *c* 10 % and *d* 100 % corona BCN-NPs with the azide-functionalized substrates; arrows designate individual NPs) (Reproduced with permission from reference [105] © 2013 The Royal Society of Chemistry)

that was designed to take place between NPs functionalized with bicyclononyne (BCN) and an azide on a silicon substrate. Subsequently, the authors evaluated whether the presence of protein corona (in medium with 10 and 100 % serum) inhibited the conjugation of BCN-NP to the azide-functionalized silicon substrates. While a high number of pristine BCN-NPs were attached to the substrate, only a few NPs incubated in serum-containing media were conjugated to the substrate.

4.4 Opportunities Provided by Protein Corona

Given that formation of a protein corona around NPs is inevitable in vivo, a paradigm shift toward rational exploitation of protein corona for drug delivery and targeting has been proposed [106, 107]. In the context of the protein corona, three

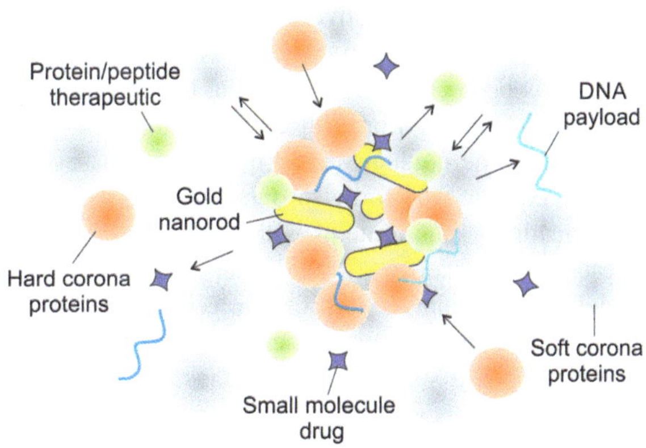

Fig. 8 The protein corona can be used to hold small and supramolecular therapeutics for subsequent release in drug delivery applications

modalities might be pursued: (i) surface engineering of NPs to create corona-free NPs [108], (ii) exploiting protein corona for drug loading and release, and (iii) exploiting the formed protein corona itself for delivery and targeting. In the first category, a routine strategy is to coat NPs with hydrophilic polymers, such as PEG, which lend non-fouling capabilities to NPs and reduce interactions with proteins [89]. Another strategy is to incorporate zwitterionic functionalities, such as amino acids and polybetaines, onto the NP surface. For instance, in a recent study, Moyano et al. [108] fabricated a series of zwitterionic non-fouling 2 nm gold core NPs of varying hydrophobicities that do not adsorb proteins when exposed to serum proteins at moderate concentrations and are devoid of hard coronas at physiological serum concentrations.

A few recent studies have focused on the exploitation of the protein corona around NPs for drug loading and release. Recently, Kah et al. [109] used the coronas of serum proteins on cetyltrimethylammonium bromide (CTAB)-coated gold nanorods for holding small molecular therapeutics (see Fig. 8). The capacity of this sponge-like reservoir for the storage of small molecules is approximately 5–10 times more than that accessed through covalent conjugation. Interestingly, the coronas were shown to hold both negatively charged (DNA oligonucleotide) and positively charged (doxorubicin) molecules. The capacity of the payload is dictated by the assembly strategy, ionic strength, and loading concentration. The payload release can be triggered by heat or through ultrafast laser excitation of the nanorods at their longitudinal surface plasmon resonance (light-triggered release). The coronas have an acceptable stabilizing effect on the nanorods in both buffer and biological environments, but leakage of therapeutic molecules from the corona remains an issue. In a similar study, authors showed that, with manipulation of the protein corona composition on NPs, the passive release rate of the DNA payload could be controlled [110]. The coronas were formed in human serum around gold nanorods, nanobones, and carbon nanotubes (CNTs). Tuning the amount of human serum albumin (HSA) in the nanorod coronas was shown to change the payload release profile.

The third goal of actively exploiting corona formation can be coupled with properties of NPs to open up the possibilities for novel futuristic applications. This theory has been established on using the corona proteins to target specific cellular targets or receptors. For such a strategy to work, several prerequisites must first be fulfilled: (i) the qualitative and quantitative assessment of the composition of the protein corona that forms around specific NPs with specific surfaces, (ii) the identification of the proteins with the highest abundance and/or the highest affinity to each NP surface, and (iii) the determination of which plasma proteins can be used to deliver a certain NP to a certain cell or tissue in vivo.

Some plasma proteins (associated in the corona composition) have the capability to direct NPs to specific target organs. For example, transferrin can enhance NP uptake in various cell lines, including wild-type rat 9L gliosarcoma cells, breast carcinoma cell lines BT20, MCF7, BT549, and HBL100, and the non-tumorigenic cell line MCF10A by a transferrin receptor-mediated process [111]. Transferrin is a plasma glycoprotein, which transports iron and binds to the Tf-receptor (TfR) in the iron-bound form, activating receptor mediated endocytosis. Since cancer cells have a higher iron demand than normal cells, transferrin conjugation or absorption enhances NP uptake in cancer cells with TfR receptor overexpression [112].

ApoA-I is a major component of high density lipoprotein (HDL) [113], which can be adsorbed to the surfaces of most hydrophobic NPs of different sizes [65, 114] through its flexible hind region [115]. ApoA-I binds to low density lipoprotein (LDL) receptors and interacts with scavenger receptors (class B type I (SR-BI)) in the brain [116], particularly in the brain capillary endothelial cells [117, 118]. This receptor mediates the transfer between very low density lipoproteins (VLDL), LDL, and HDL and cells [119]. Thus, NPs that enrich ApoA-I have a high chance of engaging in transcytosis across the blood–brain barrier (BBB) and delivering their cargo to neurons. Moreover, these NPs may also be able to target other cells with high expression of SR-BI, such as the PC3 prostate carcinoma cell line. Furthermore, SR-BI is expressed in the liver and steroidogenic glands, intestine, and placenta, and in cells such as macrophages and endothelial cells [119]. Another receptor for ApoA-I is cubilin (vitamin B12 receptor), which mediates the endocytosis of HDL [120, 121]. Cubilin is expressed in the proximal tubule in kidney and in epithelial cells in yolk sac and intestine [122, 123]. Therefore, such NPs might also be used for drug delivery applications in organs and cells rich in SR-BI or the cubilin receptor.

5 Conclusions

Taken together, protein-nanoparticle interactions, including corona formation, must be understood if such conjugates are to be utilized in protein delivery applications for cancer research and treatment. The ability to create nanosystems that interact predictably with proteins opens doors for the development of protein delivery applications with the potential to revolutionize therapeutics. This insight can guide us in generating more efficient delivery vectors by tailoring the properties of the protein corona. Our expectation is that continued research on both the fundamental and applied aspects of protein-particle interactions will enable many novel biomedical applications in the future.

References

1. Foster S, Duvall CL, Crownover EF, Hoffman AS, Stayton PS (2010) Intracellular delivery of a protein antigen with an endosomal-releasing polymer enhances CD8 T-cell production and prophylactic vaccine efficacy. Bioconjug Chem 21(12):2205–2212
2. Vasconcelos L, Pärn K, Langel Ü (2013) Therapeutic potential of cell-penetrating peptides. Ther Deliv 4(5):573–591
3. Kratz F, Elsadek B (2012) Clinical impact of serum proteins on drug delivery. J Controlled Release 161(2):429–445
4. Le Roy C, Wrana JL (2005) Clathrin-and non-clathrin-mediated endocytic regulation of cell signalling. Nat Rev Mol Cell Biol 6(2):112–126
5. Mellert K, Lamla M, Scheffzek K, Wittig R, Kaufmann D (2012) Enhancing endosomal escape of transduced proteins by photochemical internalisation. PLoS ONE 7(12):e52473
6. Yan B, Kim ST, Kim CS, Saha K, Moyano DF, Xing Y, Jiang Y, Roberts AL, Alfonso FS, Rotello VM, Vachet RW (2013) Multiplexed Imaging of Nanoparticles in Tissues Using Laser Desorption/Ionization Mass Spectrometry. J Am Chem Soc 135(34):12564–12567. doi:10.1021/ja406553f
7. Tang R, Kim CS, Solfiell DJ, Rana S, Mout R, Velázquez-Delgado EM, Chompoosor A, Jeong Y, Yan B, Zhu Z-J, Kim C, Hardy JA, Rotello VM (2013) Direct delivery of functional proteins and enzymes to the cytosol using nanoparticle-stabilized nanocapsules. ACS Nano 7 (8):6667–6673. doi:10.1021/nn402753y
8. De M, Rana S, Akpinar H, Miranda OR, Arvizo RR, Bunz UHF, Rotello VM (2009) Sensing of proteins in human serum using conjugates of nanoparticles and green fluorescent protein. Nat Chem 1(6):461–465. doi:10.1038/nchem.334
9. Saha K, Agasti SS, Kim C, Li X, Rotello VM (2012) Gold nanoparticles in chemical and biological sensing. Chem Rev 112(5):2739–2779. doi:10.1021/cr2001178
10. Yang S-T, Liu Y, Wang Y-W, Cao A (2013) Biosafety and bioapplication of nanomaterials by designing protein-nanoparticle interactions. Small 9(9–10):1635–1653. doi:10.1002/smll.201201492
11. De M, Miranda OR, Rana S, Rotello VM (2009) Size and geometry dependent protein-nanoparticle self-assembly. Chem Commun 16:2157–2159. doi:10.1039/b900552h
12. You CC, Agasti SS, De M, Knapp MJ, Rotello VM (2006) Modulation of the catalytic behavior of α-chymotrypsin at monolayer-protected nanoparticle surfaces. J Am Chem Soc 128(45):14612–14618. doi:10.1021/ja064433z
13. You CC, De M, Rotello VM (2005) Contrasting effects of exterior and interior hydrophobic moieties in the complexation of amino acid functionalized gold clusters with α-chymotrypsin. Org Lett 7(25):5685–5688. doi:10.1021/ol052367k
14. You CC, De M, Han G, Rotello VM (2005) Tunable inhibition and denaturation of α-chymotrypsin with amino acid-functionalized gold nanoparticles. J Am Chem Soc 127 (37):12873–12881. doi:10.1021/ja0512881
15. De M, Rotello VM (2008) Synthetic "chaperones": nanoparticle-mediated refolding of thermally denatured proteins. Chem Commun 30:3504–3506. doi:10.1039/b805242e
16. Kim M-S, Pinto SM, Getnet D, Nirujogi RS, Manda SS, Chaerkady R, Madugundu AK, Kelkar DS, Isserlin R, Jain S (2014) A draft map of the human proteome. Nature 509 (7502):575–581

17. Chiti F, Dobson CM (2006) Protein misfolding, functional amyloid, and human disease. Annu Rev Biochem 75:333–366
18. Laurent S, Saei AA, Behzadi S, Panahifar A, Mahmoudi M (2014) Superparamagnetic iron oxide nanoparticles for delivery of therapeutic agents: opportunities and challenges. Expert Opin Drug Deliv 0:1–22
19. Zhang Y, Meibohm B (2012) Pharmacokinetics and pharmacodynamics of therapeutic peptides and proteins. In: Pharmaceutical biotechnology: drug discovery and clinical applications, 2nd edn. Wiley, Weinheim, pp 337–367
20. Molema G, de Leij LF, Meijer DK (1997) Tumor vascular endothelium: barrier or target in tumor directed drug delivery and immunotherapy. Pharm Res 14(1):2–10
21. Netti PA, Berk DA, Swartz MA, Grodzinsky AJ, Jain RK (2000) Role of extracellular matrix assembly in interstitial transport in solid tumors. Cancer Res 60(9):2497–2503
22. Matsumura Y, Maeda H (1986) A new concept for macromolecular therapeutics in cancer-chemotherapy—mechanism of tumoritropic accumulation of proteins and the antitumor agent Smancs. Cancer Res 46(12):6387–6392
23. Dexter DL, Kowalski HM, Blazar BA, Fligiel Z, Vogel R, Heppner GH (1978) Heterogeneity of tumor cells from a single mouse mammary tumor. Cancer Res 38(10): 3174–3181
24. Cheever MA, Higano CS (2011) Provenge (Sipuleucel-T) in prostate cancer: the first FDA-approved therapeutic cancer vaccine. Clin Cancer Res 17(11):3520–3526
25. Ghosh P, Yang X, Arvizo R, Zhu Z-J, Agasti SS, Mo Z, Rotello VM (2010) Intracellular delivery of a membrane-impermeable enzyme in active form using functionalized gold nanoparticles. J Am Chem Soc 132(8):2642–2645
26. Yan M, Du J, Gu Z, Liang M, Hu Y, Zhang W, Priceman S, Wu L, Zhou ZH, Liu Z (2009) A novel intracellular protein delivery platform based on single-protein nanocapsules. Nat Nanotechnol 5(1):48–53
27. Folkes LK, Wardman P (2001) Oxidative activation of indole-3-acetic acids to cytotoxic species—a potential new role for plant auxins in cancer therapy. Biochem Pharmacol 61(2): 129–136
28. de Melo MP, de Lima TM, Pithon-Curi TC, Curi R (2004) The mechanism of indole acetic acid cytotoxicity. Toxicol Lett 148(1):103–111
29. Yeh YC, Tang R, Mout R, Jeong Y, Rotello VM (2014) Fabrication of multiresponsive bioactive nanocapsules through orthogonal self-assembly. Angew Chem Int Edit 53(20): 5137–5141
30. Allen TM, Cullis PR (2013) Liposomal drug delivery systems: from concept to clinical applications. Adv Drug Deliver Rev 65 (1):36–48. doi:http://dx.doi.org/10.1016/j.addr.2012.09.037
31. Anderson PM, Hanson DC, Hasz DE, Halet MR, Blazar BR, Ochoa AC (1994) Cytokines in liposomes: preliminary studies with IL-1, IL-2, IL-6. GM-CSF and interferon-γ. Cytokine 6 (1):92–101
32. Meyer J, Whitcomb L, Collins D (1994) Efficient encapsulation of proteins within liposomes for slow release in vivo. Biochem Biophys Res Commun 199(2):433–438
33. Ye Q, Asherman J, Stevenson M, Brownson E, Katre NV (2000) DepoFoam™ technology: a vehicle for controlled delivery of protein and peptide drugs. J Controlled Release 64(1): 155–166
34. Liguori L, Marques B, Villegas-Mendez A, Rothe R, Lenormand J-L (2008) Liposomes-mediated delivery of pro-apoptotic therapeutic membrane proteins. J Controlled Release 126 (3):217–227
35. Gao J, Zhong W, He J, Li H, Zhang H, Zhou G, Li B, Lu Y, Zou H, Kou G (2009) Tumor-targeted PE38KDEL delivery via PEGylated anti-HER2 immunoliposomes. Int J Pharm 374(1):145–152
36. Yoshikawa T, Okada N, Oda A, Matsuo K, Matsuo K, Mukai Y, Yoshioka Y, Akagi T, Akashi M, Nakagawa S (2008) Development of amphiphilic γ-PGA-nanoparticle based

tumor vaccine: potential of the nanoparticulate cytosolic protein delivery carrier. Biochem Biophys Res Commun 366(2):408–413

37. Bramwell VW, Eyles JE, Oya Alpar H (2005) Particulate delivery systems for biodefense subunit vaccines. Adv Drug Deliver Rev 57(9):1247–1265
38. Kaczmarczyk SJ, Sitaraman K, Young HA, Hughes SH, Chatterjee DK (2011) Protein delivery using engineered virus-like particles. Proc Natl Acad Sci USA 108(41): 16998–17003
39. Harris JM, Chess RB (2003) Effect of pegylation on pharmaceuticals. Nat Rev Drug Discov 2(3):214–221
40. Torchilin VP, Lukyanov AN (2003) Peptide and protein drug delivery to and into tumors: challenges and solutions. Drug Discov Today 8(6):259–266
41. Carswell E, Old LJ, Kassel R, Green S, Fiore N, Williamson B (1975) An endotoxin-induced serum factor that causes necrosis of tumors. Proc Natl Acad Sci USA 72(9):3666–3670
42. Lejeune F (1995) High dose recombinant tumour necrosis factor (rTNFα) administered by isolation perfusion for advanced tumours of the limbs: a model for biochemotherapy of cancer. Eur J Cancer 31(6):1009–1016
43. Liénard D, Lejeune FJ, Ewalenko P (1992) In transit metastases of malignant melanoma treated by high dose rTNFα in combination with interferon-γ and melphalan in isolation perfusion. World J Surg 16(2):234–240
44. Paciotti GF, Myer L, Weinreich D, Goia D, Pavel N, McLaughlin RE, Tamarkin L (2004) Colloidal gold: a novel nanoparticle vector for tumor directed drug delivery. Drug Deliv 11 (3):169–183
45. Paciotti GF, Kingston DG, Tamarkin L (2006) Colloidal gold nanoparticles: a novel nanoparticle platform for developing multifunctional tumor-targeted drug delivery vectors. Drug Develop Res 67(1):47–54
46. Libutti S, Paciotti G, Myer L, Haynes R, Gannon W, Walker M, Seidel G, Byrnes A, Yuldasheva N, Tamarkin L (2009) Results of a completed phase I clinical trial of CYT-6091: A pegylated colloidal gold-TNF nanomedicine. J Clin Oncol (Meeting Abstracts) 15S:3586
47. Briñas RP, Sundgren A, Sahoo P, Morey S, Rittenhouse-Olson K, Wilding GE, Deng W, Barchi Jr JJ (2012) Design and synthesis of multifunctional gold nanoparticles bearing tumor-associated glycopeptide antigens as potential cancer vaccines. Bioconjug Chem 23(8): 1513–1523
48. Mandal M, Lee K-D (2002) Listeriolysin O-liposome-mediated cytosolic delivery of macromolecule antigen in vivo: enhancement of antigen-specific cytotoxic T lymphocyte frequency, activity, and tumor protection. BBA-Biomembr 1563(1):7–17
49. Provoda CJ, Stier EM, Lee K-D (2003) Tumor cell killing enabled by listeriolysin O-liposome-mediated delivery of the protein toxin gelonin. J Biol Chem 278(37):35102–35108
50. Gaspar M, Perez-Soler R, Cruz M (1996) Biological characterization of L-asparaginase liposomal formulations. Cancer Chemother Pharmacol 38(4):373–377
51. Jorge JC, Perez-Soler R, Morais JG, Cruz MEM (1994) Liposomal palmitoyl-L-asparaginase: characterization and biological activity. Cancer Chemoth Pharm 34(3): 230–234
52. Kanaoka E, Takahashi K, Yoshikawa T, Jizomoto H, Nishihara Y, Uchida N, Maekawa R, Hirano K (2002) A significant enhancement of therapeutic effect against hepatic metastases of M5076 in mice by a liposomal interleukin-2 (mixture). J Controlled Release 82(2): 183–187
53. Wakita D, Chamoto K, Zhang Y, Narita Y, Noguchi D, Ohnishi H, Iguchi T, Sakai T, Ikeda H, Nishimura T (2006) An indispensable role of type-1 IFNs for inducing CTL-mediated complete eradication of established tumor tissue by CpG-liposome co-encapsulated with model tumor antigen. Int Immunol 18(3):425–434
54. Daftarian P, Mansour M, Benoit AC, Pohajdak B, Hoskin DW, Brown RG, Kast WM (2006) Eradication of established HPV 16-expressing tumors by a single administration of a vaccine

composed of a liposome-encapsulated CTL-T helper fusion peptide in a water-in-oil emulsion. Vaccine 24(24):5235–5244
55. Kwak LW, Pennington R, Boni L, Ochoa AC, Robb RJ, Popescu MC (1998) Cutting edge: liposomal formulation of a self lymphoma antigen induces potent protective antitumor immunity. J Immunol 160(8):3637–3641
56. Neelapu SS, Baskar S, Gause BL, Kobrin CB, Watson TM, Frye AR, Pennington R, Harvey L, Jaffe ES, Robb RJ (2004) Human autologous tumor-specific T-cell responses induced by liposomal delivery of a lymphoma antigen. Clin Cancer Res 10(24):8309–8317
57. Kim J-H, Kim Y-S, Park K, Kang E, Lee S, Nam HY, Kim K, Park JH, Chi DY, Park R-W (2008) Self-assembled glycol chitosan nanoparticles for the sustained and prolonged delivery of antiangiogenic small peptide drugs in cancer therapy. Biomaterials 29(12):1920–1930
58. Lim SM, Kim TH, Jiang HH, Park CW, Lee S, Chen X, Lee KC (2011) Improved biological half-life and anti-tumor activity of TNF-related apoptosis-inducing ligand (TRAIL) using PEG-exposed nanoparticles. Biomaterials 32(13):3538–3546
59. Gao J, Kou G, Chen H, Wang H, Li B, Lu Y, Zhang D, Wang S, Hou S, Qian W (2008) Treatment of hepatocellular carcinoma in mice with PE38KDEL type I mutant-loaded poly (lactic-co-glycolic acid) nanoparticles conjugated with humanized SM5-1 F (ab′) fragments. Mol Cancer Ther 7(10):3399–3407
60. Li YP, Pei YY, Zhou ZH, Zhang XY, Gu ZH, Ding J, Zhou JJ, Gao XJ, Zhu JH (2001) Stealth polycyanoacrylate nanoparticles as tumor necrosis factor-ALPHA. Carriers: pharmacokinetics and anti-tumor effects. Biol Pharm Bull 24(6):662–665
61. Tenzer S, Docter D, Kuharev J, Musyanovych A, Fetz V, Hecht R, Schlenk F, Fischer D, Kiouptsi K, Reinhardt C (2013) Rapid formation of plasma protein corona critically affects nanoparticle pathophysiology. Nat Nanotechnol 8(10):772–781
62. Gagner JE, Lopez MD, Dordick JS, Siegel RW (2011) Effect of gold nanoparticle morphology on adsorbed protein structure and function. Biomaterials 32(29):7241–7252
63. Mahmoudi M, Shokrgozar MA, Behzadi S (2013) Slight temperature changes affect protein affinity and cellular uptake/toxicity of nanoparticles. Nanoscale 5(8):3240–3244
64. Mahmoudi M, Serpooshan V (2011) Large protein absorptions from small changes on the surface of nanoparticles. J Phys Chem C 115(37):18275–18283
65. Cedervall T, Lynch I, Foy M, Berggård T, Donnelly SC, Cagney G, Linse S, Dawson KA (2007) Detailed identification of plasma proteins adsorbed on copolymer nanoparticles. Angew Chem Int Ed 46(30):5754–5756
66. Yeh Y-C, Rana S, Mout R, Yan B, Alfonso FS, Rotello VM (2014) Supramolecular tailoring of protein–nanoparticle interactions using cucurbituril mediators. Chem Commun 50(42): 5565–5568
67. Zhu ZJ, Posati T, Moyano DF, Tang R, Yan B, Vachet RW, Rotello VM (2012) The interplay of monolayer structure and serum protein interactions on the cellular uptake of gold nanoparticles. Small 8(17):2659–2663
68. Mahmoudi M, Abdelmonem AM, Behzadi S, Clement JH, Dutz S, Ejtehadi MR, Hartmann R, Kantner K, Linne U, Maffre P (2013) Temperature: the "ignored" factor at the nanobio interface. ACS Nano 7(8):6555–6562
69. Maiorano G, Sabella S, Sorce B, Brunetti V, Malvindi MA, Cingolani R, Pompa PP (2010) Effects of cell culture media on the dynamic formation of protein—nanoparticle complexes and influence on the cellular response. ACS Nano 4(12):7481–7491
70. Laurent S, Burtea C, Thirifays C, Rezaee F, Mahmoudi M (2013) Significance of cell "observer" and protein source in nanobiosciences. J Colloid Interface Sci 392:431–445
71. Casals E, Pfaller T, Duschl A, Oostingh GJ, Puntes V (2010) Time evolution of the nanoparticle protein corona. ACS Nano 4(7):3623–3632
72. Monopoli MP, Walczyk D, Campbell A, Elia G, Lynch I, Baldelli Bombelli F, Dawson KA (2011) Physical—chemical aspects of protein corona: relevance to in vitro and in vivo biological impacts of nanoparticles. J Am Chem Soc 133(8):2525–2534

73. Caracciolo G, Pozzi D, Capriotti AL, Cavaliere C, Foglia P, Amenitsch H, Laganà A (2011) Evolution of the protein corona of lipid gene vectors as a function of plasma concentration. Langmuir 27(24):15048–15053
74. Ghavami M, Saffar S, Emamy BA, Peirovi A, Shokrgozar MA, Serpooshan V, Mahmoudi M (2013) Plasma concentration gradient influences the protein corona decoration on nanoparticles. RSC Adv 3(4):1119–1126
75. Nakanishi K, Sakiyama T, Imamura K (2001) On the adsorption of proteins on solid surfaces, a common but very complicated phenomenon. J Biosci Bioeng 91(3):233–244
76. Gessner A, Lieske A, Paulke BR, Müller RH (2003) Functional groups on polystyrene model nanoparticles: influence on protein adsorption. J Biomed Mater Res, Part A 65(3):319–326
77. Choi CHJ, Hao L, Narayan SP, Auyeung E, Mirkin CA (2013) Mechanism for the endocytosis of spherical nucleic acid nanoparticle conjugates. Proc Natl Acad Sci USA 110 (19):7625–7630
78. Mahmoudi M, Lynch I, Ejtehadi MR, Monopoli MP, Bombelli FB, Laurent S (2011) Protein-nanoparticle interactions: opportunities and challenges. Chem Rev 111(9):5610–5637
79. Lundqvist M, Stigler J, Elia G, Lynch I, Cedervall T, Dawson KA (2008) Nanoparticle size and surface properties determine the protein corona with possible implications for biological impacts. Proc Natl Acad Sci USA 105(38):14265–14270
80. Vroman L (1962) Effect of Adsorbed Proteins on the Wettability of Hydrophilic and Hydrophobic Solids. Nature 196(4853):476–477
81. Hajipour MJ, Laurent S, Aghaie A, Rezaee F, Mahmoudi M (2014) Personalized protein coronas: a "key" factor at the nanobiointerface. Biomater Sci 2(9):1210–1221. doi:10.1039/c4bm00131a
82. Nagayama S, K-i Ogawara, Fukuoka Y, Higaki K, Kimura T (2007) Time-dependent changes in opsonin amount associated on nanoparticles alter their hepatic uptake characteristics. Int J Pharm 342(1):215–221
83. Lunov O, Syrovets T, Loos C, Beil J, Delacher M, Tron K, Nienhaus GU, Musyanovych A, Mailander V, Landfester K (2011) Differential uptake of functionalized polystyrene nanoparticles by human macrophages and a monocytic cell line. ACS Nano 5(3):1657–1669
84. Kobzik L (1995) Lung macrophage uptake of unopsonized environmental particulates role of scavenger-type receptors. J Immunol 155(1):367–376
85. Hamilton RF, Thakur SA, Mayfair JK, Holian A (2006) MARCO mediates silica uptake and toxicity in alveolar macrophages from C57BL/6 mice. J Biol Chem 281(45):34218–34226
86. Walkey CD, Olsen JB, Guo H, Emili A, Chan WC (2012) Nanoparticle size and surface chemistry determine serum protein adsorption and macrophage uptake. J Am Chem Soc 134 (4):2139–2147
87. Deng ZJ, Liang M, Monteiro M, Toth I, Minchin RF (2011) Nanoparticle-induced unfolding of fibrinogen promotes Mac-1 receptor activation and inflammation. Nat Nanotechnol 6(1): 39–44
88. Decuzzi P, Godin B, Tanaka T, Lee S-Y, Chiappini C, Liu X, Ferrari M (2010) Size and shape effects in the biodistribution of intravascularly injected particles. J Controlled Release 141(3):320–327
89. Hamad I, Al-Hanbali O, Hunter AC, Rutt KJ, Andresen TL, Moghimi SM (2010) Distinct polymer architecture mediates switching of complement activation pathways at the nanosphere – serum interface: implications for stealth nanoparticle engineering. ACS Nano 4(11):6629–6638
90. Rubel C, Fernández GC, Dran G, Bompadre MB, Isturiz MA, Palermo MS (2001) Fibrinogen promotes neutrophil activation and delays apoptosis. J Immunol 166(3): 2002–2010
91. Sitrin RG, Pan PM, Srikanth S, Todd RF (1998) Fibrinogen activates NF-κB transcription factors in mononuclear phagocytes. J Immunol 161(3):1462–1470
92. McNally A, Anderson J (1994) Complement C3 participation in monocyte adhesion to different surfaces. Proc Natl Acad Sci USA 91(21):10119–10123

93. Tandia B-M, Vandenbranden M, Wattiez R, Lakhdar Z, Ruysschaert J-M, Elouahabi A (2003) Identification of human plasma proteins that bind to cationic lipid/DNA complex and analysis of their effects on transfection efficiency: implications for intravenous gene transfer. Mol Ther 8(2):264–273
94. Paula AJ, Araujo Júnior RT, Martinez DST, Paredes-Gamero EJ, Nader HB, Durán N, Justo GZ, Alves OL (2013) Influence of protein corona on the transport of molecules into cells by mesoporous silica nanoparticles. ACS Appl Mater Inter 5(17):8387–8393
95. Lesniak A, Fenaroli F, Monopoli MP, Åberg C, Dawson KA, Salvati A (2012) Effects of the presence or absence of a protein corona on silica nanoparticle uptake and impact on cells. ACS Nano 6(7):5845–5857
96. Jiang X, Weise S, Hafner M, Röcker C, Zhang F, Parak WJ, Nienhaus GU (2009) Quantitative analysis of the protein corona on FePt nanoparticles formed by transferrin binding. J R Soc Interface. doi:10.1098/rsif.2009.0272.focus
97. Zhu Y, Li W, Li Q, Li Y, Li Y, Zhang X, Huang Q (2009) Effects of serum proteins on intracellular uptake and cytotoxicity of carbon nanoparticles. Carbon 47(5):1351–1358
98. JosepháJerry D (2009) Stability, toxicity and differential cellular uptake of protein passivated-Fe3O4 nanoparticles. J Mater Chem 19(35):6328–6331
99. Guarnieri D, Guaccio A, Fusco S, Netti PA (2011) Effect of serum proteins on polystyrene nanoparticle uptake and intracellular trafficking in endothelial cells. J Nanopart Res 13(9): 4295–4309
100. Ruge CA, Kirch J, Cañadas O, Schneider M, Perez-Gil J, Schaefer UF, Casals C, Lehr C-M (2011) Uptake of nanoparticles by alveolar macrophages is triggered by surfactant protein A. Nanomed-Nanotechnol 7(6):690–693
101. Wang Z, Tiruppathi C, Minshall RD, Malik AB (2009) Size and dynamics of caveolae studied using nanoparticles in living endothelial cells. ACS Nano 3(12):4110–4116
102. Caracciolo G, Callipo L, De Sanctis SC, Cavaliere C, Pozzi D, Laganà A (2010) Surface adsorption of protein corona controls the cell internalization mechanism of DC-Chol–DOPE/DNA lipoplexes in serum. BBA-Biomembr 1798(3):536–543
103. Pitek AS, O'Connell D, Mahon E, Monopoli MP, Bombelli FB, Dawson KA (2012) Transferrin coated nanoparticles: study of the bionano interface in human plasma. PLoS ONE 7(7):e40685
104. Salvati A, Pitek AS, Monopoli MP, Prapainop K, Bombelli FB, Hristov DR, Kelly PM, Åberg C, Mahon E, Dawson KA (2013) Transferrin-functionalized nanoparticles lose their targeting capabilities when a biomolecule corona adsorbs on the surface. Nat Nanotechnol 8(2):137–143
105. Mirshafiee V, Mahmoudi M, Lou K, Cheng J, Kraft ML (2013) Protein corona significantly reduces active targeting yield. Chem Commun 49(25):2557–2559
106. Mahon E, Salvati A, Baldelli Bombelli F, Lynch I, Dawson KA (2012) Designing the nanoparticle–biomolecule interface for "targeting and therapeutic delivery". J Controlled Release 161(2):164–174
107. Caracciolo G (2012) The protein corona effect for targeted drug delivery. Bioinspired Biomimetic Nanobiomaterials 2(1):54–57
108. Moyano DF, Saha K, Prakash G, Yan B, Kong H, Yazdani M, Rotello VM (2014) Fabrication of corona-free nanoparticles with tunable hydrophobicity. ACS Nano 8(7): 6748–6755. doi:10.1021/nn5006478
109. Kah JCY, Chen J, Zubieta A, Hamad-Schifferli K (2012) Exploiting the protein corona around gold nanorods for loading and triggered release. ACS Nano 6(8):6730–6740
110. Cifuentes-Rius A, de Puig H, Kah JCY, Borros S, Hamad-Schifferli K (2013) Optimizing the properties of the protein corona surrounding nanoparticles for tuning payload release. ACS Nano 7(11):10066–10074
111. Högemann-Savellano D, Bos E, Blondet C, Sato F, Abe T, Josephson L, Weissleder R, Gaudet J, Sgroi D, Peters PJ (2003) The transferrin receptor: a potential molecular imaging marker for human cancer. Neoplasia (New York, NY) 5(6):495

112. Daniels TR, Delgado T, Helguera G, Penichet ML (2006) The transferrin receptor part II: targeted delivery of therapeutic agents into cancer cells. Clin Immunol 121(2):159–176
113. Berggård T, Arrigoni G, Olsson O, Fex M, Linse S, James P (2006) 140 mouse brain proteins identified by Ca2+-calmodulin affinity chromatography and tandem mass spectrometry. J Proteome Res 5(3):669–687
114. Labarre D, Vauthier C, Chauvierre C, Petri B, Müller R, Chehimi MM (2005) Interactions of blood proteins with poly (isobutylcyanoacrylate) nanoparticles decorated with a polysaccharidic brush. Biomaterials 26(24):5075–5084
115. Cushley RJ, Okon M (2002) NMR studies of lipoprotein structure. Annu Rev Biophys Biomol Struct 31(1):177–206
116. Srivastava RAK (2003) Scavenger receptor class B type I expression in murine brain and regulation by estrogen and dietary cholesterol. J Neurol Sci 210(1):11–18
117. Goti D, Hrzenjak A, Levak-Frank S, Frank S, Van Der Westhuyzen DR, Malle E, Sattler W (2001) Scavenger receptor class B, type I is expressed in porcine brain capillary endothelial cells and contributes to selective uptake of HDL-associated vitamin E. J Neurochem 76(2): 498–508
118. Panzenboeck U, Balazs Z, Sovic A, Hrzenjak A, Levak-Frank S, Wintersperger A, Malle E, Sattler W (2002) ABCA1 and scavenger receptor class B, type I, are modulators of reverse sterol transport at an in vitro blood-brain barrier constituted of porcine brain capillary endothelial cells. J Biol Chem 277(45):42781–42789
119. Zannis VI, Chroni A, Krieger M (2006) Role of apoA-I, ABCA1, LCAT, and SR-BI in the biogenesis of HDL. J Mol Med 84(4):276–294
120. Hammad SM, Stefansson S, Twal WO, Drake CJ, Fleming P, Remaley A, Brewer HB, Argraves WS (1999) Cubilin, the endocytic receptor for intrinsic factor-vitamin B12 complex, mediates high-density lipoprotein holoparticle endocytosis. Proc Natl Acad Sci USA 96(18):10158–10163
121. Kozyraki R, Fyfe J, Kristiansen M, Gerdes C, Jacobsen C, Cui S, Christensen EI, Aminoff M, de la Chapelle A, Krahe R (1999) The intrinsic factor–vitamin B12 receptor, cubilin, is a high-affinity apolipoprotein AI receptor facilitating endocytosis of high-density lipoprotein. Nat Med 5(6):656–661
122. Sahali D, Mulliez N, Chatelet F, Dupuis R, Ronco P, Verroust P (1988) Characterization of a 280-kD protein restricted to the coated pits of the renal brush border and the epithelial cells of the yolk sac. Teratogenic effect of the specific monoclonal antibodies. J Exp Med 167(1): 213–218
123. Seetharam B, Christensen EI, Moestrup SK, Hammond TG, Verroust PJ (1997) Identification of rat yolk sac target protein of teratogenic antibodies, gp280, as intrinsic factor-cobalamin receptor. J Clin Investig 99(10):2317

Calibration-Quality Cancer Nanotherapeutics

Jillian L. Perry, Marc P. Kai, Kevin G. Reuter, Charles Bowerman, J. Christopher Luft and Joseph M. DeSimone

Abstract

Nanoparticle properties such as size, shape, deformability, and surface chemistry all play a role in nanomedicine drug delivery in cancer. While many studies address the behavior of particle systems in a biological setting, revealing how these properties work together presents unique challenges on the nanoscale. "Calibration-quality" control over such properties is needed to draw adequate conclusions that are independent of parameter variability. Furthermore, active

J.L. Perry · C. Bowerman · J. Christopher Luft · J.M. DeSimone
Lineberger Comprehensive Cancer Center, University of North Carolina at Chapel Hill, North Carolina, USA
e-mail: perryjl@email.unc.edu

C. Bowerman
e-mail: cbowerma@unc.edu

J. Christopher Luft
e-mail: jluft@email.unc.edu

M.P. Kai · J.M. DeSimone
Department of Chemical and Biomolecular Engineering, North Carolina State University, North Carolina, USA
e-mail: mpkai@ncsu.edu

K.G. Reuter · J.M. DeSimone
Department of Chemistry, University of North Carolina at Chapel Hill, North Carolina, USA
e-mail: kreuter@email.unc.edu

J. Christopher Luft · J.M. DeSimone
Department of Molecular Pharmaceutics, Eshelman School of Pharmacy, University of North Carolina at Chapel Hill, North Carolina, USA

J.M. DeSimone
Carolina Center of Cancer Nanotechnology Excellence, University of North Carolina at Chapel Hill, North Carolina, USA

C.A. Mirkin et al. (eds.), *Nanotechnology-Based Precision Tools for the Detection and Treatment of Cancer*, Cancer Treatment and Research 166,
DOI 10.1007/978-3-319-16555-4_12

targeting and drug loading strategies introduce even greater complexities via their potential to alter particle pharmacokinetics. Ultimately, the investigation and optimization of particle properties should be carried out in the appropriate preclinical tumor model. In doing so, translational efficacy improves as clinical tumor properties increase. Looking forward, the field of nanomedicine will continue to have significant clinical impacts as the capabilities of nanoparticulate drug delivery are further enhanced.

Keywords

Engineered cancer therapeutics · Personalized medicine · Tunable nanoparticle parameters · Pro-drug · Industrialized process

Contents

1 Introduction

The utilization of nanocarriers for the delivery of therapeutics has led to a significant decrease in the toxicity of chemotherapeutics and point to the potential of nanoparticle-based therapies to improve cancer treatment. Unlike small molecule drugs which distribute nonspecifically throughout the body, nanoparticles remain in the blood stream and can passively accumulate within tumors through the enhanced

J.M. DeSimone
Institute for Nanomedicine, University of North Carolina at Chapel Hill, North Carolina, USA

J.M. DeSimone
Institute for Advanced Materials, University of North Carolina at Chapel Hill, North Carolina, USA

J.M. DeSimone (✉)
Sloan-Kettering Institute for Cancer Research, Memorial Sloan-Kettering Cancer Center, New York, USA
e-mail: desimone@email.unc.edu

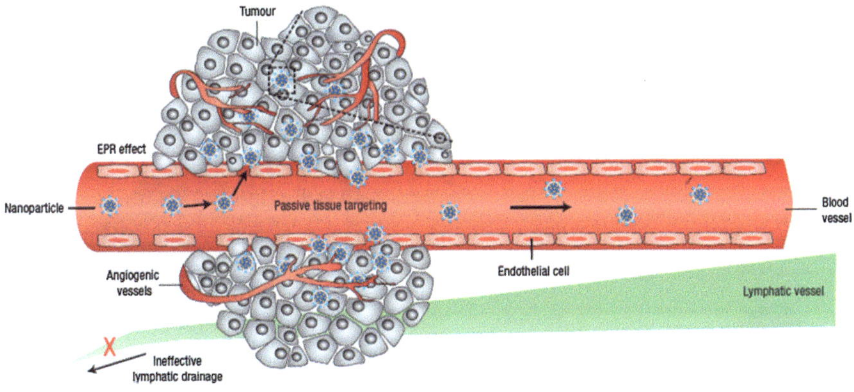

Fig. 1 Illustration of the EPR effect depicting the discontinuous epithelium lining blood vessel walls of a tumor with concurrent poor lymphatic drainage compared to normal tissue. Nanoparticles can passively accumulate via gaps in blood vessels supplying tumor tissue. Adapted with permission from Ref. [4] Copyright (2007) Nature Publishing Group

permeability and retention (EPR) effect [1, 2]. Tumors are characterized by permeable and unorganized vasculature, lack of functional lymphatics, and elevated interstitial fluid pressure (IFP), illustrated in Fig. 1 [3, 4]. The high permeability of the tumor vasculature allows macromolecules and nanoparticles to enter the tumor, while the lack of effective lymphatic drainage and elevated IFP prevents them from being removed. It is important to note that there is significant heterogeneity in the EPR as a function of tumor size, type, and location [5]. Regardless of the heterogeneity, in order to capitalize on the EPR effect, a nanoparticle therapeutic needs to stay in circulation long enough to accumulate within the tumor, which, depending on tumor type, can range from hours to days.

In general, upon injection, nanoparticles can become coated with plasma proteins (opsonization), which mark them for rapid clearance by the mononuclear phagocyte system (MPS); the MPS is comprised of bone marrow progenitors, blood monocytes, and tissue macrophages (residing in both the liver and spleen) [6]. Since extended blood retention goes hand-in-hand with tumor accumulation, much research has been devoted to the development of "stealthy" long-circulating nanoparticles that avoid the MPS. To that end, there is a general set of guidelines for engineering long-circulating particles, which is depicted in Fig. 2. Briefly, particles should be either neutral or negatively charged, within a hydrodynamic diameter size range of 8–200 nm, and surface coated with a stealthing agent, such as polyetheylene glycol (PEG) [7]. Other particle parameters, such as shape and modulus, can also greatly influence the circulation profiles of nanomaterials [8–10], and other particle attributes can have highly variable effects. Targeting moieties and therapeutic cargo can alter in vivo behavior, including serum-biomolecule interactions, pharmacokinetic (PK) profiles, and biodistribution [11–13]. Inherent in the variability of all these properties is an optimal combination that may depend on the platform or fabrication technology. The interplay between particle parameters is further

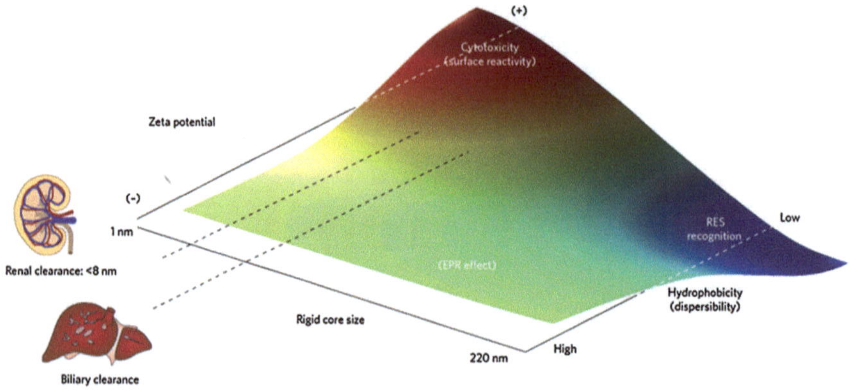

Fig. 2 Qualitative trends in biological behavior and in vivo biocompatibility of nanoparticles based on their physical characteristics. Particles with neutral or negative surface charge, between 8 and 200 nm in size, and hydrophilic (well-dispersed) properties tend to promote enhanced permeation and retention (EPR) effects. Adapted with permission from Ref. [7], Copyright (2009) Nature Publishing Group

complicated in a biological setting, where the interaction between a nanoparticle and the immune system adds a significant layer of complexity. Ultimately, anecdotal evidence is not sufficient to predict the behavior of a given nanoparticle with complete confidence. Consequently, to increase the chance of successful translation, preclinical efforts should seek to understand particle behavior in the most relevant animal model available; traditional efficacy and toxicity assays should also be used.

While many studies have investigated the behavior of particle systems in biological settings, revealing how particle properties work together presents unique challenges on the nanoscale. The major roadblock to reveal this interdependence is the previous lack of a suitable particle fabrication technique. Further, most studies on nanoparticles have focused on spherical shapes, due to their accessibility and narrow size distribution; such particles can be obtained by fabrication methods, such as dispersion, emulsion, and suspension polymerization [8, 14]. However, these systems lack sufficient control over size and shape. Particle replication in nonwetting templates (PRINT®) is one molding technique that allows for fabrication of "calibration quality" micro- and nanoparticles with independent control over their physical parameters (e.g., size, shape, surface chemistry, and modulus) [15]. The PRINT process is a continuous particle fabrication technology (Fig. 3) that begins by casting a film of pre-particle solution (this can vary from thermoplastic to thermoset materials) onto a high-surface energy delivery sheet (Fig. 3a), which is then laminated to the low-surface energy mold through a pressurized nip (Fig. 3b). Consequently, a reversible seal is formed between the mold and the substrate, allowing the liquid to be either confined to the cavities in the mold or forced out due to the low-surface energy of the interfacial area. Delamination of the mold from the delivery sheet results in PPS confined within the mold cavities, while excess PPS remains on the delivery sheet (Fig. 3c) The PRINT particles are then harvested from

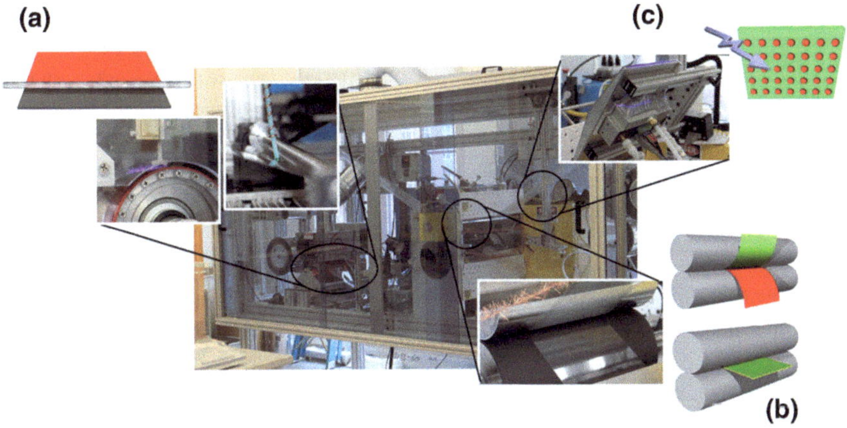

Fig. 3 Scale-up of the PRINT process to a roll-to-roll machine. PET and mold webs were mounted on the *left side* of the machine (*unwind side*), and threaded through various flexible unit operations before collecting on the *right side* (*rewind side*). **a** Film casting was achieved by corona treatment of the PET delivery sheet followed by film deposition by continuous dispensing of pre-particle solution (PPS) via a syringe pump behind a mayer rod. **b** Mold filling was accomplished by laminating the mold to the delivery sheet using a pressurized nip (heated for thermoplastics). The mold is then separated from the delivery sheet upon exiting the nip and the particles are formed within the mold either by **c** curing the PPS with a UV-LED oven (for thermosets) or by cooling the filled mold to room temperature (for thermoplastics)

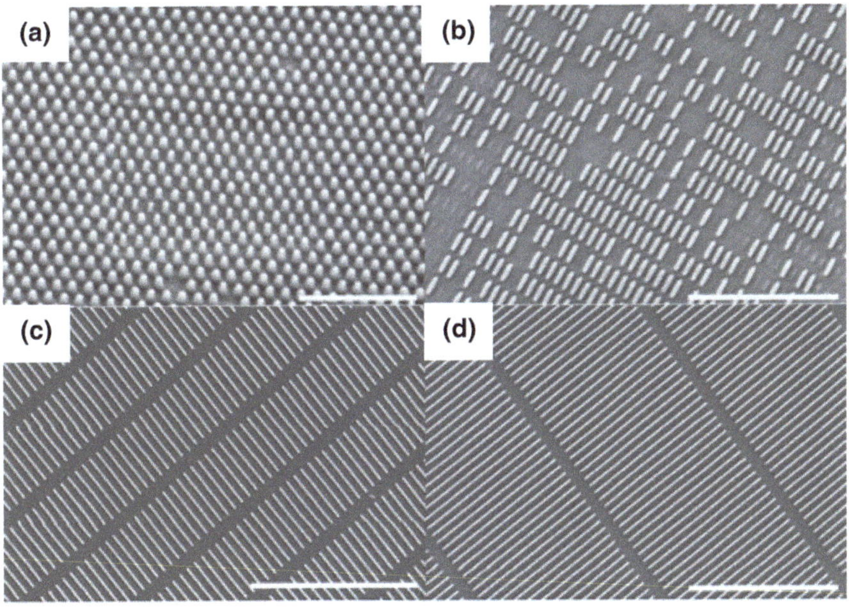

Fig. 4 Scanning electron micrographs of PRINT particles fabricated from **a** 80 nm × 180 nm, **b** 80 nm × 320 nm, **c** 80 nm × 2000 nm, and **d** 80 nm × 5000 nm molds. *Scale bar* represents **a** 1 μm, **b** 3 μm, **c** and **d** 5 μm. Figure adapted with permission from Ref. [10], Copyright (2012) American Chemical Society

the mold as isolated, "scum-free" objects as depicted in Fig. 4. In this chapter, we will focus on the use of "calibration-quality" nanotools, such as PRINT-based particles, for studying the effect of various particle parameters—size, shape, modulus, drug loading, and surface chemistry—on the biodistribution, clearance, and delivery of therapeutics.

2 Size/Shape

Engineering a long-circulating nanoparticle for therapeutic delivery is absolutely paramount for tumor accumulation. In order to harness the EPR effect, particles need to stay in circulation long enough to passively accumulate at the target site (tumor) and then release their cargo. Size and shape are key parameters in designing a long-circulating particle. Much work has been done to lay the foundation for the appropriate particle size ranges; [16, 17] however, the body of work that has explored the effect of particle shape is smaller due to the limited availability of techniques to produce nonspherical particles. The recent advances in nanoparticle synthesis and fabrication technologies resulting in shape specific particles have started to elucidate the influence that particle shape can have over circulation profiles, biodistribution, and cellular entry kinetics once the particle has reached its target site [8, 18–25].

In terms of the particle circulation profiles, margination (defined as the movement of particles in flow toward the walls of the blood vessel) enhances the particles chances to extravasate through the leaky tumor vasculature and enter the tumor. Work—both empirical and theoretical—has been conducted in this area, comparing margination rates of spherical and nonspherical micro- and nanoparticles [20, 22, 26, 27]. In general, nonspherical particles with higher aspect ratios marginate more readily than spherical particles. Furthermore, high-aspect ratio particles were found to reduce macrophage uptake in vitro [19, 28, 29] translating to decreased liver and splenic filtration and increased circulation persistence of high-aspect ratio particles as compared to spherical particles [23, 28, 30].

To be effective, nanotherapeutics must avoid clearance by the liver, spleen, and macrophages. Once long circulation is achieved, particles need to accumulate at the target site. The passive accumulation of nanoparticles within tumors is contingent on their ability to extravasate through the leaky tumor vasculature, which is highly variable across tumor models. Even with this high degree of heterogeneity, there is a general consensus across literature that decreasing nanoparticle size can improve transport within tumors; [31–33] however, there is limited data concerning nonspherical particles. Recent work comparing nanospheres to nanorods of equivalent hydrodynamic size concluded that the nanorods could penetrate tumors four times as rapidly as nanospheres [34]. After particles have extravasated into the tumor bed, they then need to deliver their therapeutic cargo. Depending on the therapeutic, this may occur via extracellular or intracellular means (or a combination of both). If intracellular delivery is necessary, particle internalization can be enhanced either by modifying particle shape or incorporating active targeting to encourage receptor-mediated endocytosis (a topic discussed in a later section). Studies have shown that shape plays an important role in the way cells interact with particles. We have demonstrated that higher aspect ratio particles enhanced HeLa cell uptake kinetics in vitro [24], thus making it more likely for rod-shaped particles to be internalized by target cells.

Additionally, increasing the aspect ratio of the nanoparticle from 1 to 2.5 resulted in the more efficient delivery of chemotherapeutics to tumors. We evaluated the delivery properties of 200 nm × 200 nm and 80 nm × 320 nm PRINT poly (lactic-co-glycolic) acid (PLGA) nanoparticles loaded with similar concentrations of docetaxel [35]. Not surprisingly, both PRINT nanotherapeutics outperformed the small molecule clinical control, Taxotere. However, the higher aspect ratio 80 nm × 320 nm particles exhibited better PK profiles than the 200 nm × 200 nm particles [35]. Specifically, the rod-like particles exhibited enhanced blood and tumor retention with lower splenic and liver accumulation. This is translated to successful efficacy studies using the 80 nm × 320 nm geometry [36].

Studies by Discher and colleagues also clearly demonstrated the therapeutic benefits of using nonspherical drug delivery vehicles. They concluded that flexible, worm-like nanocarriers were more effective than spherical particles at delivering drugs and reducing tumor volume [37]. Work by Sailor et al. revealed similar results with spherical and high-aspect ratio magnetic particles and concluded that, when only one dimension was elongated, the size of the particles could be increased without sacrificing circulation times [38]. It is important to note that as aspect ratio increases, deformability becomes a crucial parameter to effectively navigate the biological barriers within the body [10, 14].

3 Modulus/Mechanical Properties

It is well documented that particles with a hydrodynamic diameter between 8 and 200 nm are large enough to avoid renal filtration and small enough to avoid liver and splenic filtration [7]. In addition to size and shape, the modulus of particles plays a role in their ability to avoid mechanical filtration by the liver and spleen, and has the ability to circumvent size limitations above the 200 nm cutoff. For instance, red blood cells are approximately 8 μm in diameter and, due to their extraordinary deformability, they can pass through splenic slits 2–3 μm in width [39]. However, as the red blood cells age and become stiffer, they can no longer deform and are removed from the blood stream by the spleen. We have developed polymeric, micron-sized, PRINT-based red blood cell mimics (RBCMs), whose mechanical properties can be controlled based upon their cross-link densities [40]. The biodistribution and circulation profiles of these particles were evaluated in vivo, and the results showed that the softer RBCMs were able to circulate longer than their stiffer counterparts [40]. Furthermore, at the nanoscale, we observed that PRINT-based, polymeric, 80 nm × 5000 nm filamentous particles that were fabricated with lower cross-link density were able to extravasate through 200 nm porous membranes that mimicked the pores found in the liver and spleen, whereas stiffer particles were unable to efficiently extravasate [10]. Extravasation of shorter filamentous particles (80 nm × 320 nm, and 80 nm × 180 nm) through the 200 nm porous membranes was independent of modulus [10].

Shaoyi Jiang's lab reached similar conclusions in their investigations of the modulus of long-circulating zwitterionic spherical nanoparticles. Spherical particles (d = 250 nm) with a range of cross-link densities were fabricated and were passed through 220 nm porous membranes [41]. While all the "soft" particles were able to pass through the filter, the majority of the "hard" particles (with the highest cross-link densities) were stopped by the filter [41]. These trends were also predictive of in vivo particle behavior. As the cross-linker density of the particle was decreased, splenic accumulation was minimized, resulting in extended circulation persistence [41]. With a better understanding of the relevant particle parameters necessary for extending particle circulation, the focus shifts to drug loading and release.

4 Drug Loading/Drug Release Rates

Promising drug candidates are often hydrophobic compounds with poor aqueous solubility. To increase their solubility, they are typically formulated in mixtures of low molecular weight surfactants and organic solvents, which have their own nonspecific toxicity. Particulate formulations of these drugs can not only eliminate the necessity of organic solvents and surfactants, but can also alter the pharmacokinetics of the delivered therapeutic. Since small molecule drugs can extravasate through the tight endothelial junctions lining the blood vessel walls, they can freely permeate throughout the body, interacting with both healthy and diseased tissue indiscriminately and resulting in dose-limiting systemic toxicities. Formulating these drugs within a nanoparticle can keep the drug within the blood compartment (reducing the volume of distribution and interactions with healthy tissues) and allow a greater portion of the injected dose to potentially reach the target cancer cells. Maintaining controlled release of the drug from the nanotherapeutic platform is also crucial for improved efficacy and reduced toxicity. Ideally, the carrier should allow minimal premature release of drug before it reaches the intended site, but sufficient release after it reaches the target. Drug release from the particle carrier can be dictated by particle erosion and diffusion (as with biodegradable particle matrices) or by external (including light, ultrasound, and magnetic fields) or physiological (pH gradients and redox states) stimuli [42–46].

Biodegradable PLGA (poly lactic-co-glycolic acid), a thermoplastic material, is one example of a noncovalent entrapment system that is commonly used in nanoformulations [47]. For this particle matrix, generally hydrophobic drugs are noncovalently loaded by physical entrapment or weak intermolecular interactions, and release rates are based upon drug loading, matrix degradation, and passive diffusion. Typically, these particles suffer from an immediate burst release, followed by sustained drug release [47]. Polymer or lipid coatings can be used to overcome

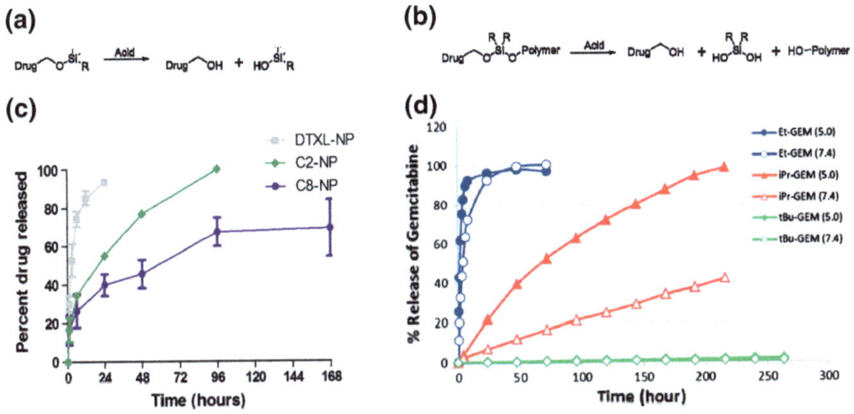

Fig. 5 Prodrug approaches for incorporating drugs into nanoparticles. Two types of silyl ether prodrugs, **a** small molecule monofunctional silyl ether and **b** polymeric asymmetric bifunctional silyl ether prodrug, for encapsulation into PRINT PLGA or PEG hydrogel particles, respectively. **c** Release kinetics of PLGA particles loaded with unmodified Docetaxel (DTXL-NP) or monfunctional silyl ether prodrugs with R equal to ethyl (C2-NP) or octyl (C8-NP). **d** Release kinetics of hydrogel particles loaded with bifunctional silyl ether Gemcitabine, with either to ethyl, isopropyl, or t-butyl R groups. Figures adapted with permission from Refs. [50] and [52], Copyright (2012 and 2014) American Chemical Society

this burst release profile. Sethi et al. utilized a cross-linkable lipid shell to decrease the release rate of wortmannin and docetaxel from PLGA particles [48]. The slower release rate translated to enhanced efficacy; while nanoparticles without a cross-linked lipid shell showed no benefit over the free drug [48].

Another method to alter drug release from the particle is to modify the drugs' affinity to the particle matrix. This can be done through a prodrug approach. Prodrugs are derivatives of drugs that undergo stimuli responsive triggering to release the active drug, offering protection from inactivation and toxic side effects [49]. Using this approach, docetaxel was lipidated through an acid-sensitive, silyl ether linkage and encapsulated within a PRINT-based PLGA nanoparticle (PRINT-C2), depicted in Fig. 5a [50]. The addition of the lipid tale enabled drug release to be extended from 24 h (for the unmodified drug) to 96 h (Fig. 5c). Furthermore, once the prodrug was released from the particles, it underwent complete conversion to the active form of the drug in plasma after 24 h. PK analysis showed that the PRINT-C2 nanoparticle had greater persistence in the plasma than both PRINT-docetaxel and Taxotere, and this can most likely be attributed to the slower release rate from the nanoparticle. A significant finding of this study was that the prodrug preferentially converted to the active drug in the tumor, most likely due to the acidic nature of the tumor tissue as compared to healthy tissue [51]. This adds another layer of control for mitigating unwanted side effects.

Encapsulation, however, is poorly suited to intrinsically porous lowly cross-linked hydrogel particles. For drugs that are difficult to retain in these types of particles we have developed a silyl ether-based prodrug approach that utilizes covalent linkages, depicted in Fig. 5b [52]. The prodrug contains an acrylate group and can therefore be directly polymerized into thermoset particles. The drug release kinetics were found to be pH dependent, and the rates could be controlled by the bulkiness of the silicon substituents (Fig. 5d). In vitro experiments demonstrated that particles could be fabricated to release drugs, such as gemcitabine and camptothecin, with comparable toxicities to the free drug. Using a similar approach, Wang et al., developed a pH sensitive gemcitabine-poly(methyl methacrylate) prodrug (GEM-PMAA), which self-assembles in water to form nanoparticles [53]. Drug release rates were found to be pH dependent and in vivo assessment proved that the particles could efficiently inhibit tumor growth and alleviate drug-associated side effects [53]. Drug release from these pH responsive linkers can occur in the acidic conditions of the tumor tissue or intracellular compartments. If intracellular delivery is necessary, nanocarriers can be modified with targeting ligands to improve cellular internalization.

5 Surface Functionalization for Passive and Active Targeting

Current methods of nanoparticle delivery to tumors are based mainly upon either passive or active targeting. Passively targeted particles are typically surface functionalized with stealthing agents, to allow for long circulation and thus accumulation through the EPR effect. Actively targeted particles are modified with targeting ligands used to recognize overexpressed receptors on the tumor cell surface. For either delivery mechanism, a common feature is that the nanoparticles in the bloodstream must first marginate and then extravasate into the interstitium of the tumor. Passive targeting relies heavily on extending particle circulation half-life, thus increasing their chances of reaching their target site. This is commonly achieved by conjugating, grafting, or adsorbing polyethylene glycol (PEG) to the surface of nanoparticles (PEGylation). Altogether, effects of PEGylation are highly dependent on two interrelated parameters: the molecular weight (MW) and surface density of the PEG coating. In general, PEG ranging from 2 to 10 k displays adequate resistance to protein adsorption and subsequent phagocytic uptake [54]. The density of PEG necessary to promote protein resistance and extend blood circulation varies drastically for different nanoparticle types. Generally, metallic and other highly immunogenic particle types require significantly greater surface PEG densities as compared to more inert matrices [55]. We have demonstrated that our cross-linked PEG matrix required a much lower surface density of PEG (less than 0.1 PEG molecules per nm^2) to enhance protein resistance and extend circulation half-life (from 0.89 to 19.5 h) [55]. Other polymers for surface passivation, such as zwitterionic coatings and sugar-based moieties, have shown promise [56, 57].

While passivation remains the most utilized surface modification, active targeting has become a fixture in particle platforms. The "holy grail" in drug delivery is to create a system that can be used to deliver a highly potent therapeutic to the diseased, cancerous tissue, while completely eliminating exposure to off-target, healthy cells [58]. To achieve targeted drug delivery, nanoparticles can be coated with ligands that bind specifically to particular overexpressed receptors on the diseased cell surface. Commonly targeted receptors implicated in various cancers include the epidermal growth factor receptor (EGFR), the vascular endothelial growth factor (VEGF), and the human epidermal growth factor receptor (HER-2, in the EGFR family), to name only a few [11, 59]. By designing a drug carrier that targets these receptors, preferential cell interactions can occur with the diseased tissue, leading to reduced exposure in healthy tissue. In reality, off-targeting remains since target-receptor expression in healthy cells persists, albeit to a lesser extent.

The diversity of targeting agents is just as expansive as the receptors they are directed toward. A myriad of ligand classes have been developed; antibodies, affibodies (small, single-domain proteins with high affinity and specificity for target), small molecules, peptides, and aptamers are just a sampling that have been coupled to particle platforms with varying levels of success [12, 59, 60]. Functionalizing particles with an active ligand is not a trivial matter. Dramatic shifts in PK profile, biodistribution, and efficacy are routinely observed, and as with PEGylation, the surface density of targeting ligands plays a role in these systematic changes [60, 61]. Wang et al. discovered unprecedented toxicity toward Ramos cells with otherwise inert transferrin-targeted PRINT particles [62]. In these studies, the cellular toxicity increased as a function of transferrin surface density, a commonly used targeting ligand that does not innately kill cells. These researchers found that while enhancing the multivalency, binding affinity associated with multiple ligand-receptor binding sites, unprecedented cell death occurred for reasons not completely understood [62].

Other research has revealed that along with ligand density, the shape of the particle can also influence targeting avidity as shown in Fig. 6. As mentioned in previous sections, a particle's shape can greatly affect its ability to marginate to the vascular surface and extravasate into the tumor bed. Theoretical work by Decuzzi and Ferrari et al. concluded that rod-shaped particles have higher adhesion/contact probability than spherical particles [63]. Further results from a computational model created by Liu et al., showed an increase in binding probability with increasing aspect ratio [64, 65]. Confirming these theoretical finds, Mitragotri reported that rod-shaped particles exhibit higher avidity and selectivity toward their target than their spherical counterparts in vivo [66]. Similarly, Sailor and colleagues showed that targeted nanoworms had enhanced multivalent interactions with cell receptors, which amplified their passive accumulation in vivo over spherical nanoparticle controls [38].

It is clear that the scientific community has done a remarkable job utilizing nanoparticles and specific particle parameters to improve drug safety and efficacy. Also, having the greatest flexibility to engineer, design, and exquisitely control the

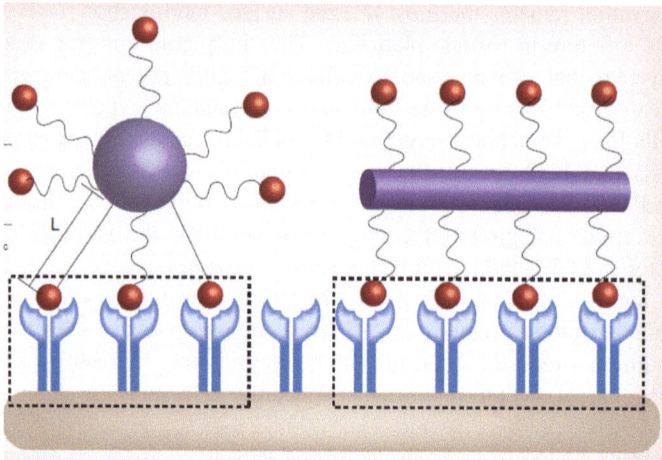

Fig. 6 Particle shape and linker length play a role in the targeting ability of nanoparticles. Figures adapted with permission from Ref. [25], Copyright (2014) Future Medicine

physiochemical properties of the particle is essential in order to manipulate the biophysiochemical barriers that dictate the performance/behavior of particles in a biological system. Continued funding (both public and private) of nanomedicine research will allow the community to build upon "lessons learned", focus on essential parameters, and investigate/identify the next parameter(s) deemed necessary to ultimately manufacture a new class of nanotherapeutics.

6 Looking Forward

The world's first nanomedicine, Doxil, was approved by the US FDA in 1995 for treating Kaposi's sarcoma, a cancer often associated with AIDS. Doxil—a liposome encapsulating doxorubicin—was a relatively simplistic nanomedicine, yet was undeniably a breakthrough [67]. In recent years, enormous advances have been made in fundamental research involving novel methods and techniques for batch and continuous processes, leading to the formation of complex nanosystems. Currently, nanomedicines advancing into the clinic are becoming much more "engineered" in that surface functionalization incorporating stealthing agents and targeting ligands are utilized to evade the immune system and enhance specificity, respectively. A challenge facing the design of a nanoparticle delivery vehicle is incorporating elegance and yet minimizing complexity for potential translation to the clinic. Often early design parameters are efficacious in vitro and/or in vivo, but the capability to translate clinically is limited. A significant reason behind this limitation is poor understanding of which tumor models are clinically relevant—and of these, which are likely to respond to a nanomedicine.

The majority of preclinical research utilizes xenograft models, generated by implanting a well-established monoclonal tumor cell line into immunocompromised mice. Alternatively, patient-derived xenografts (PDX) offer more relevance, since tumor fragments are directly transplanted from patients into immunocompromised mice. PDX models can accurately imitate the heterogeneity of the human tumor that is not always evident in cell line-based xenografts. The major limitations of these explant models are their inability to fully replicate the complex relationship between the tumor and its microenvironment. To bridge this gap, genetically engineered mouse models (GEMMs), are often used for recapitulating the genetics and heterogeneity of human tumors [68–70]. However, these models are costly and require a long time for tumors to form. Another option is the allograft model which gives the genetic relevance of GEMMs and the research utility of xenografts. Considering the variability of current preclinical tumor and mouse models, research effort is warranted for identifying the appropriate preclinical platform to capitalize on the benefits of a nanoparticle-based delivery system. Correlating preclinical results with positive clinical responses will aid in determining which tumor characteristics will be amenable to effective treatment with nanoparticle therapies. In addition to clinical translatability of the disease, it is essential that the nanofabrication system is easily manufactured at clinically relevant scales.

The successful scale-up and manufacturing of a nanomedicine presents unique challenges in pharmaceutical development. Since most nanoparticles are complex multicomponent products that require specific arrangement of the components, a full understanding of each component and their interactions is essential to define the key characteristics of the final product. Developing methods to monitor these interactions throughout critical process steps and develop or optimize analytical characterization techniques that ensure reproducibility of the product must be addressed. Furthermore, it is of paramount importance that the structural integrity and physicochemical properties of nanoparticles be preserved throughout the formulation process to the finished product. Therefore, during early development, intentional effort to assembly interdisciplinary scientific teams that engage expertise from a variety of disciplines and specialties (i.e., Material Science, Engineering, Medicine, Biology, Chemistry, and the Pharmaceutical Sciences) is essential in addressing quality, efficiency, scalability, reliability, safety, and affordability issues that are relevant to nanomanufacturing.

The path to approval for any new small molecule drug involves significant time and enormous costs. Due to the complexity and multicomponent nature of nanomedicines, additional variables may substantially increase the level of difficulty in controlling production and performance of the nanomedicine, which add to the time and cost for final health agency approval. Additionally, genetic variability among the patient population may further segment responsive subpopulations from non-responders as drug exposure is dependent upon the particle delivery system's ability to access the tumor site. This leads to the opportunity to embrace a personalized medicine approach when developing a therapeutic strategy best suited to be efficacious based upon the characteristics of the individual's cancer. Therefore, as the scientific community continues to investigate and learn about the multiple biological, spatial, and temporal interactions and the multiple communication and feedback mechanisms that are involved in the highly complex and dynamic system that is cancer, so too will the nanomanufacturing community investigate and enhance the multiple processes involved to seamlessly manufacture robust, reproducible, and cost-effective multicomponent nanosystems that will, in the near future, bring the promise and potential of nanomedicine to reality.

References

1. Jain RK, Stylianopoulos T (2010) Delivering nanomedicine to solid tumors. Nat Rev Clin Oncol 7(11):653–664
2. Moghimi SM, Hunter AC, Murray JC (2001) Long-circulating and target-specific nanoparticles: theory to practice. Pharmacol Rev 53(2):283–318
3. Matsumura Y, Maeda H (1986) A new concept for macromolecular therapeutics in cancer-chemotherapy—mechanism of tumoritropic accumulation of proteins and the antitumor agent smancs. Cancer Res 46(12):6387–6392
4. Peer D et al (2007) Nanocarriers as an emerging platform for cancer therapy. Nat Nano 2 (12):751–760
5. Marusyk A, Almendro V, Polyak K (2012) Intra-tumour heterogeneity: a looking glass for cancer? Nat Rev Cancer 12(5):323–334
6. Van Furth R (1982) Current view on the mononuclear phagocyte system. Immunobiology 161 (3–4):178–185
7. Nel AE et al (2009) Understanding biophysicochemical interactions at the nano-bio interface. Nat Mater 8(7):543–557
8. Champion JA, Katare YK, Mitragotri S (2007) Particle shape: a new design parameter for micro- and nanoscale drug delivery carriers. J Controlled Release 121(1–2):3–9
9. Euliss LE et al (2006) Imparting size, shape, and composition control of materials for nanomedicine. Chem Soc Rev 35(11):1095–1104
10. Kersey FR et al (2012) Effect of aspect ratio and deformability on nanoparticle extravasation through nanopores. Langmuir 28(23):8773–8781
11. Byrne JD, Betancourt T, Brannon-Peppas L (2008) Active targeting schemes for nanoparticle systems in cancer therapeutics. Adv Drug Deliv Rev 60(15):1615–1626
12. Benhabbour SR et al (2012) In vitro and in vivo assessment of targeting lipid-based nanoparticles to the epidermal growth factor-receptor (EGFR) using a novel Heptameric Z (EGFR) domain. J Controlled Release 158(1):63–71
13. Mahon E et al (2012) Designing the nanoparticle-biomolecule interface for targeting and therapeutic delivery. J Controlled Release 161(2):164–174
14. Liu Y et al (2012) The shape of things to come: importance of design in nanotechnology for drug delivery. Ther deliv 3(2):181–194
15. Rolland JP et al (2005) Direct fabrication and harvesting of monodisperse, shape-specific nanobiomaterials. J Am Chem Soc 127(28):10096–10100
16. Alexis F et al (2008) Factors affecting the clearance and biodistribution of polymeric nanoparticles. Mol Pharm 5(4):505–515
17. Albanese A, Tang PS, Chan WCW (2012) The effect of nanoparticle size, shape, and surface chemistry on biological systems. Annu Rev Biomed Eng 14(14):1–16
18. Champion JA, Katare YK, Mitragotri S (2007) Making polymeric micro- and nanoparticles of complex shapes. Proc Nat Acad Sci USA 104(29):11901–11904
19. Champion JA, Mitragotri S (2009) Shape induced inhibition of phagocytosis of polymer particles. Pharm Res 26(1):244–249

20. Carboni E et al (2014) Particle margination and its implications on intravenous anticancer drug delivery. Aaps Pharmscitech 15(3):762–771
21. Decuzzi P et al (2010) Size and shape effects in the biodistribution of intravascularly injected particles. J Controlled Release 141(3):320–327
22. Gentile F et al (2008) The effect of shape on the margination dynamics of non-neutrally buoyant particles in two-dimensional shear flows. J Biomech 41(10):2312–2318
23. Geng Y et al (2007) Shape effects of filaments versus spherical particles in flow and drug delivery. Nat Nanotechnol 2(4):249–255
24. Gratton SEA et al (2008) The effect of particle design on cellular internalization pathways. Proc Nat Acad Sci USA 105(33):11613–11618
25. Toy R et al (2014) Shaping cancer nanomedicine: the effect of particle shape on the in vivo journey of nanoparticles. Nanomedicine 9(1):121–134
26. Decuzzi P et al (2005) A theoretical model for the margination of particles within blood vessels. Ann Biomed Eng 33(2):179–190
27. Decuzzi P et al (2009) Intravascular delivery of particulate systems: does geometry really matter? Pharm Res 26(1):235–243
28. Champion JA, Mitragotri S (2006) Role of target geometry in phagocytosis. Proc Nat Acad Sci USA 103(13):4930–4934
29. Sharma G et al (2010) Polymer particle shape independently influences binding and internalization by macrophages. J Controlled Release 147(3):408–412
30. Arnida et al (2011) Geometry and surface characteristics of gold nanoparticles influence their biodistribution and uptake by macrophages. Eur J Pharm Biopharm 77(3):417–423
31. Perrault SD et al (2009) Mediating tumor targeting efficiency of nanoparticles through design. Nano Lett 9(5):1909–1915
32. Dreher MR et al (2006) Tumor vascular permeability, accumulation, and penetration of macromolecular drug carriers. J Natl Cancer Inst 98(5):335–344
33. Cabral H et al (2011) Accumulation of sub-100 nm polymeric micelles in poorly permeable tumours depends on size. Nat Nanotechnol 6(12):815–823
34. Chauhan VP et al (2011) Fluorescent nanorods and nanospheres for real-time in vivo probing of nanoparticle shape-dependent tumor penetration. Angew Chem Int Ed 50(48):11417–11420
35. Chu KS et al (2013) Plasma, tumor and tissue pharmacokinetics of Docetaxel delivered via nanoparticles of different sizes and shapes in mice bearing SKOV-3 human ovarian carcinoma xenograft. Nanomed Nanotechnol Biol Med 9(5):686–693
36. Chu KS et al (2013) Nanoparticle drug loading as a design parameter to improve docetaxel pharmacokinetics and efficacy. Biomaterials 34(33):8424–8429
37. Christian DA et al (2009) Flexible filaments for in vivo imaging and delivery: persistent circulation of filomicelles opens the dosage window for sustained tumor shrinkage. Mol Pharm 6(5):1343–1352
38. Park J-H et al (2008) Magnetic iron oxide nanoworms for tumor targeting and imaging. Adv Mater 20(9):1630–1635
39. Sarin H (2010) Physiologic upper limits of pore size of different blood capillary types and another perspective on the dual pore theory of microvascular permeability. J Angiogenesis Res 2:14
40. Merkel TJ et al (2011) Using mechanobiological mimicry of red blood cells to extend circulation times of hydrogel microparticles. Proc Nat Acad Sci USA 108(2):586–591
41. Zhang L et al (2012) Softer zwitterionic nanogels for longer circulation and lower splenic accumulation. ACS Nano 6(8):6681–6686
42. Bae Y et al (2003) Design of environment-sensitive supramolecular assemblies for intracellular drug delivery: polymeric micelles that are responsive to intracellular pH change. Angew Chem Int Ed 42(38):4640–4643
43. Bohmer MR et al (2009) Ultrasound triggered image-guided drug delivery. Eur J Radiol 70 (2):242–253

44. Caldorera-Moore M et al (2010) Designer nanoparticles: incorporating size, shape and triggered release into nanoscale drug carriers. Expert Opin Drug Deliv 7(4):479–495
45. Doshi N, Mitragotri S (2009) Designer biomaterials for nanomedicine. Adv Funct Mater 19 (24):3843–3854
46. Ganta S et al (2008) A review of stimuli-responsive nanocarriers for drug and gene delivery. J Controlled Release 126(3):187–204
47. Makadia HK, Siegel SJ (2011) Poly Lactic-co-Glycolic Acid (PLGA) as biodegradable controlled drug delivery carrier. Polymers 3(3):1377–1397
48. Sethi M et al (2014) Effect of drug release kinetics on nanoparticle therapeutic efficacy and toxicity. Nanoscale 6(4):2321–2327
49. Kratz F et al (2008) Prodrug strategies in anticancer chemotherapy. ChemMedChem 3(1):20–53
50. Chu KS et al (2014) Particle replication in nonwetting templates nanoparticles with tumor selective alkyl silyl ether docetaxel prodrug reduces toxicity. Nano Lett 14:1472–1476
51. Tannock IF, Rotin D (1989) Acid PH in tumors and its potential for therapeutic exploitation. Cancer Res 49(16):4373–4384
52. Parrott MC et al (2012) Incorporation and controlled release of silyl ether prodrugs from PRINT nanoparticles. J Am Chem Soc 134(18):7978–7982
53. Wang W et al (2014) Tailor-made gemcitabine prodrug nanoparticles from well-defined drug-polymer amphiphiles prepared by controlled living radical polymerization for cancer chemotherapy. J Mater Chem B 2(13):1891–1901
54. Gref R et al (2000) Stealth corona-core nanoparticles surface modified by polyethylene glycol (PEG): influences of the corona (PEG chain length and surface density) and of the core composition on phagocytic uptake and plasma protein adsorption. Colloids Surf, B 18(3–4):301–313
55. Perry JL et al. (2012) PEGylated PRINT nanoparticles: the impact of PEG density on protein binding, macrophage association, biodistribution, and pharmacokinetics. Nano Lett 12 (10):5304–5310
56. García I, Marradi M, Penadés S (2010) Glyconanoparticles: multifunctional nanomaterials for biomedical applications. Nanomedicine 5(5):777–792
57. García KP et al (2014) Zwitterionic-coated stealth nanoparticles for biomedical applications: recent advances in countering biomolecular corona formation and uptake by the mononuclear phagocyte system. Small 10(13):2516–2529
58. Pirollo KF, Chang EH (2008) Does a targeting ligand influence nanoparticle tumor localization or uptake? Trends Biotechnol 26(10):552–558
59. Lammers T et al (2012) Drug targeting to tumors: Principles, pitfalls and (pre-) clinical progress. J Controlled Release 161(2):175–187
60. Kamaly N et al (2012) Targeted polymeric therapeutic nanoparticles: design, development and clinical translation. Chem Soc Rev 41(7):2971–3010
61. Hrkach J et al (2012) Preclinical development and clinical translation of a PSMA-targeted docetaxel nanoparticle with a differentiated pharmacological profile. Science Translational Medicine 4(128)
62. Wang J et al (2010) The complex role of multivalency in nanoparticles targeting the transferrin receptor for cancer therapies. J Am Chem Soc 132(32):11306–11313
63. Decuzzi P, Ferrari M (2006) The adhesive strength of non-spherical particles mediated by specific interactions. Biomaterials 27(30):5307–5314
64. Shah S et al (2011) Modeling particle shape-dependent dynamics in nanomedicine. J Nanosci Nanotechnol 11(2):919–928
65. Liu Y, Shah S, Tan J (2012) Computational modeling of nanoparticle targeted drug delivery. Rev Nanosci Nanotechnol 1(1):66–83
66. Kolhar P et al (2013) Using shape effects to target antibody-coated nanoparticles to lung and brain endothelium. Proc Nat Acad Sci USA 110(26):10753–10758

67. Morigi V (2012) Nanotechnology in medicine: from inception to market domination. J drug delivery 2012:389485
68. Combest AJ et al (2012) Genetically engineered cancer models, but not xenografts, faithfully predict anticancer drug exposure in melanoma tumors. Oncologist 17(10):1303–1316
69. Usary J et al (2013) Predicting drug responsiveness in human cancers using genetically engineered mice. Clin Cancer Res 19(17):4889–4899
70. Herschkowitz JI et al (2007) Identification of conserved gene expression features between murine mammary carcinoma models and human breast tumors. Genome Biol 8(5):R76

Cancer Nanotherapeutics in Clinical Trials

Abigail K.R. Lytton-Jean, Kevin J. Kauffman, James C. Kaczmarek and Robert Langer

Abstract

To be legally sold in the United States, all drugs must go through the FDA approval process. This chapter introduces the FDA approval process and describes the clinical trials required for a drug to gain approval. We then look at the different cancer nanotherapeutics and in vivo diagnostics that are currently in clinical trials or have already received approval. These nanotechnologies are catagorized and described based on the delivery vehicle: liposomes, polymer micelles, albumin-bound chemotherapeutics, polymer-bound chemotherapeutics, and inorganic particles.

Dr. Abigail K.R. Lytton-Jean, Kevin J. Kauffman and James C. Kaczmarek contributed equally to this work

A.K.R. Lytton-Jean · K.J. Kauffman · J.C. Kaczmarek · R. Langer (✉)
David H. Koch Institute for Integrative Cancer Research,
Massachusetts Institute of Technology, Cambridge, MA 02139, USA
e-mail: rlanger@mit.edu

R. Langer
Harvard-MIT Division of Health Sciences and Technology, Massachusetts Institute of Technology, Cambridge, MA 02139, USA

R. Langer
Institute for Medical Engineering and Science, Massachusetts Institute of Technology, Cambridge, MA 02139, USA

K.J. Kauffman · J.C. Kaczmarek · R. Langer
Department of Chemical Engineering, Massachusetts Institute of Technology, Cambridge, MA 02139, USA

C.A. Mirkin et al. (eds.), *Nanotechnology-Based Precision Tools for the Detection and Treatment of Cancer*, Cancer Treatment and Research 166,
DOI 10.1007/978-3-319-16555-4_13

Keywords
Cancer therapeutics · Nanotechnology · Clinical trials · Liposomes · Polymer micelles · Albumin-bound · Chemotherapeutics · Polymer-bound chemotherapeutics · Inorganic nanoparticles

Contents

1 Introduction

For researchers in the fast-paced and always-changing field of cancer nanotechnology, it can be difficult to stay up-to-date with the progress of cancer therapies in clinical trials. After all, due to the lengthy FDA-approval process, drugs approved today are often the technology of research started over a decade ago. Furthermore, once drugs have completed preclinical testing—for which peer-reviewed publications in the literature are often prevalent—and entered clinical trials (typically in a private commercial realm), the status of these drugs is less frequently published. Because innovative and promising drugs in the literature can sometimes be "lost" in the poorly-reported and slow-moving clinical trials phase, we seek to review various examples of cancer nanotechnology currently undergoing clinical trials as of July 2014. In this chapter we will explore the development of representative classes of cancer nanotechnology therapies from preclinical to clinical stages to ascertain the state of the field and gain insights into the FDA drug approval process.

Nanotechnology is an attractive strategy for addressing certain challenges in cancer therapy. These challenges include drug stability, solubility, and tumor localization/targeting. Nanoparticles, polymers, and proteins are often used to improve the bioavailability of cancer-fighting drugs through encapsulation or conjugation, typically by improving the often hydrophobic drug's solubility in the aqueous physiological conditions. Also, it is common for nanoparticles to passively accumulate in tumors more prevalently than normal healthy tissues through a mechanism called the "enhanced permeability and retention" (EPR) effect. It is speculated that the larger-than-average fenestrations in endothelial cell layers of tumors caused by erratic angiogenesis facilitates entry of nanoparticles into the cell ("enhanced permeability"), while the lack of proper lymphatic drainage in tumor cells hinders nanoparticle clearance ("enhanced retention") [1]. The EPR effect causes preferential nanoparticle accumulation in tumors, which can lead to improved efficacy of the drug, reduction of systemic toxicity from the drug, and reduction in side effects typically associated with chemotherapy experienced by the patient. The EPR effect is the primary mechanism of action for many of the nanoparticles discussed in this chapter, but there are a few that are actively targeted.

In this chapter, we will begin with a brief description of clinical trials and the FDA approval process. Then, different types of nanotechnology-based cancer therapies or in vivo diagnostics will be introduced. Most of the nanotechnologies in the cancer field lend themselves to characterization by a drug delivery vehicle: liposomes, polymer micelles, albumin-bound chemotherapeutics, polymer-bound chemotherapeutics, or inorganic particles. For this reason, we will describe the fundamental principles of each class and review their clinical development separately. Although we have limited our discussion to these classes for the sake of brevity, they are hardly definitive; we are particularly excited by recent advances in antibody therapy [2] and immunotherapy [3] for cancer treatment, which have been reviewed elsewhere in greater detail. Table 1 provides a list of the clinical trials discussed in this book chapter. Finally, we conclude with a perspective on the field and future work.

2 Intro to Clinical Trials

The FDA as we know it today was born out of a 1937 tragedy in which 107 people died after taking the marketed "elixir" drug sulfanilamide, which prompted the government to enact more strict codes for food and drugs that could be legally sold in the United States [4]. Since then, the FDA approval process has continued to evolve in response to changes in drug technology, particularly in the realm of cancer therapy. Currently, there are three mandatory clinical trial phases through which a drug must go to be marketed in the United States, as well as a Phase 0 and a Phase IV which are sometimes required (Fig. 1). Given the lengthy, somewhat complex nature of clinical trials, as well as the wealth of cancer-treating nanoparticles spanning multiple phases, knowledge of the FDA approval process can provide a helpful context for evaluating the progress of a given treatment.

Before a therapy can reach clinical trials, it must be discovered, created, and tested both in vitro and in vivo. This period of invention and development is referred to as the "preclinical" phase and on average takes 6.5 years [5]. The ultimate goal of the preclinical period is to receive approval for an Investigational New Drug (IND) Application from the FDA. A successful IND lists the drug's manufacturing information, clinical protocols, investigator information, and must

Table 1 List of clinical trials discussed in this book chapter, including the phase, year initiated, cancer indication, and ClinicalTrials.gov identifier

Phase	Year initiated	Description	Cancer indication	ClinicalTrials.gov identifier
0	2014	Silica imaging nanoparticle	Head and neck, melanoma, prostate, cervical, uterine	NCT02106598
0	2014	Magnetic nanoparticle thermoablation	Prostate	NCT02033447
0	2013	Iron oxide imaging nanoparticle	Head and neck	NCT01895829
1	2012	Anti-EGFR immunoliposome	Solid tumors	NCT01702129
1	2009	siRNA-loaded liposome	Solid tumors	NCT00882180
1	2011	siRNA-loaded liposome	Colorectal, pancreatic, gastric, breast, ovarian with hepatic metastases	NCT01437007
1	2009	siRNA-loaded liposome	Advanced solid tumors	NCT00938574
1	2012	siRNA-loaded liposome	Advanced cancers	NCT01591356
1	2014	siRNA-loaded liposome	Solid tumors, multiple myeloma, non-Hodgkins lymphoma	NCT02110563
1	2008	Iron oxide imaging nanoparticle	Brain neoplasms	NCT00769093
1	2014	Paclitaxel-loaded polymer micelle	Breast	NCT02064829
1	2011	Prostate-targeted docetaxel-loaded polymeric nanoparticle	Metastatic, solid tumors	NCT01300533
1	2008	Albumin-bound paclitaxel nanoparticle	Head and neck	NCT00736619
1	2007	Chemotherapeutic-polymer conjugate	Small cell lung, non-small cell lung	NCT00455052
1	2011	Magnetic biopsy needle and nanoparticles	Leukemia	NCT01411904
1	2011	Hafnium nanoparticles for targeted radiation	Adult soft tissue sarcoma	NCT01433068
1	2013	Hafnium nanoparticles for targeted radiation	Head and neck	NCT01946867
2	1999	Liposomal cisplatin	Ovarian	NCT00004083
2	2010	siRNA-loaded liposome	Neuroendocrine, adrenocortical	NCT01262235
2	2013	siRNA-loaded liposome	Pancreatic	NCT01808638
2	2011	Paclitaxel-loaded polymer micelle	Bladder, ureter	NCT01426126
2	2004	Albumin-bound paclitaxel nanoparticle	Breast	NCT00093145
2	2008	Chemotherapeutic-polymer conjugate	Breast	NCT00802945

(continued)

Table 1 (continued)

Phase	Year initiated	Description	Cancer indication	ClinicalTrials.gov identifier
2	2006	Chemotherapeutic-polymer conjugate	Solid tumors	NCT00333502
2	2011	Chemotherapeutic-polymer conjugate	Non-small cell lung	NCT01380769
2	2001	Chemotherapeutic-polymer conjugate	Fallopian tube, ovarian, primary peritoneal cavity	NCT00017017
2	2011	Chemotherapeutic-polymer conjugate	Glioblastoma multiforme	NCT01402063
2	2013	Prostate-targeted docetaxel-loaded polymeric nanoparticle	Prostate	NCT01812746
3	2012	Paclitaxel-loaded polymer micelle	Breast	NCT01644890
3	2008	Albumin-bound paclitaxel nanoparticle	Breast	NCT00785291
3	2007	Chemotherapeutic-polymer conjugate	Non-small cell lung	NCT00576225
3	2011	Chemotherapeutic-polymer conjugate	Breast	NCT01492101
4	2008	Liposomal paclitaxel	Solid tumors	NCT00606515
4	2009	Iron oxide imaging nanoparticle	Pancreatic	NCT00920023

also show a wealth of animal data suggesting positive pharmacological effects and tolerable toxicity [4]. In short, the transition from preclinical work to clinical trials requires a novel therapeutic that shows sufficient promise to be effective and safe in treating human disease. However, due to the diversity of nano-platforms, as well as the low number of nanoparticle therapeutics that have made it all the way to clinical approval, determining what constitutes a "promising" nanoparticle drug can be difficult; to this end, the National Cancer Institute developed the Nanotechnology Characterization Laboratory (NCL) in 2004 to address the standards for preclinical nanoparticle data [5, 6]. Researchers may apply via the NCL website (http://ncl.cancer.gov) to have their preclinical nanomaterials and devices extensively characterized in an "assay cascade," that is, a battery of tests to determine physical properties, toxicology, pharmacology, efficacy, and many other parameters. The NCL has the infrastructure to perform this crucial preclinical testing on a large scale, allowing researchers to characterize their nanotechnology far more rigorously than they could alone. Moreover, this process can provide an accurate prediction of whether or not a technology will likely face significant efficacy or safety issues when it reaches clinical trials.

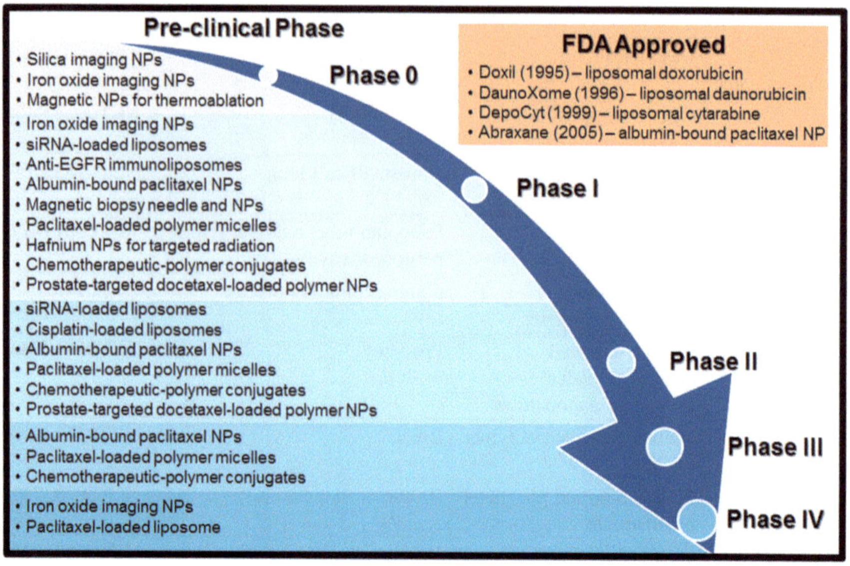

Fig. 1 A schematic of the FDA approval process with a list of clinical trials discussed in this book chapter

While many therapeutics will move directly from a successful IND to Phase I trials, some will first go through Phase 0 trials. Phase 0 trials, also developed by the National Cancer Institute, generally involve microdosing (dosing at sub therapeutic levels) of the drug in order to obtain relevant pharmacokinetic data; this process was developed largely as a means of streamlining the FDA approval process by serving as an early benchmark of the translatability of an IND [7]. As with the NCL, Phase 0 trials provide more clarity in the matter of which therapeutics are ready to take the leap from preclinical studies to the arduous FDA clinical trials.

The bulk of the FDA approval process, both in terms of time and capital, lies in Phases I–III. Phase I evaluates the safety of the new drug as its primary endpoint. Generally, 20–100 volunteers are enrolled, and the drug dose administered is gradually increased in order to observe patient response. These volunteers are usually healthy, though exceptions can be made for especially life-threatening illnesses [4]. Phase I generally lasts anywhere from 6 to 18 months, and about 2/3 of all INDs will progress to Phase II [5]. Phase II trials look at the efficacy of the drug. 100–300 patients are typically enrolled and the effective dose method of delivery, optimal dosing interval, and further safety concerns are all evaluated [4]. Phase II can last anywhere from 6 months to 2 years, and is typically where the majority of drugs fail, given that it is the first stage to really evaluate the effectiveness of the new drug in humans [5]. Indeed, since 2003 only about 3 in 10 drugs have made it through Phase II testing [8]. If a drug is deemed effective and suitably safe after Phase II, it then moves on to Phase III. Phase III is the largest (and most expensive) of all the phases, with 1000–5000 patients from all over the United

States often enrolled in randomized, placebo-controlled experiments [5]. This phase can take up to 10 years to complete, and roughly 10 % of the drugs that make it through Phase II will still fail in Phase III [4]. While less common than failure in Phases I and II, failure in Phase III can be caused due to safety or efficacy shortcomings that only manifest themselves in a large-scale pool of patients, or could simply be a result of the sponsoring institution running out of capital to fund the trial [9]. Success in Phase III depends largely on a drug's efficacy or safety relative to a "gold standard" (i.e., it has to have some advantage over a market treatment currently used) [5].

If a treatment makes it through Phase III testing, the sponsor can file for a New Drug Application. The FDA will take 6–12 months to review this application; at the end of the review process, they can approve the drug if the Phase III data suggests increased efficacy relative to the gold standard, reject the drug if they believe it to have insufficient safety or to not represent a significant improvement over current treatments, or ask for more data on the drug [4]. An approved NDA means that the new drug can be marketed in the US, after which many drugs will undergo Phase IV trials. Phase IV studies take numerous forms, such as gathering data from consumers and double-blind tests in hospitals that are more similar to Phases II and III testing; these trials are largely concerned with drug safety, especially in populations such as children, which are rarely a part of premarket studies [10]. On the whole, it is estimated that the approval rate for all new oncology drugs entering clinical trials is somewhere around 5 % [7], with oncology drugs having a historically lower than average rate of making it through Phase III [8]. Moreover, an average drug requires somewhere around $1 billion total cost to make it through clinical trials [11].

3 Different Treatments Used with Nanotechnology Cancer Therapeutics

The nanoparticle cancer therapies currently undergoing clinical trials primarily fall into the following three classes of therapeutic mechanisms: chemotherapeutics, short interfering RNA, and imaging agents.

3.1 Chemotherapeutics

While "chemotherapy" is sometimes used as a blanket term to describe any pharmaceutical intervention for cancer, chemotherapeutics are most often considered to be small-molecule drugs that act as purposefully cytotoxic agents. Chemotherapeutics are chosen or designed such that they primarily affect dividing cells, making them aggressive in suppression of the unrestrained cell growth characteristic of cancer [12]. The first chemotherapeutics discovered and employed in human studies were alkylating agents derived from mustard gas used in World War I. Alkylating agents are able to form covalent bonds with DNA, thereby disrupting the mitotic cycle [13]. Since then, many other chemotherapies with alternate mechanisms of action have been developed. Anti-microtubule agents that disrupt the necessary organization of microtubules during cell mitosis, ultimately leading to a disruption in cell division, have been effective at retarding cancer growth. For instance, paclitaxel, which belongs to a class of anti-microtubule agents known as taxanes, inhibits cell growth by stabilizing microtubules, which prevents them from performing the reorganizations necessary for mitosis [14]. Other examples of anti-microtubule agents include dolostatins and rhizoxin [15]. Additionally, cytotoxic antibiotics have been developed as chemotherapeutic agents; while such drugs certainly possess antibacterial properties, they prevent cancer growth by inhibiting cell growth through a variety of mechanisms. Doxorubicin, one of the most commonly used antibacterial chemotherapeutics, binds to several DNA-associated enzymes, thereby damaging DNA and disrupting its replication [16].

Although chemotherapeutics act by impeding cellular division, this effect is not specific to cancer cells. For this reason, chemotherapeutics are associated with many well-known adverse side effects resulting from systemic toxicity. Cancer nanotechnology, therefore, seeks to take advantage of the potent cytotoxicity of chemotherapeutics while limiting the off-target effects that compromise patient well-being. Strategies for doing so include encapsulation of chemotherapeutics in order to improve biodistribution, drug stability, and retention within tumors (e.g., through the EPR effect); complexation with nanoparticles to improve solubility; formulation into nanoparticles with targeting ligands that specifically bind to tumors; and others which are discussed throughout this chapter.

3.2 Short Interfering RNA

Following fundamental studies of RNA interference (RNAi) in *C. elegans* [17], as well as the studies of the underlying mechanism of RNAi [18], short interfering RNA platforms have been studied intensively as therapeutic agents. Short interfering RNAs, also known as siRNAs, are 21–23 nucleotide double-stranded RNAs that can be synthetically designed to target specific cellular mRNA (Fig. 2). Once in the cytoplasm, these siRNAs are incorporated into an enzyme complex called RISC (RNAi silencing complex), which targets and degrades mRNA complementary to the antisense strand of the siRNA [19]. Properly designed siRNA therapies result in decreased protein expression caused by the mRNA degradation and lead to knockdown of a specific phenotype.

Because mRNA degradation is sequence-specific, siRNA has the potential to silence a wide range of target proteins, many of which are "undrugable" and cannot be targeted by small molecule chemotherapeutic agents. Moreover, siRNA is modular: the same base siRNA delivery technology can be applied to different protein targets by changing only the sequence of the siRNA strand. This is especially important for a complex disease like cancer, which can dynamically

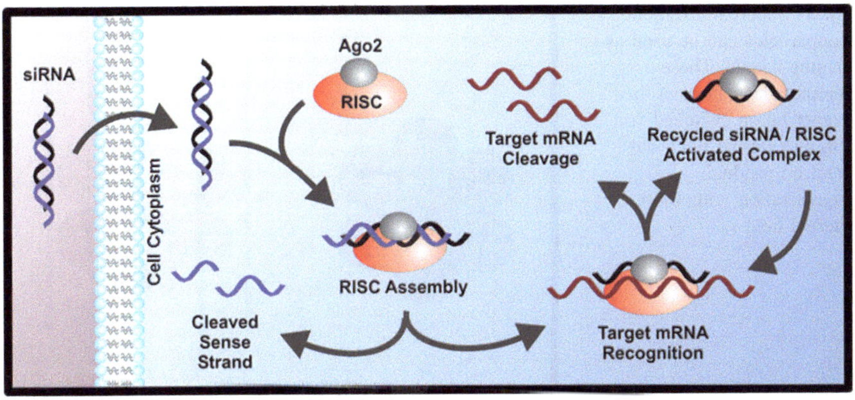

Fig. 2 siRNAs can be used to silence expression of cancer-related proteins through the RNA interference mechanism

upregulate multiple oncogenic proteins at once (providing multiple target proteins) and has been known to acquire resistance to some conventional chemotherapeutics [20].

The main hurdle in translation of siRNA lies in the mechanism of delivery; naked siRNA delivered systemically is quickly degraded by various nucleases and/ or quickly cleared through the kidneys, making its biodistribution unfavorable and its bioavailability low. Additionally, siRNA is too large and negatively charged to cross cell membranes on its own [21]. To this end, numerous nanoparticle delivery systems that aid in siRNA delivery for treatment of a variety of diseases have been and continue to be studied and developed [22]. Similar to nanoparticles for chemotherapeutics, such particles are intended to enhance siRNA stability, aid in cellular entry, and improve biodistribution. Indeed, siRNA nanoparticle delivery for cancer treatment has been heavily researched, with a few promising technologies having reached clinical trials [23].

3.3 Imaging Agents

In addition to cancer treatments, nanoparticles are also advantageous in diagnosis and imaging of the disease. Given their unique properties, magnetic nanoparticles have the potential to play an important role in the realm of cancer imaging. Superparamagnetic nanoparticles are smaller than the size of a single magnetic domain and, without an applied field, retain no residual magnetization [24]. Thus, these superparamagnetic particles can have their magnetization switched entirely "on" or "off" at will (Fig. 3). When used with magnetic resonance imaging, these nanoparticles confer greater contrast, signal strength, and less aggregation than other imaging agents [25]. As such, several clinical trials utilize nanoparticles as

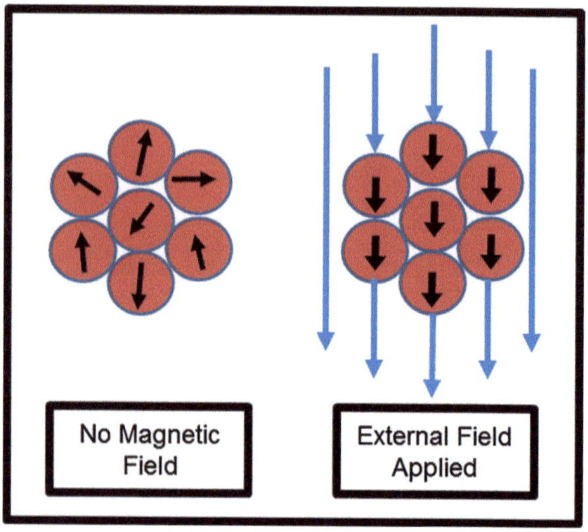

Fig. 3 Superparamagnetic nanoparticles can be used as imaging agents. These nanoparticles can have their magnetization switched "on" with an external field, but retain no residual magnetization without an external field

MRI contrast agents. Nanoparticles designed to accumulate in tumors can also serve as optical probes, with fluorescence in the near infrared range used as an alternative to MRI [26].

4 Current Clinical Trials

There are currently many FDA clinical trials involving nanoparticle therapies for cancer, along with a few examples of nanoparticles that have gained FDA approval. The different nanotechnologies that have successfully translated from the lab to clinical trials will be discussed in the following categories: liposomes, polymeric nanoparticles, protein-bound chemotherapeutics, polymer-bound chemotherapeutics, and inorganic nanoparticles.

4.1 Liposomes

Liposomes are spherical nanoparticles made from amphipathic lipids, a class of molecules which have a polar head group and a nonpolar tail group. In aqueous conditions, these lipids naturally self-assemble into a lipid bilayer which forms the outer layer of the liposome (Fig. 4). Drugs can be either entrapped inside the aqueous core or in the hydrophobic membrane, depending on whether the drug is hydrophilic or hydrophobic. Because the encapsulation of the drug in the liposome can reduce drug degradation, limit potential off-target toxicity of the drug, and increase the concentration of the drug at tumors due to the EPR effect, conventionally toxic chemotherapeutics are a natural choice for liposomal formulations. In fact, one of the earliest studies to show the utility of liposomal drugs in vivo used a

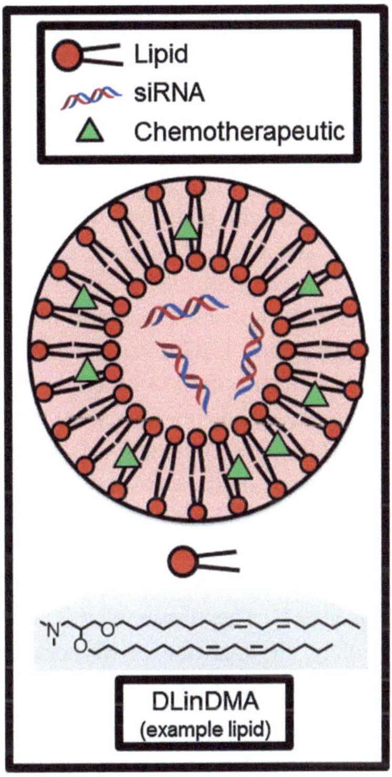

Fig. 4 Liposomes can encapsulate and deliver hydrophobic small molecule chemotherapeutics and/or siRNAs. DLinDMA is an example of an amphiphilic lipid molecule which comprises the bilayer membrane of the nanoparticle

chemotherapeutic: in 1977, Kobayashi et al. demonstrated significantly enhanced survival of leukemic mice with liposomal ara-C (DNA synthesis inhibitor) compared to ara-C alone [27]. In the decades that followed, liposome technology progressed further with advances such as targeting ligands and triggered release, and several liposomal formulations have made it to the market with hundreds of millions of dollars of sales per year [28].

Arguably the most influential liposome technology developed early on was Doxil, a liposomal formulation of the chemotherapeutic doxorubicin. The importance of Doxil to the field of cancer nanotechnology cannot be overstated; as the first FDA-approved nanoparticle, it paved the way for all future generations of liposomal and polymeric particles. For an excellent, firsthand account of the development of Doxil from concept to the clinic, we refer the reader to the review of Barenholz [29]. Briefly, the preclinical development of Doxil was catalyzed by several discoveries, including that (1) PEGylation of the nanoparticle surface increased circulation time and reduced non-specific uptake by macrophages [30] and (2) the use of an ammonium sulfate gradient during drug loading leads to crystallization of the drug and a marked increase in encapsulation efficiency [31]. Pharmacokinetic studies in rats and dogs were conducted [32], and a pilot clinical trial in Israel demonstrated that Doxil localized to the tumor and was overall well-tolerated in humans [33]. Doxil received FDA approval in 1995 for AIDS-related

Kaposi's sarcoma, in 1999 for ovarian cancer, and in 2003 for breast cancer [28]. Doxil is also prolific in clinical trials for other indications or in combination with other therapies, which are too numerous to go into detail here: a search of ClinicalTrials.gov for "Doxil" finds over 1500 registered ongoing or completed clinical trials. Suffice it to say that Doxil's impact on the field of cancer technology over 20 years after its FDA approval is still enduring today.

After Doxil, many other chemotherapeutics have been incorporated into liposomes and have undergone clinical trials, such as liposomal paclitaxel [34] and liposomal cisplatin [35]; several have also received FDA approval, including DaunoXome (liposomal daunorubicin) and DepoCyt (liposomal cytarabine). Additionally, a non-PEGylated version of liposomal doxorubicin called Myocet has been developed to combat one of the side effects of Doxil, a condition called palmar-plantar erythrodysaesthesia or "hand-foot syndrome" which is characterized by a dermatologic toxic reaction in the hands and feet [36]. There are currently various Myocet formulations in all stages of clinical trials; the most clinically advanced is a Phase III combination therapy for HER2-positive metastatic breast cancer with Mycoet, paclitaxel, and Trastuzumab (Herceptin), an antibody which has been independently shown to improve survival in women with HER2-positive breast cancer [37]. The results of this Phase III trial were published in 2014, and unfortunately, the addition of Myocet was found to not significantly improve the progression-free survival over the standard first-line therapy of paclitaxel and Trastuzumab alone [38]. However, the authors note that Myocet may have benefit and warrants further study in patients whose tumors display a particular subset of biomarkers [38]; because of the great heterogeneity of cancer, it is not uncommon for clinical trials to discover that certain classes of patients respond better to treatment than others.

As an example of a promising new liposome technology, liposomes have recently been developed which are actively targeted to the tumor instead of merely passively accumulating at the site due to the EPR effect. The surface of these liposomes contains antibodies against epidermal growth factor receptor (EGFR), a protein that is overexpressed on the surface of many types of cancer cells [39]. These so-called immunoliposomes delivered cytotoxic drugs more efficiently to EGFR-expressing cells in vitro than liposomes without antibody [40], and this effect was confirmed in vivo with mice [41]. The antibody used on these immunoliposomes was also capable of binding to cells expressing EGFRvIII, a deletion mutant of EGFR often associated with breast carcinoma, non-small cell lung carcinoma, and high-grade glioma. In 2012, the anti-EGFR immunoliposomes became the first targeted nanoparticle delivering chemotherapeutics to undergo a Phase I clinical trial [42]. The doxorubicin-loaded immunoliposomes were administered to patients with EGFR-overexpressing tumors (including pancreatic, head/neck, colorectal, urothelial, and several other cancer indications) in escalating doses, and the recommended dose was determined to be 50 mg/m^2 for future Phase II trials; notably, no patients developed palmar-plantar erythrodysaesthesia or cardiotoxicity, which can be associated with Doxil [36, 43].

Liposomes can also be used as a drug delivery vehicle for siRNA therapeutics. Although the most clinically advanced siRNA liposome formulations treat heredity liver disorders (such as Alnylam Pharmaceutical's patisiran, currently enrolling in a Phase III clinical trial [44]), several siRNA liposome formulations for cancer therapy have recently began or completed Phase I clinical trials. The siRNA entrapped in the aqueous core of these liposomes are designed to silence the expression of genes associated with cancer, including oncogenes (e.g. *EphA2*) [45] and angiogenesis-promoting genes (e.g., *PKN3*) [46]. In these lipid nanoparticle formulations, cationic or ionizable lipids are often used because their positive charge (1) improves the entrapment efficiency of the negatively charged siRNA, (2) increases cellular uptake, and (3) facilitates escape of the siRNA from the endosome [22]. The chemical structure of these lipids, which are often composed of an amine-containing polar headgroup and nonpolar hydrocarbon tails, is strongly associated with the potency of the resulting siRNA liposome formulation. Accordingly, thousands of lipid and lipid-like molecules have been synthesized over the past decade using both rational design approaches [47–49] and the synthesis of large combinatorial libraries [50–52] with only a select few of these lipids having sufficient potency and non-toxicity to reach clinical trials. For this reason, liposomes used to encapsulate siRNA are often called "lipid nanoparticles" (LNPs) to highlight the importance of the lipid itself.

Reported in the mid-2000s, one of the first lipids to be developed for use in siRNA-LNP formulations was DLinDMA (Fig. 3) [47]. DLinDMA and siRNA can be formulated into nanoparticles along with phospholipid (to increase efficacy), cholesterol (to increase stability), and lipid-anchored PEG (to increase LNP circulation time, reduce aggregation, and reduce nonspecific uptake by macrophages) [22]. Because it contains these additional excipients, the nanoparticle is classified as a SNALP (stable nucleic acid lipid nanoparticle); SNALPs have been shown to typically distribute to the liver in vivo and increase efficacy to hepatocytes over simple lipid-siRNA lipoplexes [53]. In 2006, DLinDMA SNALPs were used to demonstrate the first RNAi-mediated gene silencing in non-human primates [54], and this technology was then translated to the clinic by Alnlyam Pharmaceuticals, which in 2011 completed a Phase I dose-escalation clinical trial [55]. The DLinDMA SNALP was called ALN-VSP02 and incorporated siRNAs targeting both vascular endothelial growth factor (VEGF, promotes angiogensis) and kinesin spindle protein (KSP, promotes cellular proliferation) for the treatment of solid tumors. Published results from this Phase I clinical trial indicate that intravenously-administered ALN-VSP02 was generally well-tolerated at doses ranging from 0.1 to 1.5 mg/kg with no clinically significant liver toxicity in 40 of 41 patients; additionally, one patient with nodal and extensive liver metastases from endometrial cancer achieved a complete response to the therapy and three patients (with both hepatic and extrahepatic metastases) achieved stable disease for 1–1.5 years [56].

Over the past decade, the chemical structures of the lipids have been optimized by different researchers to reduce toxicity, change the biodistribution, and reduce the IC_{50} of the siRNA-LNPs. (A lower IC_{50} is desirable because it reduces the amount of lipid required and thus may alleviate concerns of any potential long-term toxicity of the lipid.) A lipid called DLin-MC3-DMA from Tekmira was incorporated into a formulation called TKM-PLK1 designed to treat primary or secondary liver cancer, and it has recently completed Phase I clinical trials [48, 57]. The siRNA encapsulated in TKM-PLK1 targets polo-like kinase 1 (PLK1), a protein frequently overexpressed in tumors and associated with carcinogenesis [58]. A cationic lipid called AtuFECT01 from Silence Therapeutics which targets the endothelium [59] was incorporated into a formulation called Atu027. Atu027 contained siRNA targeting protein kinase 3 (PKN3, promotes angiogenesis) which retarded tumor growth in preclinical mouse studies [46], was well-tolerated in Phase I trials for patients with advanced solid tumors with some anti-metastatic activity observed [60, 61], and is currently enrolling in a Phase1b/IIa study in patients with pancreatic adenocarcinoma [62].

TKM-PLK1, ALN-VSP02, and Atu027 have completed Phase I clinical trials, but there are additional siRNA-LNP formulations for cancer therapy on the horizon. A Phase I clinical trial has been registered which uses the neutrally-charged lipid 1,2-dioleoyl-*sn*-3-phosphatidylcholine (DOPC) to encapsulate siRNA targeting epithelial cell kinase (EphA2), an oncoprotein overexpressed in many types of human tumors and especially ovarian tumors [45, 63]. Moreover, Dicerna Pharmaceuticals is currently enrolling patients for a Phase I clinical trial for the treatment of solid tumors, multiple myeloma, and non-Hodgkins lymphoma with their LNP formulation called DCR-MYC [64]. DCR-MYC is made from a proprietary lipid called EnCore and contains 27-nucleotide siRNA targeting the oncogene *myc*, which is frequently overexpressed in cancers [65–67].

4.2 Polymeric Nanoparticles

In contrast with LNPs, polymers can also be used to form spherical polymeric micelles to encapsulate and deliver chemotherapeutics. These nanoparticles are distinguished from polymer-bound chemotherapeutic conjugates (discussed in a later section) because the drug here is encapsulated in the particle rather than covalently bonded to the polymer, although it should be noted that some conjugates will spontaneously form nanoparticle-like shapes in aqueous conditions [68]. Polymer micelles are made when an amphiphilic block copolymer self-assembles in an aqueous solution, trapping the hydrophobic drug in the hydrophobic inner micelle core (Fig. 5).

The formulation called Genexol PM (also called cynviloq or IG-001) from Sorrento Therapeutics uses a micelle made from a PEG (hydrophilic) and poly(D,L-lactic acid) (hydrophobic) block copolymer to encapsulate paclitaxel and release the drug in a controlled fashion [69]. Genexol PM is approved for clinical use in South Korea to treat metastatic breast cancer and non-small cell lung cancer (NSCLC), is undergoing Phase II clinical trials in the United States for various other indications like urothelial cancer [70], and is currently underway in a bioequivalency crossover clinical trial with albumin-bound paclitaxel as the control alternative therapy [71].

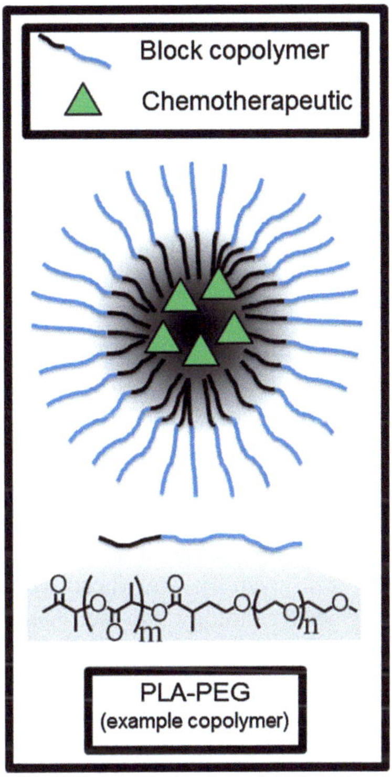

Fig. 5 Polymer micelles can encapsulate and deliver hydrophobic small molecule chemotherapeutics in the hydrophobic core. PLA-PEG is an example of an amphiphilic block copolymer which self-assembles into the micelle in aqueous conditions

According to a Sorrento press release [72], Genexol PM may be approved by the FDA under the 505(b) [46] pathway for a new drug application if it demonstrates equivalent pharmacokinetics to albumin-bound paxlitaxel. Several other polymeric micelle-based nanoparticles are in Phase III clinical trials. For example, NK105, made from a PEG/modified polyaspartate block copolymer and encapsulating paclitaxel [73], is now in a Phase III clinical trial for the treatment of breast cancer [74]. NK105 was also well-tolerated with 14/56 patients having a clinical response in a Phase II trial for gastric cancer [75].

Targeting moieties are also being introduced to the polymeric nanoparticle surface to deliver the particle to tumor cells with better specificity. One example of this concept is Bind Therapeutic's BIND-014, a nanoparticle made from a PEG/poly(lactic-co-glycoclic acid) block copolymer encapsulating docetaxel. The surface of BIND-014 is decorated with a small molecule called ACUPA which has been shown to have specificity for prostate specific membrane antigen (PSMA), a receptor overexpressed on prostate cancer cells and solid tumor vasculature [76]. In preclinical development, BIND-014 was selected and optimized from a large combinatorial library of nanoparticles with varying sizes, polymer composition, ACUPA targeting molecule density, and other parameters [77]. It was reported that 2/12 patients had tumor shrinkage in a Phase I dose escalation clinical trial [77, 78], and BIND-014 is currently being studied in two Phase II clinical trials for prostate cancer [79] and NSCLC [80].

Polymer nanoparticles have predominately been used to deliver small molecule chemotherapeutics but can also be used to deliver siRNA for cancer treatment. First described in 1999 by Davis and coworkers, β-cyclodextrin is a sugar-derived, linear, cationic polymer which can form nanoparticle complexes with negatively charged siRNA [81]. The nanoparticle formulation called CALAA-01 is made of β-cyclodextrin, PEG, and siRNA targeting ribonucleotide reductase subunit M2 (RRM2, an enzyme involved in DNA replication); furthermore, the surface of CALAA-01 nanoparticles is decorated with the targeting moiety transferrin, a protein which binds to transferrin receptors overexpressed on cancer cells [82]. In 2008, a Phase I clinical trial for CALAA-01 began, the first clinical trial for targeted siRNA nanoparticles [83]. An interim publication released in 2010 investigated the tumor biopsies of some of the patients and discovered the first evidence of siRNA-mediated mRNA degradation in humans [84], an important proof-of-concept for the mechanism of RNAi and a milestone in the gene delivery field. The Phase I trial was terminated in 2013, and to our knowledge no explanation for this termination has been given by Arrowhead Research Corporation. However, the β-cyclodextrin polymer technology has seen better success in polymer-bound chemotherapeutics as discussed in a subsequent section.

4.3 Albumin-Bound Chemotherapeutics

By numbers alone, the most common types of cancer-treating nanoparticle currently undergoing clinical trials are albumin-bound chemotherapeutics. Specifically, these platforms utilize albumin to form a nanoparticle carrier for chemotherapeutic drugs. Because albumin is a natural carrier of hydrophobic materials, it can be used to overcome solvation issues with highly hydrophobic drugs that would otherwise need to be solubilized and delivered using harmful solvents. Additionally, albumin is known to interact with the cellular GP60 receptor, which improves the trafficking of the bound drug into the cell [85]. These nanoparticle albumin bound (*nab*) platforms therefore overcome the dose restrictions on potent chemotherapeutic agents (such as taxanes) dissolved in cytotoxic solvents.

The first albumin-bound chemotherapeutic to be approved by the FDA was Abraxane. Abraxane (delivered intravenously) is a nanoparticle composed of the taxane chemotherapeutic paclitaxel bound to albumin (generally referred to as a *nab*-paclitaxel platform), and is prepared by high-pressure homogenization of paclitaxel in the presence of human serum albumin [86]. During formulation, each albumin molecule noncovalently binds several hydrophobic paclitaxel molecules, and these individual albumin/paclitaxel particles combine and stabilize into nanoparticles of ~130 nm in diameter (Fig. 6) [87]. Up until the development of Abraxane, paclitaxel was only available as Taxol; due to paclitaxel's poor water solubility, the solvent used in Taxol was Cremophor, a mixture of castor oil and

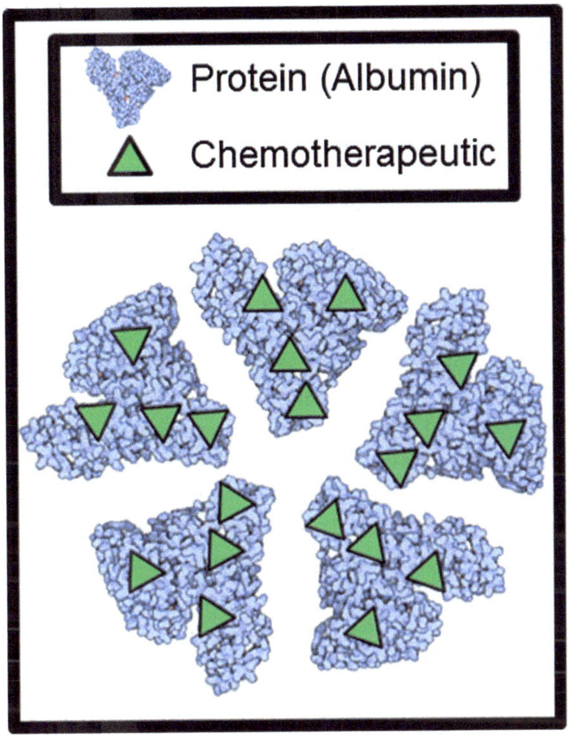

Fig. 6 Proteins, typically albumin, can solvate hydrophobic small molecule chemotherapeutics. These proteins associate into nanoparticles, which can deliver the encapsulated drug to cells. (Albumin protein cartoon from protein data bank at http://www.rcsb.org/)

ethylene oxide. Despite paclitaxel's antitumor activity [14], the Cremophor solvent is associated with a host of adverse effects, including neurotoxicity and clinical acute hypersensitivity reaction [88]. The dose-limiting toxicity of the Cremophor solvent motivated the search for alternative means of paclitaxel delivery, and ultimately resulted in the development of Abraxane. In 2000, Abraxis Bioscience (the company that developed Abraxane) presented preclinical data which showed that paclitaxel bound to albumin nanoparticles had equivalent efficacy to Taxol in nude mice, and significantly reduced toxicity in rats [89]. In clinical trials, Abraxane was shown to have a tolerated dose 50 % larger than that of Taxol, which likely contributed to its improvement in patient tumor response over Taxol [90]. These clinical successes led to the drug's approval in January of 2005 for breast cancer in patients that failed combination therapy [85]. Since then, it has been approved for NSCLC and pancreatic cancer in 2012 and 2013, respectively [91].

In the wake of Abraxane's success, there are currently several Phase I–III clinical studies investigating *nab*-Paclitaxel efficacy in treating different cancers (or cancer stages) and as part of various combination therapies. As an example, there is currently a Phase I trial testing the effectiveness of *nab*-paclitaxel with the antibody cetuximab and intensity-modulated radiation therapy for patients with head and neck squamous cell carcinoma [92]. To date, results have been published demonstrating that the *nab*-paclitaxel platform as a part of this particular combination therapy has a mean tolerated dose up to 2.5x higher than that of traditional paclitaxel treatments [93]. Likewise in a Phase II trial, *nab*-paclitaxel is being tested alongside carboplatin (an alkylating chemotherapeutic) and trastuzumab (an antibody) to see if the combination is effective in treating metastatic breast cancer [94]. Once more, results showed that use of *nab*-paclitaxel was potent in its anticancer properties with less associated toxicity than Cremophor-based paclitaxel, with no patients experiencing high grades of neuropathy [95]. A Phase III clinical trial testing a combination of the antibody bevacizumab along with *nab*-paclitaxel for stage IIIC and IV breast cancer is currently ongoing [96].

While there are several similar trials that use the same *nab*-paclitaxel platform, there is also a clinical trial investigating a *nab*-rapamycin drug. Rapamycin is an antibiotic which inhibits signals from mTOR, an important intracellular kinase involved in cell proliferation and survival [97]. Like paclitaxel, rapamycin suffered from poor solubility, so it was formulated with albumin in hopes of improving its efficacy [85]. Indeed, early work by Abraxis BioScience demonstrated that the *nab*-rapamycin platform as a single agent could greatly inhibit tumor growth in a murine breast tumor xenograph model, suggesting it held similar promise to the improvements shown using Abraxane [98]. This drug (ABI-009) is currently in Phase I/II clinical trials [99].

4.4 Polymer-Bound Chemotherapeutics

Similar to albumin, polymers can also be conjugated to chemotherapeutics to increase the drug's solubility, bioavailability, and tolerability (Fig. 7). Whereas albumin binds to the chemotherapeutic through hydrophobic interactions, the polymers must be covalently conjugated to the drug through an enzymatically-cleavable bond so that the drug can be released; furthermore, polymers used in this application are typically linear and must be water-soluble and biocompatible [100]. One of the first polymers used for this application was *N*-(2-Hydroxypropyl) methacrylamide (HPMA), a hydrophilic and non-immunogenic polymer first developed in the 1970s by Jindřich Kopeček and coworkers [101]. In 1999, the results of the first clinical trial of HPMA-conjugated doxorubicin (PK1) were published [102], but further progress in clinical trials has been slow over the past two decades with several HPMA conjugates failing Phase I trials and only a handful of others reaching Phase II trials [103].

One disadvantage of HPMA is that it is non-biodegradable and therefore may result in long-term toxicity accumulation. Accordingly, the biodegradable polymer poly-L-glutamic acid was conjugated to paclitaxel to make the drug called CT-2103 (paclitaxel poliglumex, also called Opaxio) from Cell Therapeutics, which had promising preclinical anti-cancer activity and was well tolerated in Phase I clinical trials [104]. In a Phase III trial, results indicated that CT-2013 did not extend survival in patients with NSCLC compared to free chemotherapeutic; however, it was observed that there was a strong but not significant increase in survival for

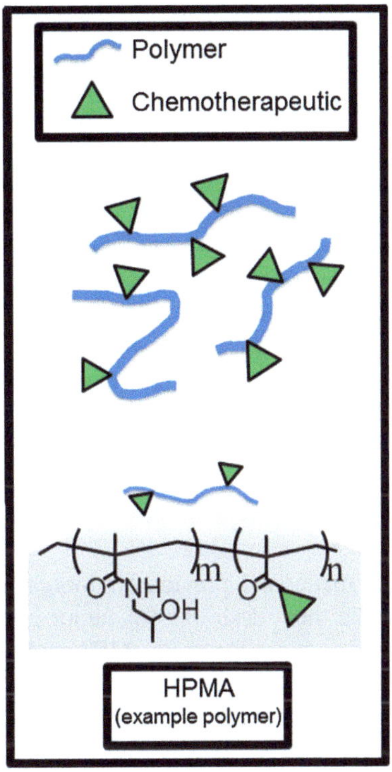

Fig. 7 Polymers can be chemically ligated to chemotherapeutics to increase their solubility. HPMA is an example of a copolymer that has traditionally been used in this application

women over men [105]. This observation, coupled with preclinical data indicating that CT-2103 efficacy is enhanced by estradiol (female sex hormone), encouraged two separate follow-up clinical trials: first, a Phase III clinical trial [106] is currently underway studying the efficacy of CT-2103 in women with NSCLC who have baseline estradiol concentrations above 25 pg/mL; second, a Phase II clinical trial [107] was conducted in men with prostate cancer who were co-treated with estradiol and CT-2013, but this trial was terminated. Despite its prevalence in clinical trials (including a Phase III trial for ovarian cancer [108], a Phase II trial for glioblastoma [109], and many others), after nearly 20 years in the clinic it has not yet been published that CT-2013 has increased survival over free drug for any indication, highlighting just how difficult it is for drugs to surmount Phase III trials.

Many other strategies for polymer-bound chemotherapeutics have been employed. First reported in 2004, XMT-1001 is camptothecin conjugated to a biodegradable, hydrophilic polyacetal polymer [110] and was shown to have efficacy in mice bearing HT-29 human colon carcinoma xenografts [111]. XMT-1001 is currently in Phase I clinical trials for the treatment of lung cancers [112], and preliminary results indicated that the drug was well tolerated with prolonged stable disease in 9/54 patients [113]. Another example is CRLX-101, which is camptothecin conjugated to β-cyclodextrin [68]. CRLX-101 had efficacy in mice bearing xenograft tumors for six different cancer types [114], and in 2013 completed its first-in-human Phase I/IIa clinical trial in advanced solid tumors [115]. Stable disease was achieved in 28/44 patients treated at the maximum tolerable dose and 16/22 of the patients with NSCLC [116]. A Phase II study with CRLX-101 for NSCLC is currently underway [117].

As an example of another polymer-bound chemotherapeutic which has progressed further in the clinic, NKTR-102 (etirinotecan pegol) from Nektar Therapeutics is a PEGylated topoisomerase I-inhibitor. NKTR-102 is most clinically advanced for the treatment of breast cancer, but it is also being studied in the treatment of ovarian, colorectal, lung, and brain cancers. In a Phase II trial [118] to determine the safety and efficacy of NKTR-102 in incurable metastatic breast cancers, the drug was generally well tolerated and 20/70 patients had complete or partial response to the treatment [119]. NKTR-102 is currently being studied in a Phase III trial [120] with over 800 patients which is anticipated to have results in early 2015.

4.5 Inorganic Nanoparticles

As mentioned previously, inorganic magnetic nanoparticles have properties that make them desirable for tumor imaging applications. The potential of iron oxide nanoparticles for use as MRI contrast agents was realized in the 1990s, and studies began in the hopes of moving to clinical trials [121]. Ferumoxtran-10, a dextran-coated iron oxide nanoparticle, was the first generation of metallic nanoparticles to seek FDA approval as an MRI contrast agent. The iron oxide core of the particle provided the image contrast, while the dextran coating minimized reaction with serum proteins and opsonization (Fig. 8) [121]. Early experiments showed that the

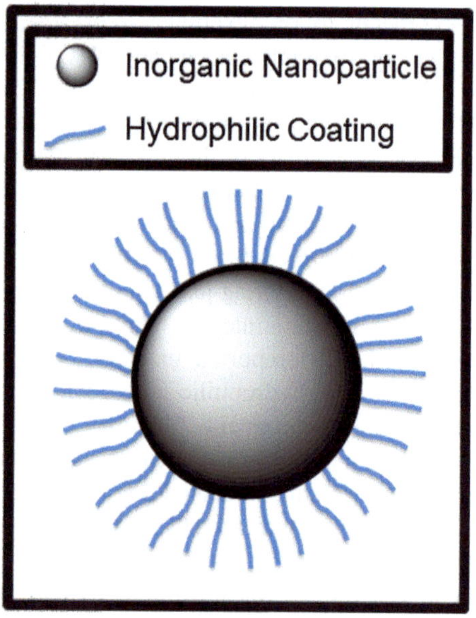

Fig. 8 Inorganic nanoparticles, comprising an inorganic core with hydrophilic polymer coatings, have both imaging and therapeutic applications

nanoparticles were taken up mainly by macrophages after circulation, which are much more present in non-cancerous tissue than in tumors [122]. Reticuloendothelial uptake reduces the postcontrast magnetic resonance of the particles and it was therefore hypothesized that a strong magnetic signal should remain in tumors during post-imaging. Thus, the nanoparticles, known as Combidex in the United States, began clinical trials for use as a lymph node imaging tool [123]. However, this trial was terminated early, and the development of Combidex in the US was discontinued. It is possible that this discontinuation was due to the relatively high rate of false positives involved with the use of ferumoxtran-10 [124].

Since the discontinuation of ferumoxtran-10, a second-generation iron oxide nanoparticle, ferumoxytol (also known as Feraheme), has begun clinical evaluations. Ferumoxytol is an iron oxide nanoparticle coated with a semisynthetic carbohydrate instead of dextran. This novel hydrophilic coating was designed to reduce the immune response associated with dextran-coated iron oxide particles and to also reduce the free iron concentration in vivo, thus decreasing toxicity [125]. Ferumoxytol carries the extra advantage of being previously clinically approved to treat iron deficiency resulting from chronic kidney disease in adults. Thus, in addition to imaging advantages of magnetic nanoparticles already discussed, ferumoxytol has been shown safe for chronic kidney disease and angiographic imaging, which confers an advantage over the commonly used Gadolinium-based MRI contrast agents that can cause nephrogenic systemic fibrosis in patients [126]. There are several clinical studies that involve ferumoxytol as an MRI contrast agent for different cancers, spanning the entire range of trials: a Phase 0 trial is testing it as a general MRI-enhancement agent [127], and a Phase IV trial is investigating ferumoxytol as a contrast agent in lymph nodes associated with pancreatic cancer [128]. Ferumoxytol is also involved in a clinical trial for imaging cancer in thyroid lymph nodes [129], as well as in pediatric and adult soft tissue sarcoma [130]. The nanoparticle is even being used to image the differential drug effects of doxorubicin (a chemotherapeutic) and bevacizumab (an antibody) on brain tumors [131].

In addition to iron oxide nanoparticles, there are currently clinical trials underway that use silica nanoparticles as cancer imaging agents. Unlike other nanoparticles, such as iron oxide nanoparticles, the chemistry of silica nanoparticles is such that various functional groups can easily be attached to the surface [132]. Thus, these particles can be tagged for versatile imaging applications (such as fluorescence and radioactive tags) and can also be functionalized to target receptors that are known to be overexpressed in cancer cells, giving them an active targeting aspect to complement the passive EPR effect. Based on these advantages, silica nanoparticles for tumor imaging were developed and shown to have high binding affinity for specific cancer-related receptors and promising imaging capabilities in a preclinical model of human melanoma [133]. The first clinical trials using these particles in 2010 tagged these silica nanoparticles with radioactive iodine-124, which allowed for PET imaging for melanoma and brain cancer [134]. Additionally, a Phase 0 clinical trial that seeks to use silica nanoparticles labeled with the fluorescent tag Cy5.5 to aid in surgical intervention for head and neck melanoma is currently recruiting participants; the nanoparticles are injected into the neck at the time of surgery and a handheld camera is used to track their movement and eventual accumulation [135].

As an alternative to imaging, there is a clinical trial investigating the use of a magnetic biopsy needle in concert with iron oxide nanoparticles for detection of tumors [136]. The technology, known as MagProbe, relies on attachment of iron oxide nanoparticles coated with anti-CD34 antibodies which target the overexpressed CD34 receptor on leukemia cells in the bone marrow. Once these particles are endocytosed by cells, a magnetic biopsy needle is inserted into the bone marrow and collects the cells tagged with the magnetic nanoparticles. These cells can then be analyzed further, aiding in the diagnosis of leukemia patients [137].

Although inorganic nanoparticles for cancer treatment in clinical trials are largely used for diagnostic purposes, a few therapeutic applications have made their way to clinical trials. A technique called thermoablation, which involves the heating of magnetic nanoparticles by exposing them to an alternating current in a magnetic field, is a potential means of selectively killing tumor cells; the magnetic nanoparticles should preferentially accumulate in tumor cells through the EPR effect so that it is mainly the tumor cells that will be killed upon the current-induced heating of the particles [138]. This technology is currently being assessed in a Phase 0 clinical trial studying the biodistribution and clearance of the iron oxide nanoparticles [139]. Nanobiotix has developed their own therapeutic metallic nanoparticle, NBTXR3, which is a hafnium oxide crystal that emits large amounts of electrons upon x-ray radiation [140]. Similar to the thermoablation technique, it is hypothesized that the preferential accumulation of NBTXR3 in tumor cells will localize the effects of the radiation, thereby reducing the off-target effects of conventional radiation therapy. Two Phase I trials involving NBTXR3 are currently ongoing: one for adult soft tissue sarcoma [141] and another for head and neck cancer [142].

5 Perspective

Advances in cancer nanotechnology are the culmination of advances made in many areas of research. They are built on understanding the fundamentals of cell biology, cancer biology and the identification of cellular mechanisms and morphologies that present potential therapeutic pathways. The therapeutics discussed in this chapter have relied upon chemistry for the synthesis of new materials, nanoparticle formation for the stabilization of drugs and new materials, and an understanding of how the human body traffics and metabolizes these nanoparticles. All areas are interwoven to create the field of cancer nanotechnology and therefore rely on advancements in all research areas before new drug outcomes can be achieved. As a consequence, nanotechnology-based cancer therapeutics is limited by our fundamental understanding of cancer. Recently developed nanotechnology-based cancer therapies have had difficulty showing improved efficacy over the respective gold standards of their fields, but nanotechnology has played an important factor in reducing chemotherapeutic toxicity over the years. It is important not to understate how reducing chemotherapeutic side effects can dramatically improve the quality of life for millions of cancer patients. Moreover, it is encouraging that the knowledge gleamed from decades of clinical trials in cancer nanotechnology can instruct us with future directions for therapeutic development.

One area with great potential to advance cancer nanotechnology is gene sequencing. The majority of therapies that have passed through clinical trials up until now were initiated prior to readily accessible sequencing technology. To some degree, cancer researchers have been "shooting in the dark," unable to identify the differences (and similarities) between cancers on a genetic level. As sequencing has become more routine and affordable, it has become possible to dissect cancer into subsets with specific mutations. Gene sequencing, with the advancements of bioinformatics to process the large amounts of data, will play an important role in the next wave of cancer therapeutics. We may see new life brought back to previously unsuccessful therapies or find therapies that are specific to particular cancer mutations. We have seen several examples in this chapter where cancer therapies failed in an overall trial but were beneficial to a specific subset of the cancer patients. This will undoubtedly improve our understanding of cancer and help direct the right therapies to the right patients. Furthermore, with an improved understanding of cancer genetics and therapeutic mechanisms, we will likely discover combination therapies that can improve cancer treatment where the individual therapies were not effective alone.

Active targeting is another area that could significantly improve therapeutics. Currently, all FDA-approved liposomes and albumin-bound chemotherapeutics rely solely on the EPR effect to passively accumulate in cancerous tissues. Active targeting could increase efficacy by improving the tumor localization of chemotherapeutics and decrease side effects by reducing the amount of drugs in healthy tissues. Some targeted therapies such as anti-EGFR immunoliposomes and PSMA-targeting BIND-014 micelles (discussed in this chapter) are already undergoing clinical trials with promising results. Actively-targeted nanoparticles, however, can only be as good as their targeting ligands. Although the literature is full of many different small molecule-, peptide-, protein-, aptamer-, and antibody-based targeting ligands which work great in vitro, it is often difficult to see improved targeting in far more complicated in vivo conditions. As targeting ligands improve in both binding and specificity, we anticipate more targeted therapies will translate to the clinic.

A third area likely to contribute to the advancement of cancer nanotechnology comes from the field of gene delivery, specifically RNAi therapies (including siRNA). An siRNA liposome formulation for a hereditary liver disorder is already in Phase III clinical trials [44] in 2014, just 4 years after its novel lipid component was first reported in the literature [49]. Many more siRNA therapies are progressing through the pipelines at multiple pharmaceutical companies across the world, but cancer-treating siRNAs have lagged behind those of genetic disorders. Because these siRNA-liposomes often transfect hepatocytes of the liver, it was a natural choice to first target liver diseases, but the development of new types of liposomes and targeted liposomes should open up avenues for exploring other cancerous organs. Furthermore, most cancer-targeting siRNA liposomes only contain one or two different types of siRNA; since cancer is so heterogeneous and can upregulate many oncoproteins, the encapsulation of multiple types of siRNA may be beneficial and straightforward due to the modular nature of siRNA. One may even envision tailoring the siRNA sequences to the specific oncoproteins expressed by individual patients.

In conclusion, the FDA-approval process is a thorough and necessary procedure for drugs to reach the clinic. Over the past several decades, small molecule chemotherapeutics, siRNAs, and imaging agents have been formulated with different nanotechnologies, such as liposomes, polymeric micelles, albumin-bound systems, polymer-bound systems, and inorganic molecules. A majority of the successful drugs in the clinic and those described in this chapter were developed over a decade ago and are variations on a theme: chemotherapeutic-loaded liposomes and albumin-bound chemotherapeutics. In the coming years, however, we hope to see some of the new drugs and nanotechnologies described in this chapter reaching the clinic and improving the lives of cancer patients.

References

1. Fang J, Nakamura H, Maeda H (2011) The EPR effect: unique features of tumor blood vessels for drug delivery, factors involved, and limitations and augmentation of the effect. Adv Drug Deliv Rev 63(3):136–151
2. Scott AM, Wolchok JD, Old LJ (2012) Antibody therapy of cancer. Nat Rev Cancer 12 (4):278–287
3. Vanneman M, Dranoff G (2012) Combining immunotherapy and targeted therapies in cancer treatment. Nat Rev Cancer 12(4):237–251
4. Lipsky MS, Sharp LK (2001) From idea to market: the drug approval process. J Am Board Fam Med 14(5):362–367
5. Eifler AC, Thaxton CS (2011) Nanoparticle therapeutics: FDA approval, clinical trials, regulatory pathways, and case study. In: Hurst SJ (ed) Methods in molecular biology (Clifton NJ), vol 726. pp 325–38
6. NCI (2014) Nanotechnology characterization laboratory. http://ncl.cancer.gov
7. Kinders R et al (2007) Phase 0 clinical trials in cancer drug development: from FDA guidance to clinical practice. Mol Interv 7(6):325–334
8. Hay M et al (2014) Clinical development success rates for investigational drugs. Nat Biotechnol 32(1):40–51
9. DiMasi J (2001) Risks in new drug development: approval success rates for investigational drugs. Clin Pharmacol Ther 69(5):297–307
10. Glasser SP, Salas M, Delzell E (2007) Importance and challenges of studying marketed drugs: what is a phase IV study? Common clinical research designs, registries, and self-reporting systems. J Clin Pharmacol 47(9):1074–1086
11. DiMasi JA, Hansen RW, Grabowski HG (2003) The price of innovation: new estimates of drug development costs. J Health Econ 22(2):151–185
12. Lundqvist EÅ (2012) Principles of chemotherapy. Int J Gynecol Obstet 119 (Suppl(M)): S151–S154
13. Siddik ZH (2002) Mechanisms of action of cancer chemotherapeutic agents : DNA-interactive alkylating agents and antitumour platinum-based drugs
14. Rowinsky EK, Donehower RC (1995) Paclitaxel (taxol). New Engl J Med 332:1004–1014
15. Rowinsky EK, Donehower RC (1991) The clinical pharmacology and use of antimicrotubule agents in cancer chemotherapeutics. Pharmacol Ther 52(1):35–84

16. Tacar O, Sriamornsak P, Dass CR (2013) Doxorubicin: an update on anticancer molecular action, toxicity and novel drug delivery systems. J Pharm Pharmacol 65(2):157–170
17. Fire A et al (1998) Potent and specific genetic interference by double-stranded RNA in Caenorhabditis elegans. Nature 391:806–811
18. Zamore PD et al (2000) RNAi: double-stranded RNA directs the ATP-dependent cleavage of mRNA at 21–23 nucleotide intervals. Cell 101(1):25–33
19. McManus MT, Sharp PA (2002) Gene silencing in mammals by small interfering RNAs. Nat Rev Genet 3(10):737–747
20. Gottesman M (2002) Mechanisms of cancer drug resistance. Annu Rev Med 53:615–627
21. Whitehead KA, Langer R, Anderson DG (2009) Knocking down barriers: advances in siRNA delivery. Nat Rev Drug Discov 8(2):129–138
22. Kanasty R et al (2013) Delivery materials for siRNA therapeutics. Nat Mater 12(11):967–977
23. Shen H, Sun T, Ferrari M (2012) Nanovector delivery of siRNA for cancer therapy. Cancer Gene Ther 19(6):367–373
24. Frey N et al (2009) Magnetic nanoparticles: synthesis, functionalization, and applications in bioimaging and magnetic energy storage. Chem Soc Rev 38(9):2532–2542
25. Rosen JE et al (2012) Iron oxide nanoparticles for targeted cancer imaging and diagnostics. Nanomed Nanotechnol Biol Med 8(3):275–290
26. Josephson L et al (2002) Near-infrared fluorescent nanoparticles as combined MR/optical imaging probes. Bioconjug Chem 13(3):554–560
27. Kobayashi T et al (1977) Enhancement of anti-tumor activity of 1-B-D-Arabinofuranosylcytosine by encapsulation in liposomes. Int J Cancer 20:581–587
28. Allen TM, Cullis PR (2013) Liposomal drug delivery systems: from concept to clinical applications. Adv Drug Deliv Rev 65(1):36–48
29. Barenholz Y (2012) Doxil®–the first FDA-approved nano-drug: lessons learned. J Controlled Release 160(2):117–134
30. Allen TM, Chonn A (1987) Large unilamellar liposomes with low uptake into the reticuloendothelial system. FEBS Lett 223(1):42–46
31. Haran G et al (1993) Transmembrane ammonium sulfate gradients in liposomes produce efficient and stable entrapment of amphipathic weak bases. Biochim Biophys Acta 1151 (2):201–215
32. Gabizon AA, Barenholz Y, Bialer M (1993) Prolongation of the circulation time of doxorubicin encapsulated in liposomes containing polyethylene glycol-derivatized phospholipid: pharmacokinetic studies in rodents and dogs. Pharm Res 5:703–708
33. Gabizon A et al (1994) Prolonged circulation time and enhanced accumulation in malignant exudates of doxorubicin encapsulated in polyethylene-glycol coated liposomes prolonged circulation time and enhanced accumulation in malignant exudates of doxorubicin encapsulated in polyet. Cancer Res 54:987–992
34. ClinicalTrials.gov (2008) NCT00606515: Pharmacokinetics study of liposomal paclitaxel in humans (LPS-PK-H). http://www.clinicaltrials.gov/show/NCT00606515
35. ClinicalTrials.gov (1999) NCT00004083: liposomal cisplatin in treating patients with recurrent ovarian cancer. http://www.clinicaltrials.gov/show/NCT00004083
36. Lorusso D et al (2007) Pegylated liposomal doxorubicin-related palmar-plantar erythrodysesthesia ('hand-foot' syndrome). Ann Oncol: Off J Eur Soc Med Oncol/ESMO 18(7):1159–1164
37. Arteaga CL et al (2012) Treatment of HER2-positive breast cancer: current status and future perspectives. Nat Rev Clin Oncol 9(1):16–32
38. Baselga J et al (2014) Phase III trial of nonpegylated liposomal doxorubicin in combination with trastuzumab and paclitaxel in HER2-positive metastatic breast cancer. Ann Oncol: Off J Eur Soc Med Oncol/ESMO 25(3):592–598
39. Nicholson R, Gee J, Harper M (2001) EGFR and cancer prognosis. Eur J Cancer 37:9–15

40. Mamot C et al (2003) Epidermal growth factor receptor (EGFR)—targeted immunoliposomes mediate specific and efficient drug delivery to EGFR- and EGFRvIII-overexpressing tumor cells. Cancer Res 63:3154–3161
41. Mamot C et al (2005) Epidermal growth factor receptor-targeted immunoliposomes significantly enhance the efficacy of multiple anticancer drugs in vivo. Cancer Res 65 (24):11631–11638
42. ClinicalTrials.gov (2012) NCT01702129: anti-EGFR immunoliposomes in solid tumors. http://www.clinicaltrials.gov/show/NCT01702129
43. Mamot C et al (2012) Tolerability, safety, pharmacokinetics, and efficacy of doxorubicin-loaded anti-EGFR immunoliposomes in advanced solid tumours: a phase 1 dose-escalation study. Lancet Oncol 13(12):1234–1241
44. ClinicalTrials.gov (2013) NCT01960348: the study of an investigational drug, ALN-TTR02, for the treatment of transthyretin (TTR)-mediated amyloidosis. http://www.clinicaltrials.gov/show/NCT01960348
45. Thaker PH et al (2004) EphA2 expression is associated with aggressive features in ovarian carcinoma. Clin Cancer Res 10:5145–5150
46. Aleku M et al (2008) Atu027, a liposomal small interfering RNA formulation targeting protein kinase N3, inhibits cancer progression. Cancer Res 68(23):9788–9798
47. Heyes J et al (2005) Cationic lipid saturation influences intracellular delivery of encapsulated nucleic acids. J Controlled Release 107(2):276–287
48. Jayaraman M et al (2012) Maximizing the potency of siRNA lipid nanoparticles for hepatic gene silencing in vivo. Angew Chem Int Ed Engl 51(34):8529–8533
49. Semple SC et al (2010) Rational design of cationic lipids for siRNA delivery. Nat Biotechnol 28(2):172–176
50. Akinc A et al (2008) A combinatorial library of lipid-like materials for delivery of RNAi therapeutics. Nat Biotechnol 26(5):561–569
51. Dong Y et al (2014) Lipopeptide nanoparticles for potent and selective siRNA delivery in rodents and nonhuman primates. Proc Natl Acad Sci 111(11):3955–3960
52. Love KT et al (2010) Lipid-like materials for low-dose, in vivo gene silencing. In: Proceedings of the national academy of sciences of the united states of America, vol 107, issue no 5, pp 1864–1869
53. Whitehead KA et al (2012) In vitro-In vivo translation of lipid nanoparticles for hepatocellular siRNA delivery. ACS Nano 6:6922–6929
54. Zimmermann TS et al (2006) RNAi-mediated gene silencing in non-human primates. Nature 441(7089):111–114
55. ClinicalTrials.gov (2009) NCT00882180: dose escalation trial to evaluate the safety, tolerability, pharmacokinetics and pharmacodynamics of intravenous ALN-VSP02 in patients with advanced solid tumors with liver involvement. http://www.clinicaltrials.gov/show/NCT00882180
56. Tabernero J et al (2013) First-in-humans trial of an RNA interference therapeutic targeting VEGF and KSP in cancer patients with liver involvement. Cancer Discov 3(4):406–417
57. ClinicalTrials.gov (2011) NCT01437007: TKM 080301 for primary or secondary liver cancer. http://www.clinicaltrials.gov/show/NCT01437007
58. Strebhardt K, Ullrich A (2006) Targeting polo-like kinase 1 for cancer therapy. Nat Rev Cancer 6:321–330
59. Santel A et al (2006) A novel siRNA-lipoplex technology for RNA interference in the mouse vascular endothelium. Gene Ther 13(16):1222–1234
60. ClinicalTrials.gov (2009) NCT00938574: study with Atu027 in patients with advanced solid cancer. http://clinicaltrials.gov/show/NCT00938574
61. Strumberg D et al (2012) Antimetastatic activity of Atu027, a liposomal small interfering RNA formulation, targeting protein kinase N3 (PKN3): final results of a phase I study in patients with advanced solid tumors. J Clin Oncol (Meeting Abstracts) 30:e13597

62. ClinicalTrials.gov (2013) NCT01808638: Atu027 plus gemcitabine in advanced or metastatic pancreatic cancer (Atu027-I-02). http://www.clinicaltrials.gov/show/NCT01808638
63. ClinicalTrials.gov (2012) NCT01591356: EphA2 gene targeting using neutral liposomal small interfering RNA delivery. http://www.clinicaltrials.gov/show/NCT01591356
64. ClinicalTrials.gov (2014) NCT02110563: phase I, multicenter, dose escalation study of DCR-MYC in patients with solid tumors, multiple myeloma, or lymphoma. http://www.clinicaltrials.gov/show/NCT02110563
65. Kim D-H et al (2005) Synthetic dsRNA Dicer substrates enhance RNAi potency and efficacy. Nat Biotechnol 23(2):222–226
66. Dudek H et al (2014) Knockdown of β-catenin with dicer-substrate siRNAs reduces liver tumor burden in vivo. Mol Ther 22(1):92–101
67. Wong DH et al (2012) MYC dicer substrate siRNA formulated in EnCore lipid nanoparticle reduces tumor burden in the Hep3B orthotopic hepatocellular carcinoma model. In: International liver cancer association annual conference
68. Davis ME (2009) Design and development of IT-101, a cyclodextrin-containing polymer conjugate of camptothecin. Adv Drug Deliv Rev 61(13):1189–1192
69. Kim SC et al (2001) In vivo evaluation of polymeric micellar paclitaxel formulation: toxicity and efficacy. J Controlled Release 72(1–3):191–202
70. ClinicalTrials.gov (2011) NCT01426126: Study of genexol-PM in patients with advanced urothelial cancer previously treated with gemcitabine and platinum. http://www.clinicaltrials.gov/show/NCT01426126
71. ClinicalTrials.gov (2014) NCT02064829: bioequivalence study of IG-001 versus abraxane in metastatic or locally recurrent breast cancer. http://www.clinicaltrials.gov/show/NCT02064829
72. Sorrento (2014) Sorrento announces first patient dosed in registration trial to evaluate bioequivalence between cynviloq and abraxane. http://www.sorrentotherapeutics.com/sorrento-announces-first-patient-dosed-in-registration-trial-to-evaluate-bioequivalence-between-cynviloq-and-abraxane/
73. Hamaguchi T et al (2005) NK105, a paclitaxel-incorporating micellar nanoparticle formulation, can extend in vivo antitumour activity and reduce the neurotoxicity of paclitaxel. Br J Cancer 92(7):1240–1246
74. ClinicalTrials.gov (2012) NCT01644890: a phase III study of NK105 in patients with breast cancer. http://www.clinicaltrials.gov/show/NCT01644890
75. Kato K et al (2012) Phase II study of NK105, a paclitaxel incorporating micellar nanoparticle, for previously treated advanced or recurrent gastric cancer. Invest New Drugs 30(4):1621–1627
76. Ghosh A, Heston WDW (2004) Tumor target prostate specific membrane antigen (PSMA) and its regulation in prostate cancer. J Cell Biochem 91(3):528–539
77. Hrkach J et al (2012) Preclinical development and clinical translation of a PSMA-targeted docetaxel nanoparticle with a differentiated pharmacological profile. Sci Transl Med 4 (128):1–11
78. ClinicalTrials.gov (2011) NCT01300533: a study of BIND-014 given to patients with advanced or metastatic cancer. http://www.clinicaltrials.gov/show/NCT01300533
79. ClinicalTrials.gov (2013) NCT01812746: a phase 2 study to determine the safety and efficacy of BIND-014 (Docetaxel nanoparticles for injectable suspension), administered to patients with metastatic castration-resistant prostate cancer. http://www.clinicaltrials.gov/show/NCT01812746
80. ClinicalTrials.gov (2013) NCT01792479: a phase 2 study to determine the safety and efficacy of BIND-014 (Docetaxel nanoparticles for injectable suspension) as second-line therapy to patients with non-small cell lung cancer. http://www.clinicaltrials.gov/show/NCT01792479

81. Gonzalez H, Hwang SJ, Davis ME (1999) New class of polymers for the delivery of macromolecular therapeutics. Bioconjugate Chem 10(6):1068–1074
82. Bartlett DW, Davis ME (2008) Impact of tumor-specific targeting and dosing schedule on tumor growth inhibition after intravenous administration of siRNA-containing nanoparticles. Biotechnol Bioeng 99(4):975–985
83. ClinicalTrials.gov (2008) NCT00689065: safety study of CALAA-01 to treat solid tumor cancers. http://www.clinicaltrials.gov/show/NCT00689065
84. Davis ME et al (2010) Evidence of RNAi in humans from systemically administered siRNA via targeted nanoparticles. Nature 464(7291):1067–1070
85. Hawkins MJ, Soon-Shiong P, Desai N (2008) Protein nanoparticles as drug carriers in clinical medicine. Adv Drug Deliv Rev 60(8):876–885
86. Ibrahim NK et al (2002) Phase I and pharmacokinetic study of ABI-007, a of paclitaxel phase I and pharmacokinetic study of ABI-007, a cremophor-free, protein-stabilized, nanoparticle formulation of paclitaxel. Clin Cancer Res 8:1038–1044
87. Desai N (2008) Nab technology: a drug delivery platform utilising endothelial gp60 receptor-based transport and tumor derived SPARC for targeting. Drug Deliv Rep, Winter 200:37–41
88. Gelderblom H et al (2001) Cremophor EL: the drawbacks and advantages of vehicle selection for drug formulation. Eur J Cancer 37(13):1590–1598
89. Desai NP et al (2000) Protein-based nanoparticles for drug delivery of paclitaxel. In: Transactions of the sixth world biomaterials congress, Kamuela, Hawaii, USA: society for biomaterials, USA, p 199
90. Gradishar WJ et al (2005) Phase III trial of nanoparticle albumin-bound paclitaxel compared with polyethylated castor oil-based paclitaxel in women with breast cancer. J Clin Oncol 23 (31):7794–7803
91. NCI (2013) FDA approval for paclitaxel albumin-stablized nanoparticle formulation. http://www.cancer.gov/cancertopics/druginfo/fda-nanoparticle-paclitaxel
92. ClinicalTrials.gov (2008) NCT00736619: weekly nanoparticle albumin-bound paclitaxel (Abraxane) + weekly cetuximab + radiation therapy (IMRT intensity-modulated radiation therapy) in patients with stage III-IVB head and neck squamous cell carcinoma (HNSCC). http://www.clinicaltrials.gov/show/NCT00736619
93. Fury MG et al (2014) Phase I study of weekly nab-paclitaxel + weekly cetuximab + intensity-modulated radiation therapy (IMRT) in patients with stage III-IVB head and neck squamous cell carcinoma (HNSCC). Ann Oncol: Off J Eur Soc Med Oncol/ESMO 25(3):689–694
94. ClinicalTrials.gov (2004) NCT00093145: Study of albumin-bound paclitaxel (Abraxane) in combination with carboplatin and herceptin in patients with advanced breast cancer. http://www.clinicaltrials.gov/show/NCT00093145
95. Conlin AK et al (2010) Phase II trial of weekly nanoparticle albumin-bound paclitaxel with carboplatin and trastuzumab as first-line therapy for women with HER2-overexpressing metastatic breast cancer. Clin Breast Cancer 10(4):281–287
96. ClinicalTrials.gov (2008) NCT00785291: paclitaxel, paclitaxel albumin-stabilized nanoparticle formulation, or Ixabepilone with or without Bevacizumab in treating patients with stage IIIC or stage IV breast cancer. http://www.clinicaltrials.gov/show/NCT00785291
97. Hidalgo M, Rowinsky EK (2000) The rapamycin-sensitive signal transduction pathway as a target for cancer therapy. Oncogene 19(56):6680–6686
98. Desai N, D'Cruz O, Trieu V (2010) Combination regimens of nab-rapamycin (ABI-009) effective against MDA-MB-231 breast-tumor xenografts. Cancer Res 69(24 Supplement):6106
99. ClinicalTrials.gov (2013) NCT02009332: phase 1/2 study of ABI-009 in nonmuscle invasive bladder cancer. http://www.clinicaltrials.gov/show/NCT02009332
100. Duncan R (2006) Polymer conjugates as anticancer nanomedicines. Nat Rev Cancer 6 (9):688–701
101. Kopecek J, Kopecková P (2010) HPMA copolymers: origins, early developments, present, and future. Adv Drug Deliv Rev 62(2):122–149

102. Vasey PA et al (1999) Phase I clinical and pharmacokinetic study of PK1 [N—(2-Hydroxypropyl) methacrylamide copolymer doxorubicin]: first member of a new class of chemotherapeutic agents—drug-polymer conjugates. Clin Cancer Res 5:83–94
103. Duncan R, Vicent MJ (2010) Do HPMA copolymer conjugates have a future as clinically useful nanomedicines? A critical overview of current status and future opportunities. Adv Drug Deliv Rev 62(2):272–282
104. Singer JW et al (2003) Poly-(L)-Glutamic Acid-Paclitaxel (CT-2103) [XYTOTX], a bioregradable polyermic drug conjugate. In: Polymer drugs in the clinical stage. pp 81–99
105. O'Brien MER et al (2008) Randomized phase III trial comparing single-agent gemcitabine or vinorelbine for the treatment of PS 2 patients with chemotherapy-naive advanced non-small cell lung cancer. J Thorac Oncol 3(7):728–734
106. ClinicalTrials.gov (2007) NCT00576225: CT-2103/carboplatin versus paclitaxel/carboplatin for NSCLC in women with estradiol >25 pg/mL. http://www.clinicaltrials.gov/show/NCT00576225
107. ClinicalTrials.gov (2007) NCT00459810: paclitaxel poliglumex and estradiol in treating patients with stage IV prostate cancer. http://www.clinicaltrials.gov/show/NCT00459810
108. ClinicalTrials.gov (2001) NCT00017017: CT-2103 in treating patients with recurrent ovarian epithelial or fallopian tube cancer or primary peritoneal cancer. http://www.clinicaltrials.gov/show/NCT00017017
109. ClinicalTrials.gov (2011) NCT01402063: PPX and concurrent radiation for newly diagnosed glioblastoma without MGMT methylation
110. Yurkovetskiy AV et al (2004) Synthesis of a macromolecular camptothecin conjugate with dual phase drug release. Mol Pharm 1(5):375–382
111. Walsh MD et al (2012) Pharmacokinetics and antitumor efficacy of XMT-1001, a novel, polymeric topoisomerase I inhibitor, in mice bearing HT-29 human colon carcinoma xenografts. Clin Cancer Res 18(9):2591–2602
112. ClinicalTrials.gov (2007) NCT00455052: a study of intravenous XMT-1001 in patients with advanced solid tumors. http://www.clinicaltrials.gov/show/NCT00455052
113. Sausville EA et al (2010) Phase I study of XMT-1001 given IV every 3 weeks to patients with advanced solid tumors. J Clin Oncol (Meeting Abstracts) 28:e13121
114. Schluep T et al (2006) Preclinical efficacy of the camptothecin-polymer conjugate IT-101 in multiple cancer models. Clin Cancer Res 12(5):1606–1614
115. ClinicalTrials.gov (2006) NCT00333502: study of CRLX101 (formerly named IT-101) in the treatment of advanced solid tumors. http://www.clinicaltrials.gov/show/NCT00333502
116. Weiss GJ et al (2013) First-in-human phase 1/2a trial of CRLX101, a cyclodextrin-containing polymer-camptothecin nanopharmaceutical in patients with advanced solid tumor malignancies. Invest New Drugs 31(4):986–1000
117. ClinicalTrials.gov (2011) NCT01380769: a phase 2 study of CRLX101 in patients with advanced non-small cell lung cancer. http://www.clinicaltrials.gov/show/NCT01380769
118. ClinicalTrials.gov (2008) NCT00802945: study to evaluate the safety and efficacy of NKTR-102 in patients with metastatic or locally advanced breast cancer. http://www.clinicaltrials.gov/show/NCT00802945
119. Awada A et al (2013) Two schedules of etirinotecan pegol (NKTR-102) in patients with previously treated metastatic breast cancer: a randomised phase 2 study. Lancet Oncol 14(12):1216–1225
120. ClinicalTrials.gov (2011) NCT01492101: the BEACON study (breast cancer outcomes with NKTR-102)
121. Jung CW, Jacobs P (1995) Physical and chemical properties of superparamagnetic iron oxide MR contrast agents: ferumoxides, ferumoxtran, ferumoxsil. Magn Reson Imaging 13(5):661–674
122. Anzai Y, Prince M (1997) Iron oxide-enhanced MR lymphography: the evaluation of cervical lymph node metastases in head and neck cancer. J Magn Reson Imaging 7(1):75–81

123. ClinicalTrials.gov (2005) NCT00147238: a validation study of MR lymphangiography using SPIO, a new lymphotropic superparamagnetic nanoparticle contrast. http://www.clinicaltrials.gov/show/NCT00147238
124. Heesakkers RA et al (2009) Prostate cancer: detection of lymph node metastases outside the routine surgical area with purpose: methods: results: conclusion. Radiology 251(2):408–414
125. Spinowitz BS et al (2005) The safety and efficacy of ferumoxytol therapy in anemic chronic kidney disease patients. Kidney Int 68(4):1801–1807
126. Neuwelt EA et al (2009) Ultrasmall superparamagnetic iron oxides (USPIOs): a future alternative magnetic resonance (MR) contrast agent for patients at risk for nephrogenic systemic fibrosis (NSF)? Kidney Int 75(5):465–474
127. ClinicalTrials.gov (2013) NCT01895829: ferumoxytol—iron oxide nanoparticle magnetic resonance dynamic contrast enhanced MRI. http://www.clinicaltrials.gov/show/NCT01895829
128. ClinicalTrials.gov (2009) NCT00920023: pre-operative staging of pancreatic cancer using superparamagnetic iron oxide magnetic resonance imaging (SPIO MRI). http://www.clinicaltrials.gov/show/NCT00920023
129. ClinicalTrials.gov (2013) NCT01927887: pre-operative nodal staging of thyroid cancer using ultra-small superparamagnetic iron oxide magnetic resonance imaging (USPIO MRI): preliminary study. http://www.clinicaltrials.gov/show/NCT01927887
130. ClinicalTrials.gov (2012) NCT01663090: ferumoxytol-enhanced MRI in adult/pedi sarcomas. http://www.clinicaltrials.gov/show/NCT01663090
131. ClinicalTrials.gov (2008) NCT00769093: assessing dynamic magnetic resonance (MR) imaging in patients with recurrent high grade glioma receiving chemotherapy. http://www.clinicaltrials.gov/show/NCT00769093
132. Bagwe RP et al (2004) Optimization of dye-doped silica nanoparticles prepared using a reverse microemulsion method. Langmuir 20(19):8336–8342
133. Benezra M et al (2011) Multimodal silica nanoparticles are effective cancer-targeted probes in a model of human melanoma. J Clin Investig 121(7):2768–2780
134. ClinicalTrials.gov (2010) NCT01266096: PET imaging of patients with melanoma and malignant brain tumors using an 124 l-labeled cRGDY silica nanomolecular particle tracer: a microdosing study. http://www.clinicaltrials.gov/show/NCT01266096
135. ClinicalTrials.gov (2014) NCT02106598: targeted silica nanoparticles for image guided intraoperative sentinel lymph node mapping in head and neck melanoma patients. http://www.clinicaltrials.gov/show/NCT02106598
136. ClinicalTrials.gov (2011) NCT01411904: study of the detection of lymphoblasts by a novel magnetic needle and nanoparticles in patients with leukemia. http://www.clinicaltrials.gov/show/NCT01411904
137. Jaetao JE et al (2009) Enhanced leukemia cell detection using a novel magnetic needle and nanoparticles. Cancer Res 69(21):8310–8316
138. Kirby R, Eeles R, Neal D (2013) Prostate cancer UK: the blue skies forum. Trends Urol Men's Health 4(6):39–43
139. ClinicalTrials.gov (2014) NCT02033447: Magnetic nanoparticle thermoablation-retention and maintenance in the prostate: a Phase 0 study in men (MAGNABLATE I). http://www.clinicaltrials.gov/show/NCT02033447
140. Maggiorella L et al (2012) Nanoscale radiotherapy with hafnium oxide nanoparticles. Future oncol 8(9):1167–81
141. ClinicalTrials.gov (2011) NCT01433068: NBTXR3 crystalline nanoparticles and radiation therapy in treating patients with soft tissue sarcoma of the extremity. http://www.clinicaltrials.gov/show/NCT01433068
142. ClinicalTrials.gov (2013) NCT01946867: NBTXR3 crystalline nanoparticles and radiation therapy in treating patients with locally advanced squamous cell carcinoma of the oral cavity or oropharynx. http://www.clinicaltrials.gov/show/NCT01946867

MIX
Papier aus verantwortungsvollen Quellen
Paper from responsible sources
FSC® C105338
FSC
www.fsc.org

If you have any concerns about our products,
you can contact us on
ProductSafety@springernature.com

In case Publisher is established outside the EU,
the EU authorized representative is:
Springer Nature Customer Service Center GmbH
Europaplatz 3, 69115 Heidelberg, Germany

Printed by Libri Plureos GmbH
in Hamburg, Germany